SCRIBES OF SPACE

SCRIBES OF SPACE

PLACE IN MIDDLE ENGLISH LITERATURE AND LATE MEDIEVAL SCIENCE

MATTHEW BOYD GOLDIE

CORNELL UNIVERSITY PRESS
Ithaca and London

Copyright © 2019 by Cornell University

All rights reserved. Except for brief quotations in a review, this book, or parts thereof, must not be reproduced in any form without permission in writing from the publisher. For information, address Cornell University Press, Sage House, 512 East State Street, Ithaca, New York 14850. Visit our website at cornellpress.cornell.edu.

First published 2019 by Cornell University Press

Library of Congress Cataloging-in-Publication Data

Names: Goldie, Matthew Boyd, author.
Title: Scribes of space : place in Middle English literature and late medieval science / Matthew Boyd Goldie.
Description: Ithaca : Cornell University Press, 2019. | Includes bibliographical references and index.
Identifiers: LCCN 2018029904 (print) | LCCN 2018047128 (ebook) | ISBN 9781501734052 (epub/mobi) | ISBN 9781501734069 (pdf) | ISBN 9781501734045 (cloth)
Subjects: LCSH: English literature—Middle English, 1100–1500—History and criticism. | Place (Philosophy) in literature. | Geographical perception in literature. | Local color in literature. | Physics—England—History—To 1500. | Geographical perception—England—History—To 1500.
Classification: LCC PR275.P48 (ebook) | LCC PR275.P48 G65 2018 (print) | DDC 820.9/001—dc23
LC record available at https://lccn.loc.gov/2018029904

For Steve Kruger, mentor and friend

Contents

List of Illustrations   ix
Acknowledgments   xi

Introduction: Late Medieval Space   1

1. Local Space, Edges, and Contents: Chorography and Late Medieval English Maps   18

2. Local Literature: Vernacular Local Space and John Lydgate's *Siege of Thebes*   55

3. Horizontal Space: Measuring Local Area with Astrolabes, Quadrants, and *Topographia*   76

4. Horizontal and Abstracted Spaces: *The Book of Margery Kempe* and *The Book of Sir John Mandeville*   101

5. The Science of Motion: New Ideas of Impetus and Measurement   123

6. Motion in Literature: Place and Movement in the *House of Fame*   141

7. Intense Proximate Affect: Nicole Oresme's *Tractatus de configurationibus qualitatum et motuum*   168

8. Proximal Literature: Nearness and Distinction in the *Legend of Good Women*   188

Afterword: Ubiquitous Being in
the Pardoner's Prologue and Tale   209

Notes   219
Bibliography   261
Index   285

Illustrations

1	Two-page drawing of Canterbury Cathedral's grounds from the Eadwine Psalter	32
2	Single-page drawing of Canterbury Cathedral's grounds from the Eadwine Psalter	33
3	Map of Cliffe, Kent	37
4	Map of Sherwood Forest, Nottinghamshire	41
5	Small map of Inclesmoor, West Riding, Yorkshire	43
6	Large map of Inclesmoor, West Riding, Yorkshire	44
7	"Mappa Thaneti . . . Insule," Kent	51
8	Illumination of John Lydgate's *Siege of Thebes*	67
9	Quadrant with sliding latitude plate, shadow square, and plumb bob with a pearl bead	88
10	Measurements of towers using quadrants	91
11	Graphs of motion in Nicole Oresme's *Tractatus de configurationibus qualitatum et motuum*	139

Acknowledgments

My gratitude for their generous help and brunches first goes to the Saturday Medieval Group: Valerie Allen, Jennifer N. Brown, Glenn Burger, Steven F. Kruger, David Lavinsky, Michael G. Sargent, and Sylvia Tomasch. This writing group brought to the project their goodwill, attention, guidance, expertise, and patience, and I am very thankful. Thank you also to the American Geographical Society Library at the University of Wisconsin for a McColl Research Fellowship, which began the whole project in a very different form. Other libraries and librarians were of real help, especially for images: the New York Public Library; New York University Bobst Library; the manuscript room at the British Library; the New York Academy of Medicine Library; Sandy Paul at Trinity College Library, Cambridge; Wim Braakman at the University of Groningen Library; Evelien Hauwaerts at Openbare Bibliotheek, Bruges; and the librarians at Rider University Library. Rider University provided summer fellowships and reimbursements. PJ continues to keep my spirits up and offers sound advice in all matters from approach to prose. Part of the introduction builds on "An Early English Rutter: The Sea and Spatial Hermeneutics in the Fourteenth and Fifteenth Centuries," *Speculum* 90 (2015): 701–27. Some of chapter 6 is adapted in "Spatial History: Estres, Edges, and Contents," *Studies in the Age of Chaucer* 40 (2018).

SCRIBES OF SPACE

Introduction
Late Medieval Space

Scribes of Space examines thirteenth-, fourteenth-, and fifteenth-century science and literature for their insights into space, not in the sense of the stars and planets of the cosmos but for the more everyday and earthbound areas that people encounter, sense, and apprehend, and through which they move. The scientific texts include late medieval physics (especially mechanics), geographical writings, technologies of measurement in what were known as the mechanical arts, and maps. The literary texts are primarily Middle English poetry and prose, particularly Geoffrey Chaucer; most of the discussions of literature center on a sustained reading of a Chaucerian work. The period covered is a crucial time when fundamental spatial paradigms were deliberately reconsidered, often innovatively, in scientific writings. The era is also when poetry and prose explored space through engagement with scientific ideas and attention to the literary and experiential qualities of space. On occasion the literature addresses space in a conscious manner similar to scientific or technical writings, and at other times it is less intentional, its modes of expression ranging from the philosophical to the ways, for instance, narrators and characters perceive, and are altered by, the spaces around them.[1]

One underlying idea in *Scribes of Space* is that space changes throughout history. By this I mean that the defining attributes of a physical area transform historically, even in paradigmatic ways, and not merely types of space, such as boroughs, cities, parks, and gardens, or built structures like cathedrals or halls. Changes in space are neither a fact of nature nor the result of material alterations to space itself but instead are bound up with modifications in human understanding, observation, and experience. To adapt the conclusion of Brian Harley and David Woodward's seminal *History of Cartography*, space undergoes "cognitive transformations," and analysis of it reveals how a "developing picture of reality—what was actually perceived—was modified."[2] When in the thirteenth, fourteenth, and fifteenth centuries, scholastic science, the mechanical arts, and geographical writings not only accepted ancient ideas about the essential characteristics of space but also thought of crucial new ways of understanding them, these developments did not simply influence the larger culture but were bound up with concurrent changes in people's understandings of the areas around them. That is, the science did not so much have an effect on the culture and its literature in the sense of preceding them but was instead part of a shared transformation. Human apprehensions of basic physical space interacted with related epistemes of scientific knowledge, spatial perception, and presence, giving rise to new ways of thinking and feeling and being.

Connecting spatiality with fundamental aspects of human presence is not a new idea. Aristotle's *Physics*, the work of essential importance to Aristotle's ideas about nature in general, centers on this relationship. (The work is often titled *Lecture on Nature*). As is well known, the *Physics* played a central role in nearly all late medieval apprehensions of the world following its reintroduction to the Latin West in the twelfth and thirteenth centuries. In fact, it is not too much of an exaggeration to say that what is called a renaissance at this time in history was primarily a rebirth of physics. Having discussed infinite capacities in book 3, Aristotle arrives at the characteristics of place or space in book 4, ideas about which he acknowledges present "many difficulties." He urges his audience to learn about space and then comes to his key observation: "A student of nature must have knowledge about *place* [*topos*] . . . : whether it is or not, and in what way it is, and what it is. For everyone supposes that things that are are somewhere, because what is not is nowhere." He reemphasizes his point: "the nonexistent is nowhere," and he then asks rhetorically "where for instance is a goat-stag or a sphinx?" as examples of those things that do not exist and therefore are no place.[3] Aristotle's key point is to draw attention to the relationship between location and existence. To exist is to be somewhere, and conversely nonbeing is nowhere.

Aristotle's connection between space and being poses a challenge for the study of medieval literature (indeed literature of all periods) because it excludes fictional poetry, prose, and drama. When in the thirteenth century

Thomas Aquinas addresses the same Aristotelian passage in his *Expositio in VIII libros Physicorum Aristotelis*, he concurs with Aristotle but slightly alters his authority's distinction between things that are in a place and therefore exist, and things that are not anywhere and thus nonexistent. He writes: "What does not exist is nowhere, i.e., in no place, for there is no place where the goat-stag or the sphinx exist, which are certain fictions after the manner of chimeras" (Quod non est, nusquam est, idest in nullo loco est: non enim est dare ubi sit Tragelaphus aut sphinx, quae sunt quaedam fictitia sicut Chimaera).[4] Aquinas adds a different element to Aristotle, modifying the question about space from the subject of existence to *fictitia*, thereby relegating that which does not exist to the realm of fiction, the imaginary, literature. For Aquinas, literature deals in a contradiction—fictional beings—but medieval fiction in fact has little problem with violating the distinction between that which is in a place and therefore real, and the unplaced, nonexistent, and imagined. Indeed, it can deal in double nonbeings that are both fictional and in unreal places. The fiction of medieval poetry and prose takes up animals such as goat-stags, sphinxes, and chimera, as well as less exotic creatures like human characters and other made-up entities, and it works with them in imaginary locations.

Scribes of Space explores science and literature in paired chapters. The first of each pair examines scientific, technical, and historical innovations about space in the thirteenth, fourteenth, and fifteenth centuries, while the second addresses contemporary Middle English poetry and prose along with some Scottish and other literatures. These pairs are not meant to blur substantial, disciplinary, and formal differences between the science and the literature; indeed, it not only explores them but also teases out unexpected similarities between their qualities. The points of similarity and difference between the science and the literature of the time are worth lingering over. First, late medieval mechanics and Middle English literature together articulate the characteristics of space in ways that are similar to the concept of balance in Joel Kaye's *History of Balance*. Space is one of the "mediating structures . . . between environment and intellectual invention, between sensation and science" that "weave together the experiential, the intuitive, and the ideational, and that are shaped by the 'sense' of how things actually work and find order in the world."[5] Kaye's *History of Balance* and his earlier study *Economy and Nature in the Fourteenth Century* are models for what follows in some respects, but the "mediating structure" of space is complex and should not be taken as collapsing important distinctions between the scientific and the literary writings.

The greatest difference between particularly the Latin natural philosophy and Middle English poetry and prose on the topic of space—beyond the obvious and significant ones of language, genre, and context—is that the science is generally not concrete while the literature often engages with space in corporeal ways. The writings on the physics of mechanics are almost without exception abstract and theoretical, rarely appearing to draw on or be modified by observed phenomena, or applying their ideas to objects or events in the world. The scholastic thinkers of the time have justifiably been called "philosophical empiricists" and their work "empiricism without observation," appellations that sound like paradoxes to the modern ear.[6] What the phrases suggest, however, is that while science in the twelfth and thirteenth centuries turned away from metaphysics and toward what Aquinas called "sensible things" (a significant change in the history of science and technology in itself), it remains difficult to judge whether the objects in the natural world, the *res sensibiles* that the science appears to address, are observed phenomena.[7] The literature, on the other hand, addresses individuated and concrete beings, albeit through fiction and other modes of writing. Poetry's creatures may be imaginary, but they appear and act as distinct and mostly tangible phenomena.

There are, however, aspects of the scientific and literary writings that are similar to each other, or parallel, or that even connect in direct or less direct ways beyond space operating as a "mediating structure." *Scribes of Space* examines the direct historical evidence for the circulation of mechanical ideas and their explicit influence on literature, but it also considers the modes in which both were written. It is often said that one point of connection lies in the fact that medieval literature does not have the strong boundaries of genre that apply to later literature; authors in the period transition between what may be considered traditional poetic subjects and others, including science, with more ease than many authors in other periods of literature. Geoffrey Chaucer, "hier in philosophie / To Aristotle in our tonge" as Thomas Hoccleve called him, is a good example with his translation of Boethius's *Consolation of Philosophy*, *A Treatise on the Astrolabe*, and "The Complaint of Mars," as well as many scientific features that appear throughout his romances, lyrics, and other writings.[8] The argument about the porosity of literature in the period is true as far as it goes, but another aspect of the similarity between the treatises on physics and late medieval literature is form and style. The "philosophical empiricism" of the time has also been described as having been written in the "subjunctive mood," a rhetorical mode of nonobservational speculation about things.[9] That is, natural philosophy discusses possibilities, taking a principle and exploring its ramifications (usually in exhaustive detail) when applied to various situations and phenomena. As mentioned, these scientific consequences, however, do not appear to be proved in the modern sense in which a speculation is observed in specific natural objects or an experiment performed to demonstrate a result. Middle English literature is similar in that it is also often philosophical and "subjunctive" in terms of mode when it explores spatial relations; like the scientific discourses, it is speculative in its investigations into the possibilities and implications of space. Even though it might be less deliberate than science or mechanical arts—as in less direct, intentional, and consistent—it is nevertheless still deliberative.

Beyond these modal similarities, the closest parallel to the literature probably lies in the mechanical arts, particularly geometry and its related arts or sciences—architecture, surveying, navigation, commerce, agriculture, and so on—whose prominence increases in the late medieval period. I consider the *artes mechanicae* in part because they can be read as an important "mediating structure" between the philosophical and applied sciences, between theory and real-world practices. They were initially denigrated as inferior to philosophy and the liberal arts, but by the twelfth century they were valued because they contributed to understanding nature and offered practical assistance. Hugh of St. Victor, for example, describes the mechanical arts in his *Didascalicon* of about 1125 as "adulterated" (*adulterina*) not because they were inferior but because they imitated nature and were the result of human labor. He classifies them as part of philosophy.[10] At one point Hugh writes, "The mechanical sciences are the seven handmaids which Mercury received in dowry from Philology, for every human activity is servant to eloquence wed to wisdom" (hae sunt septem ancillae quas Mercurius a Philologia in dotem accepit, quia nimirum eloquentiae, cui iuncta fuerit sapientia).[11] In his mid-thirteenth-century *De ortu scientiarum*, Robert Kilwardby, who taught at Oxford and later became archbishop of Canterbury and a cardinal, went further, directly equating the speculative sciences with the practical and vice versa. He affirms that "those [parts of philosophy] which are practical are also speculative. . . . It consequently seems that each of the speculative sciences is also practical. It appears, therefore, that the speculative sciences are practical and the practical speculative" (illae quae practicae sunt sint etiam speculativae. . . . Videtur igitur quod unaquaeque dictarum speculativarum sit etiam practica. Videtur ergo quod et speculativae sint practicae et practicae speculativae).[12] Jerome Taylor writes that the mechanical arts began "to flourish in the rising centers of urban life" in the late Middle Ages, and they seem to have found a place in the universities by the fourteenth century.[13] I aim to draw attention to the mechanical arts or sciences in their own right and to reveal their significant, sometimes philosophical, understandings of space. Middle English poetry and prose seem to occupy a similar place in the culture in that they also work in the area between philosophical abstractions and material things, between theories and practices.

Literature, however, parses concrete individuations of places and spaces, especially in relation to human character, more often than natural philosophy or the mechanical arts. Middle English and other literatures explore space as described by the philosopher Jeff Malpas (whose work informs this study throughout): "To be within a place is to find oneself oriented to its currents and directions; in the fullest sense, it is to be capable of acting within it and moving through it; it is to gain a feeling for the patterns and rhythms of the place, of its own movements, of the density of the spaces within it, of the possibilities that it enables and the demands that it imposes."[14] Late medieval science does not single out human interactions, because its ideas are rigorously applied to all phenomena. If human attributes are addressed, they are considered only in the context of strict commonalities among qualities everywhere in the world. In contrast, while it is not the case that medieval literature's sole attention is on humans, it is nevertheless impossible to deny that they are a principal focus. Both disciplines, however, address many kinds of phenomena, so it may be said that late medieval physics, the mechanical arts, and literature are about "objectivity," that is, about objects, including human ones.[15]

A final difference between the science and the literature is that the poetry and prose do not simply record space in the world, nor do its places and situations correspond with real areas in a plain, conforming manner. Literary and artistic presentations alter, interpret, and compose the spaces that audiences experience in narratives and images. The literary works also do not commonly present accurate portraits of how people saw the spaces around them. So much of literature is precisely concerned with narrators and characters who misperceive, deliberately misrepresent, and otherwise diversely interpret and experience the spaces they inhabit. That literature, especially poetry, is deliberately shaped is nevertheless a strength when considering it alongside the mechanical sciences on the topic of space. It is because of, not in spite of, these complexities that literature and other forms of artistic production are revealing about space. The Middle English texts set characters and objects in specific places with which they are forced to interact, characters and things that are in turn deeply affected by the parameters, characteristics, and other qualities of the spaces. In doing so, the texts in turn reveal to their audiences the consequences of particular spaces.[16]

The literary works in *Scribes of Space* have been chosen for their insights into the scientific debates about the distinctive qualities of space, the speculative qualities they have in common with natural philosophy, their parallels with the mechanical arts, and their revelations about spatial apprehension more generally. I have not tried to be comprehensive in selecting the literature (an impossibly large task), and I am keenly aware that other works would benefit from a spatial analysis. It is modestly hoped that the attention to significant concepts in medieval mechanics within physics and other disciplines will lead to future explorations into the relations between these sciences and literature. The texts examined here are nevertheless diverse, belonging to a range of genres and produced under different circumstances: some are more literary, and others exhibit little obvious investment in poetic or tropological qualities; some are explicitly scientific while others do not appear on the surface to engage with late medieval science or technology. Chaucer is important because when he writes on space, even in his less directly scientific writings, he is philosophically experimental in ways that are often similar to his scholastic contemporaries and perhaps even more similar to the (usually anonymous) practitioners of the mechanical arts. Alexander Gabrovsky writes that "medieval physics has been largely excluded from previous studies that focus on Chaucer's representations of sublunary change," yet "Chaucer . . . recreates the sublunar world in his poetic imagination as a kind of thought experiment, which puts to test medieval theories of natural philosophy."[17] While Chaucer's direct and explicit historical connections with medieval science are important, the temperament of his engagement with natural philosophy in his poems—more than the fact of historical sources and influences—draws a study of late medieval space to his works.

The following analysis investigates how the period apprehended spatial area in terms of four main qualities: space's defining characteristics, the standpoint from which one commonly perceives an area, motion through space, and the proximity of objects within a location. My opening claim is that the area that was most meaningful to people at the time—the space that the sciences, literature, and other discourses engaged with in the most depth—was an immediately local one. The space I address is not "space" as in the cosmos or a spatial abstraction but concrete and physically smaller geographical areas. Malpas captures this category of spatial extent when he describes what it means "to be within a place," "to find oneself oriented to its currents and directions," "to be capable of acting within it and moving through it," and "to gain a feeling" of an area. In drawing attention to immediate and physical senses of place, I reveal different apprehensions and understandings of space and being than at the larger scales. Cosmological and global spaces are more reliant on written authorities, and they tend to be hierarchical in that they entail a geographical or geometrical scheme of the heavens above earthbound and individual locations. My purpose here is somewhat corrective because unlike larger geographical entities—the universe, the globe, and the nation—the local areas of common spatial experience in the late Middle Ages have received less attention in scientific, geographical, and literary studies.[18] When discrete spatial entities have been considered—cities, gardens, markets, theater areas, parks, cathedrals, monasteries and nunneries, gatehouses—they usually have special qualities or are architectural units, and the focus is not on space per se.[19]

Chapter 1 focuses on local space by examining Claudius Ptolemy's ideas about chorography—detailed description of an individual place—in his *Geography*, and it pairs his ideas with English maps of smaller areas that survive from the later Middle Ages. The chapter tries not to presume that what was local in the late medieval era is also local today or in another period; the very definition of local space, the paradigms according to which a local space might be apprehended, might be similar but also different. Ptolemy's ideas about chorography set up some of the parameters for what was considered local, and the maps further reveal how local area was apprehended. About thirty-five English local maps survive from the late Middle Ages, and they are quite diverse in terms of form and style. Historical evidence suggests that mapmaking of smaller areas was much less established as a genre or discipline within the mechanical arts than cartography of larger areas; therefore, the local maps are less conventional and conceivably more responsive to how people perceived the areas around them. By analyzing the maps, which often have accompanying documentation, one can gain insights into how their makers and their audiences recognized and understood local area. The chapter considers how the mapmakers thought about the edges, contents, and temporalities of local space. It demonstrates that the boundaries or borders of a physical or conceptual nature (as in law, the imagination, history, and so on) provided a local space with definition. These edges also served as transitional zones between one locale and another, zones that had a variety of characteristics. The contents of map spaces interact with these edges and are sometimes arranged in patterns or organized according to categories. Thus, objects within a local space could be thought of in systematic ways in that they could be classified into groups or types. Temporality also was of concern to the mapmakers, who often present current phenomena and historical objects side by side on a map; in many cases the landscape is as marked by the past as it is by the present, with two or more times interacting with each other in one place.

Chapter 2 builds on the scientific and cartographic features of local space through an examination of semantic and literary evidence about local space. A reader can gain an initial impression of how a late medieval sense of local space's qualities—its directions, patterns, densities, and possibilities—are different from the present day by considering the words used to describe spatial entities. For instance, though I use *local* to denote the common sense of space, Middle English did not use the word with a geographical meaning. Instead, writings often turn to the borrowed French word *estre* or *ester* to specify a smaller geographical area, which ranged from a space within a building to a hundred or a shire. When *estre* first entered English in the late thirteenth century, it had already begun to lose the connotations of its origin in the verb *être*, but it retained a sense of a *state* or *condition*. Middle English *estre* has within it what Aristotle, Aquinas, Malpas, and other philosophers have drawn attention to—namely, an intimate relationship between space and being. The chapter continues with analysis of John Lydgate's *Siege of Thebes* from 1421 or 1422 and the iconic manuscript portrait of the Canterbury pilgrims on the road in British Library, Royal 18 D 2 from the third quarter of the fifteenth century. Lydgate's *Siege* and the later image are similar in depicting the edges, contents, and temporalities of an estral space in particularly deliberate fashion. I posit that the *Siege* and the illumination engage in a kind of spatial realism when they present the narrator—John Lydgate—joining the company of Canterbury pilgrims. My point, however, is that the poem and the image cannot avoid a sense of incoherence in the social group depicted in the poem and the illumination because the local space they show may be systematic, but it also contains contradictions.

The second quality of space at the time, and my second main point, is that estral space is what I call *horizonal* in two senses. First, the local space of the period typically reaches out from a viewer in a planar fashion—in Middle English terms, at a given moment it "extendeth" or "streccheth" out in a horizontal zone or band near the earth. Second, estral space spreads out to the horizon and ends there; it has the potential to go on forever, but interest focuses on what is "compassed" all about within the "orisonte." Medieval estral space in its horizontality differs in important ways from earlier Roman understandings of space, which have been characterized as largely "hodological," or one-dimensional.[20] The Anglo-Saxonist Nicholas Howe has considered Pliny's geographical understanding of area and explains that Roman authors generally "locate places not by pointing to them on an illustrative map or by setting their coordinates on a grid, but by writing them in an ordered sequence that typically begins with a well-defined and isolatable site and then moves outward to such other regions as Britain." Howe discerns a similar understanding of places as "contiguous and sequential" in Anglo-Saxon writings, what he calls "narrative cartography."[21] Examples of this kind of linear apprehension and depiction of space continue in the late Middle Ages in pilgrimage narratives, selected maps, and some romances, but this mode of understanding space, though discernible and significant in cultural practices of the Middle English period, is not as common as the horizonal.[22] The prevailing apprehension of space spreads out to the horizon across an area parallel to the earth in almost a two-dimensional fashion. Nothing about this estral space is meant to imply that late medieval culture was somehow shortsighted or impoverished in terms of spatial experience. It simply had certain characteristics that are as rich as any period.

One implication of highlighting estral horizontality is that it turns discussion away from generalizations about the late medieval period as hierarchical in the sense of a chain of being extending down from the godhead, the time and place of creation taking precedent over the present, the king above his subjects, the spirit over the physical body, the head over the lower regions, and so on. As mentioned, this means temporarily bracketing medieval discussions of celestial space; after all, philosophy of the time and humoral theories tend just to discuss all space without distinguishing among scales (although a countertendency to distinguish between macrocosmic and local space already exists). A related implication of horizonality for geography specifically is to rotate orientation from a vertical alignment—as in familiar cosmological schemes of fixed stars and eight spheres stacked above the earth—to a horizontal orientation.[23] *Mappaemundi* and Macrobian maps of climes (indeed most but importantly not all cartographical presentations), some dream states, visions, and other special circumstances—all also well studied—show that people without the physical means of a wholly aerial standpoint could imagine and depict "spaceship earth" from an overhead bird's-eye and larger perspective.[24] These and other ways of perceiving and experiencing overviews of space were culturally important, and they will be taken into account especially in relation to grid-based isotropic formations on maps that emerge during this period and what they imply, but the evidence suggests that people did not habitually or frequently experience space in this way.

The overwhelming number of late medieval written and visual presentations of space largely ignore an overhead perspective and very extended views. The scientific ways of thinking and the literature instead explore and develop considerations of what was contained in the estral compass—namely, diverse objects in locations around a viewer's position. I lay out the characteristics of the horizonal in the third chapter via an examination of astrolabes, quadrants, and the treatises that describe them. Astrolabes are the best-known objects here, but I draw attention to their simplest and possibly their most frequently used features—the *limbus* and the *umbra recta*—along with the related measuring device of the quadrant, all of which were employed to measure the elevations of celestial phenomena but also the heights, depths, and breadths of earthbound objects. I also examine the treatises that describe their uses; their descriptions of interactions between the viewer and sighted objects further reveal the ways the horizonal space was apprehended. I last consider the related late medieval scientific-rhetorical trope of *topographia* for additional insight into the understanding of objects within a horizonal perspective. Writers such as Matthew of Vendôme and Gerald of Wales describe *topographia* as involving objects within an area that are notable because they arise in fitting or unfitting fashion out of the specifics of a particular place and time. These items possess different weights and characteristics; they are not flattened by an overhead perspective and are therefore not of equal value, whether that value is visual, physical, emotional, philosophical, moral, political, or thematic.

Chapter 4 explores the implications of the horizontal apprehension of space in two very different texts: *The Book of Margery Kempe* and *The Book of Sir John Mandeville*. The former describes a developing series of visions and social challenges in the life of the fifteenth-century mystic woman Margery Kempe; the latter is an encyclopedic compilation of geographical lore and travel itineraries from the Near East to the Far East, purportedly experienced and written by a fourteenth-century English knight. Some overlap exists between these two very different texts, however, in that they are both in part travel writings, which were becoming prominent at this time.[25] Margery traveled quite extensively in England and made several pilgrimages to Spain, the Low Countries, and Jerusalem. Commonalities between the two books, and the nature of each text's geographical knowledge, have begun to be examined in the critical literature. In light of chapter 3, I have a different focus from source studies—namely, an attempt to discern how Margery and Mandeville perceive the horizontal spaces around them. Chapter 4 reveals that both tend to see the spaces around them as containing diverse objects that are related to their own selves in particular ways, but both also occasionally achieve a nonhorizontal overview of an area. This more vertical kind of bird's-eye perspective is infrequent but important, and the chapter explores the conditions that made it possible and why Margery and Mandeville nevertheless return to horizontal apprehensions of space.

Nearly all scientific discussions of space in the late medieval era have to do with motion. This is due to Aristotle's influence as well as challenges to, and divergences from, his ideas. The fifth chapter focuses on this important quality. Motion was central to Aristotle's idea of space in his *Physics*, and motion was crucial to nearly all of his ideas about physical objects and about nature more generally. The *Physics* seeks to discover the principles of nature—change, infinity, place, void, and time—and motion is key in the idea of change and is the subject of much of the work. Along with other writings, the reappearance of the *Physics* in European universities and elsewhere led to a reorganization of academic disciplines, including the establishment of the faculty of arts and sciences as we know it, a "near frenzy to measure everything imaginable," and fundamental alterations in how people thought about the world.[26] By focusing on movement in Aristotle and (mainly) fourteenth-century science, I hope to turn discussion away from another important but possibly overemphasized topic in histories of medieval motion and location, namely, natural place. It is well known that the late medieval era inherited the Aristotelian theory that objects are either at rest in their natural places or not; when they are not, they have a natural inclination to move to their places. A violent or artificial force keeps an object from resting in its *locus naturalis*. Natural place therefore implied a spatial hierarchy in which it was more fitting for an object to remain there rather than be (temporarily) elsewhere. Dante and others employ these ideas, which received powerful support from theology.

As chapter 5 shows, later medieval mechanics and literature, however, tell an additional story in which new ideas about motion developed beyond Aristotle, ideas that de-emphasize the concept of natural place. In the fourteenth century, Thomas Bradwardine at Oxford University formulated a new theorem about velocity that drew attention to rate of motion at any point in time or in any place. Contemporary with his ideas, the concept of *impetus* appears, a term Jean Buridan coined to denote the way an object can embody motion rather than be drawn to a final point. This is also the time of the Merton School of Calculators at Oxford, who developed new methods for measuring motion that also focus on points within the transit of a mobile entity. And there is also the French philosopher Nicole Oresme, who is credited with being the first to graph motion (I discuss his work in greater depth in a later chapter). My claim is that these new concepts suggest that an object was not solely or significantly oriented toward one natural place either of origin or of destination, but that each location in its existence could be as significant as any other. Motion became nonteleological and, like space itself, nonhierarchical in its organization of points or moments in transit.

The conscience, war, council, the soul, the flesh, blood, advice, the sea, flames, weights, thoughts, woods, the sun, hatred, people, stars, destiny, the heart, questions, reason, sight, and time—these objects and others all move in Chaucer's writings. Chapter 6 addresses the subject of motion in literature in light of the evidence from both traditional Aristotelian mechanics and fourteenth-century developments about key elements of motion. Much of Middle English poetry and prose is traditional and even old-fashioned in following the powerful ideas about natural place, an object's inclination to stay still, and so on. But the new science and arts of mechanics also had an impact on literature so that some works register how people were thinking in new ways about movement through space. I begin the chapter with an excursus on Robert Henryson's *Orpheus and Eurydice* from the second half of the fifteenth century to examine how the poem depicts Orpheus's flight as he searches for Eurydice across the earth, in the heavens, and down in hell. The main focus of the chapter, however, is on the structure and themes of Chaucer's *House of Fame*. I read the poem as centrally concerned with a dreamer, "Geffrey," who does not know where he is. As the philosophy tells us, if one does not know where one is, one also does not know oneself, and the *House of Fame* describes the narrator's search for a purpose or at least to know which questions to ask. Beyond the well-discussed topic of the motion of sound in book 2 of the poem, Chaucer's dream vision seems elsewhere to register fourteenth-century debates about motion. I address the evidence of Chaucer's knowledge of those innovations along the way, but my examination of the poem focuses more on the fact that, while the dreamer-narrator seems to embody his movement through the spaces he encounters, the cause of his motion is obscured and, consequently, his motion has no direction that can answer his question about where he is.

Such a turning away from an end point in motion and toward spatial equivalences applies to a single mobile body, but what about more than one? What spatial relationships attain between one object and another? Chapter 7 analyzes late medieval scientific developments in thinking about physical relationships between distinct bodies. Discussions of *propinquitas* occurred in science within the context of analyses of motion, given motion's central role in mechanics, but the topic was also addressed in terms of what was called the "configuration of qualities" or the "uniformity and difformity of intensions." A particular focus of these subjects was on how the distance between attributes of proximate bodies is causal in the ability of one entity to effect changes in another. The chapter principally considers Oresme's mid-fourteenth-century *Tractatus de configurationibus qualitatum et motuum*, his book-length investigation of physical affectivity in which he graphs motion and a treatise that describes a being's "qualities" as arranged in *configurationes* and *proportiones*.[27] Oresme developed an almost flat ontological view of the world in which everything—animate and inanimate creatures, elements such as earth and fire, and intellectual and emotional faculties like the imagination and feelings—is made up of qualities that interact with one another in measurable ways so long as they are proximate. It is a kind of mechanical, or rather, dynamic, view of the world that developed at the same time as the invention of the clock, the astrolabe, the compass, and devices in optical science. Of key interest is the distance between objects as well as the nature of those objects' qualities with the caveat that they remain distinct from one another. Were two entities to merge, they would become one, and the distances and enlivening proxemics would disappear. Oresme extends his systematic discussion of proximate entities to psychosomatic and social entities, and he develops a picture of a world of "intensities" that interact in complex arrangements. I delve into his ideas, and I partly examine them in light of ideas about affectivity.

Chapter 8 explores the implications of those tensions in the *Legend of Good Women*, especially in what may be *the* story of proximity—Pyramus and Thisbe—in Chaucer's Legenda Tesbe Babilonie. The chapter offers a reading of the *Legend of Good Women* that identifies "distaunce" between entities as a theme running through the *Legend* in several narratives: the Prologue, the Legend of Cleopatra, Pyramus and Thisbe, and the Legend of Ariadne. Proximity, I argue, contributes to the structure of the poem as a whole on a deep level. In three parts, the chapter first looks at the *Legend* narrator's proximity to the God and Goddess of Love in the Prologue and the image of Cleopatra considering Anthony's death in the conclusion of their story in order to explore how the dynamics of proximity are established in the poem. The second section analyzes the lovers and the wall in the Legend of Thisbe and how Chaucer's version of the poem focuses attention on the wall as a joining and mediating structure. The third section looks at the remainder of the *Legend* and how nearness comes to be a threat to the women in the poem. I argue that proximate relations generate social and physical ambiguities in the poem that become progressively darker in tone.

The afterword opens up the subject of space. Where *Scribes of Space* intentionally keeps a tight focus on the essential qualities of space and moves out to nearby entities only in the last chapter, the afterword inverts that concentration and addresses the topic of ubiquity: the ability to be everywhere. As Aristotle and Aquinas state in relation to the mythical creatures of the goat-stag and sphinx, place is necessarily bound up with being. To be is to be in a place. Ubiquity is in an important sense the opposite of being in a place, and spatial profusion or pervasiveness suggests in turn that a being that is ubiquitous will necessarily have a different nature. Indeed, is it possible to be everywhere? The afterword pairs philosophy together with a reading of the spaces in Chaucer's Pardoner's Prologue and Tale. Chaucer brings real and allegorical characters with a very limited sense of space alongside characters that occupy no single place but instead many places, figures who are more than embodiments of *contemptus mundi*. He does so, I argue, to explore the difference between constricted and ubiquitous spaces, and to investigate the ramifications of ubiquity for human beings and being more generally.

A note on terminology appears necessary, particularly the term *space*. I have already registered my anachronistic use of the word *local*, and other words are potentially difficult when applied to medieval texts; *area*, for example, was not used in Middle English, and in Latin it most frequently applied to building sites. Using the word *space* remains complicated today. The situation of too many meanings and applications of spatial concepts that Henri Lefebvre critiqued in *The Production of Space* in 1974 persists, ironically in no small part because of Lefebvre's own work, which, among other texts, inspired the "spatial turn" in the humanities: "We are . . . confronted by an indefinite multitude of spaces, each one piled upon, or perhaps contained within, the next: geographical, economic, demographic, sociological, ecological, political, commercial, national, continental, global. Not to mention nature's (physical) space, the space of (energy) flows, and so on."[28] One geographer describes looking back on the spatial turn as "like walking into the aftermath of an academic explosion. What had once been a reasonably coherent body of thought, grounded in phenomenology and mostly the concern of humanistic geographers and environmental psychologists, seems to have flown off in all directions."[29] The meanings and implications of *space* continue to be obscure despite (or because of) the contributions of Lefebvre's and others' writings, the most frequently mentioned of which are Michel Foucault's "Of Other Spaces," Pierre Bourdieu's *The Field of Cultural Production*, and Michel de Certeau's *The Practice of Everyday Life*.[30] In fact, we are at a point where geographers, social scientists such as E. V. Walter, philosophers such as Edward Casey, and others have rejected the term *space* and instead opted for the term *place*, but such an attempt is complicated because Aristotle's word *topos* (τόπου) does not allow a distinction between the two words; I use them both.[31]

A further complication that coincided historically with the spatial turn lies in social constructivist arguments about space. The geographer Doreen Massey has described how space "was seen only as an outcome; geographical distributions as only the *results* of social processes" and "geographers simply mapping the outcomes of processes studied in other disciplines."[32] She points to "a kind of double usage" in the meaning of the word, "where space is both the great 'out there' and the term of choice for characterisations of representation, or of ideological closure."[33] Instead, in her work, Massey has selected a composite delineation in which space "includes distance, and differences in the measurement, connotations and appreciation of distance. It includes movement. It includes geographical differentiation, the notion of place and specificity, and of differences between places. And it includes the symbolism and meaning which in different societies, and in different parts of given societies, attach to all of these things."[34] I follow her in part with an "inclusive" approach rather than seeing space only as a product of ideology, but all the while I attempt to keep the focus on the essential qualities of space in the sense that Malpas and other philosophers emphasize. In many respects, the meaning of space I propose coincides with that of psychologist James J. Gibson, to whose works I will have recourse throughout this study. Gibson describes how the "basic orienting system" of terrestrial animals depends on "the detection of the stable, permanent *framework* of the environment," and he also clarifies that "[t]his is sometimes called the perception of 'space,' but that term implies something abstract and intellectual, whereas what is meant is something concrete and primitive—a dim, underlying, and ceaseless awareness of what is permanent in the world."[35] I furthermore try to avoid metaphorical uses of space, as in what I call "the space of" or "the place of" studies of entities that are not really spatial at all. Another part of my approach has been to look not only *at* the late Middle Ages through the lenses of current theoretical, philosophical, and literary ideas but also *to* the period's own scientific ways of thinking and its literature in order to respond with sensitivity to their registrations of space's complexities.

Finally, a note about historical parameters and periodization. It is difficult to tell how long the period of scientific and social innovation about space lasted into and/or beyond the fifteenth century. The broader scope of this study to include technical and literary evidence beyond the scholastic material presents additional challenges to setting parameters in the history of space. Although scholarship has answered some questions about periodization, exceptions continue. Earlier histories often characterized the sixteenth and seventeenth centuries as a period in which there was "the birth of a new physics" in contrast to the preceding centuries, and some histories continue these generalizations about the Middle Ages.[36] Also, often but not completely gone from art historical studies that address space are arguments showing the influence of prominent works such as Samuel Edgerton's 1975 *The Renaissance Rediscovery of Linear Perspective*. His acclaimed work posited a clear medieval-modern divide in which Renaissance linear perspective established a fundamental alteration in thinking and art practice, and "a definitive victory over medieval parochialism and superstition."[37]

Generalizations about a great change from late medieval to early modern thinking about space in science, geography, art history, and other disciplines have largely passed away in thoughtful study. More careful contextual analyses of shorter periods of time, more specific locales, and particular disciplines or areas of cultural production reveal occurrences of incremental change, some abrupt departures, and several continuities over time and period divisions. Perhaps it is not necessary today to keep chipping away at sharp differences between "medieval" and "renaissance" characteristics. After all, criticism as early as Pierre Duhem's foundational 1909 *Le système du monde* described sixteenth-century scientists as "seized by a strange delusion" (*saisis d'une étrange illusion*) that what they did took place "on a terrain where nothing else was standing" (*sur un terrain où rien n'était plus debout*). In fact, as Duhem (who began the modern study of medieval science) said, scientists in the sixteenth century were often *continuateurs* and sometimes *plagiaires*, or there was "a long series of partial transformations, each claiming only to retouch or to enlarge some piece of the edifice without change to the whole" (*une longue suite de transformations partielles, dont chacune prétendait seulement retoucher ou agrandir quelque pièce de l'édifice sans rien changer à l'ensemble*).[38] Linda Voigts has argued similarly that print culture of scientific books in late fifteenth-century Britain, rather than causing an abrupt revolution in thinking about spatial sciences, "had a retardative influence on the development of scientific thought," and "because of the conservative and popularising nature of publication in the first century and a half of printing, innovation was unlikely to find an audience via the book."[39] Geography is likewise a field of developing ideas about which it is difficult to pronounce large epochal change despite intermittent claims to the contrary. I discuss Claudius Ptolemy's *Geography* in the first chapter, and Patrick Gautier Dalché argues that its new translation directly from the Greek language in Florence at the end of the fourteenth century, despite including methodology for projecting a sphere on a flat surface as well as the use of coordinates, had the effect of satisfying humanist "[p]hilological and topographical curiosity" rather than the humanists seeing it as a "scientific treatise on cartography and geometry/optics."[40] Instead of the late Middle Ages being portrayed, as Gautier Dalché disclaims elsewhere, as a time of *"non objective"* and *"'théologique' ou 'moralisée'"* geography with a lack of knowledge of Ptolemy, Ptolemaic and other geographical ideas were prominent, reasonably widespread, and purposefully examined.[41]

Scribes of Space therefore joins with other medieval scholarship that has changed perceptions of the Middle Ages as a monolithic and benighted age. A more important intention is to highlight the particular and dynamic spatial ideas and characteristics of the time and, in drawing attention to the writers and thinkers who engaged with and developed them, to foster more interest in the fascinating and valuable scientific and mechanical tracts themselves. Fortunately, the work of Joel Kaye, Gillian Beer, Charlotte Sleigh, Margareth Hagen, Janine Rogers, Kellie Robertson, and others has already suggested the benefits for scholars in the humanities to study science.[42] Likewise, it is hoped that the following discussion will encourage historians of science and technology, geographers, and others to consider literature as just as deeply involved with key spatial questions. My larger goal is to contribute to ongoing analysis of space in literature, history, geography, and physics across time.

CHAPTER 1

Local Space, Edges, and Contents
Chorography and Late Medieval English Maps

It is difficult to think of how local space, that is, the very definition and extent of what is considered a local area, might change over the centuries. After all, what is local today? How large is a local area? What defines its parameters, and what are its characteristics? How could those fundamentals change over time? This chapter considers what was local in late medieval England and the paradigms according to which the local was defined. Part of the challenge is that historical alterations in space are more apparent at the larger scales, a geocentric versus heliocentric view of the cosmos being the starkest example. The ease of perceiving how people represented larger spaces is perhaps one explanation for the great many studies of universal diagrams, *mappaemundi*, and nationalist discourses in theology, geography, history, and literature. Smaller scales are more everyday, they are less often studied scientifically, and they are generally taken for granted. The evidence I examine indicates that large-scale perspectives were, however, not the only—and perhaps not the principal—ways that people recognized space. Indeed, even today we occasionally perceive the whole earth and think globally in various ways, but we more frequently encounter and experience the areas closest to us. It is possible, in fact, that people still do not think of universal or global spaces in any conscious or thorough fashion except in the physical sciences. Yet local space is at least as significant in people's understandings of physical area, in culture, and in history as space at a universal, cosmological, global, or protonational scale. Moreover, it would be a mistake to think that local space is somehow free from the vicissitudes of history. The aim of this chapter, therefore, is to bring to light the features of local space in late medieval culture. Readers will find some continuities between late medieval perceptions of local area and current ones, but they will also discover differences in how local space was presented and understood.

If we were trying to discern what people thought was local today, what information and signs would be representative? Would it include the common routes people use, an area that could fit on a screen, municipal definitions, a series of landmarks? How would the evidence be similar or different in late medieval Britain? To answer these questions about late medieval local space, I present two sources: local maps of smaller areas and what Claudius Ptolemy calls *chorography*, the detailed presentation of regional geographical features. In this chapter, I argue that surviving local maps from the time period are, in a sense, visual experiments about understanding and presenting space. Because cartographical practices were neither institutionally formalized nor standardized, the maps are witnesses to the ways mapmakers apprehended the spaces before them and worked through the challenges of forming them on a page. The maps are reproduced in R. A. Skelton and P. D. A. Harvey's *Local Maps and Plans from Medieval England*, and the editors state in their introduction that—in contrast to maps of all of England or the British Isles (such as the Matthew Paris, Gough, or John Hardyng maps), which they describe as the "products of scholarship or of philosophical speculation"—the maps in their collection appear from the evidence to be "entirely of areas known personally to their authors."[1] The evidence suggests that the mapmakers combined personal perception and more official kinds of knowledge about local areas; many of the maps were made for judicial ends and have accompanying documentation of different types. Their methods were, furthermore, embedded within social expectations; the maps were not made for their makers' own delight nor to solve methodological problems of cartographic presentation but instead were, the evidence suggests, made for others to help them understand an area, in effect to express how people already think about a local space. People's shared (conscious and unconscious) perceptions of spatial properties shaped the mapmakers' approaches and their maps. The diversity and the mundane, heuristic, and quotidian nature of the maps are their strength as evidence because they are more intimately connected to everyday apprehensions of space. The reciprocity among mapmaker, audience, and map is a hermeneutics of local space, a way of understanding local area.

Less well known than *mappaemundi*, Macrobian zonal maps, and maps of all of England or Britain, the English maps of smaller areas require introduction.

Approximately thirty-five local maps survive from the late medieval period in England, each showing some variety in land areas. One pair of Canterbury Cathedral plans depicts an area that is about a square mile, but a more typical late fifteenth-century map, such as one that survives from Lincolnshire, shows approximately fifteen square miles, while a mid-fifteenth-century map of the village of Boarstall in Buckinghamshire displays approximately thirty square miles. The largest areas depicted on these kinds of maps are of Sherwood Forest and the Isle of Thanet, both about 150 square miles.[2] While it may seem solipsistic or redundant for a discussion of local space to select what have, after all, been called "local maps," the following observations do not take map scale or objective scale (the size of a given area on the ground) at face value. Instead, the analysis includes examination of the scalar choices of their makers to discern what they were thinking of in terms of local area. After all, as the geographer Denis Cosgrove points out, "Enlarging or reducing the space generated and occupied by phenomena alters their form, their significance, their relations of meaning with other phenomena. Scale selection and manipulation is thus a powerfully imaginative and generative act which at once records and sets in train chains of meaning and association in an active process of knowing."[3]

Since there does not appear to have been established training in mapmaking, part of the interest in the maps concerns disciplinary boundaries in the sense that they were produced between the scholastic learning of the quadrivium and practical use. They would appear to reside in the field of what was called the "mechanical arts," transitional practices between the world within and outside the university. Hugh of St. Victor's twelfth-century *Didascalicon*, his influential synthesis of all types of learning that sought to illuminate what to read and why to read it, does not include cartography. Maps, however, might arguably belong within the discipline of geometry or "earth-measure" (*mensura terrae*) in the quadrivium. Planimetry and altimetry are part of geometry in Hugh's classification, and all are within the liberal arts.[4] But because of their practical origins and, presumably on occasion, application, it is nevertheless also possible to see mapmaking as worthy of inclusion within his classification of the *artes mechanicae*. As mentioned in the introduction, Hugh is credited with raising the mechanical arts to being a part of philosophy. Maps are arguably "mechanical" in that they are "human works" (*opera humana*), because that form of knowledge "supervises the occupations of this life" (*huius vitae actiones dispensat*).[5]

While it would be incorrect to restrict the maps to inside the walls of the university, it would be equally inaccurate to suggest that their "mechanical" or "occupational" functions exhaust the meanings of the walls, buildings, flora and fauna, roads, and rivers they depict. A reasonable amount of information about the maps is available because they are often accompanied by, and occasionally embedded in, written documentation. Some of the maps were made for practical and frequently forensic or administrative ends: to record irrigation systems, to resolve boundary disputes between one field and another, and so on. These legal and other uses undoubtedly affected the production of some of the maps, but it is not possible to know whether all were made for practical purposes; besides, their trees, field tenements, religious and secular buildings, walls, roads, and other items exceed and are otherwise only occasionally explainable in terms of legal or other applications. As mentioned, the items can come from one or a combination of sources: recall from longer ago, from more recent observation of a particular local area, or from another map or in a written source without the mapmaker having seen an actual area in person. In the first place, the maps are diverse in style. Some appear crudely schematic, while others are painterly and ornate. Certain maps appear to be imitative of book illuminations, the maps secondary to, or even parasitic on, established artistic practices. Artistic and textual conventions undoubtedly affected the translation of spatial information to a page, but the potentially more significant elements are point of view, level of detail in rendering objects, and arrangement. These appear to be the mapmakers' principal concerns, always with consideration of the audience. The emphasis in what follows therefore falls on how the mapmakers think through spatial problems in order to present them to their audiences, meaning that interest lies in between the map and the local space, in the interactions between page and place.

This chapter is structured around three topics: local space, edges, and contents. First, it considers what a local area is and how far it extended—indeed, whether it was thought of in terms of physical extension such as distance or other factors. The important work here is Ptolemy's *Geography*, which I explore in some detail along with its phenomenological implications. Second, the evidence suggests that late medieval culture defined local area to a certain extent by considering its edges: objective ones such as walls, tenement limits, or rivers, and cartographic ones such as the margins of the map page, diagrammatic edges, or other forms of drawn boundary. I therefore examine how edges on the ground and on the page contribute to creating different senses of a bounded local space, the nature of those edges, and their effects. I do so by studying the pair of Canterbury Cathedral maps and a map of Cliffe in Kent on the banks of the Thames. Third, the chapter addresses the contents of local spaces. Asking which things—flora, fauna, manufactured structures, and so on—were considered part of a local space allows us not simply to catalog. The maps suggest the importance of considering whether items are individuated or whether they are drawn systematically according to certain types. Also, the relations among items belonging to a local area contribute to what may be called the texture of local space, the warp and weft of interactions among local objects. The arrangement and presentation of the open fields, walls, roads, forests, and other features suggest the way viewers interact with the maps. The maps in this section of the chapter are of Sherwood Forest and of Inclesmoor in Yorkshire. The evidence about a local area's contents also indicates that the temporality of a space's features was an important consideration for late medieval culture, showing how the objects belonging to an area signify present realities and historical events. One further map of the Isle of Thanet at the eastern tip of Kent provides the evidence here because it reveals with particular clarity the temporal features of a local area's contents.

Ptolemy's *Geography* and Homogeneous versus Heterogeneous Space

What was a local area? What defined it in terms of extent, edge, and ways of thinking about its contents? And what is a useful framework to begin considering local space in these terms? Historians have not come to a consensus on the most basic question about what constituted a local area. In terms of sheer distance, Dick Harrison, in *Medieval Space: The Extent of Microspatial Knowledge in Western Europe during the Middle Ages*, points out that, among the *annalistes*, Marc Bloch proposed that "medieval human beings lived isolated lives in their own villages, rarely if ever travelling to other places" while Jacques Le Goff "maintained the opposite—that medieval men and women were extremely mobile."[6] Harrison's point is that the size of a local space "varies according to cultural situation, social structure, age, sex, personal habits, etc." He employed historical methods to plot how far people habitually traveled. It turns out this was quite far; for example, his study of Somerset in the first half of the fourteenth century shows people regularly journeyed twenty-five or thirty miles.[7] A focus on miles and on human mobility, however, is not the best place to begin. Motion was a complex topic at this point in history, one that extended beyond the human sphere and that was addressed explicitly in scientific writings; I return to it in a later chapter, but it seems to lead away from local space rather than answering what it was.

Geographical writings directly address the topic of local space. Ptolemy's *Geography* is commonly associated with the transition from medieval to early modern ideas about cartography and space more generally, but the late medieval era was quite aware of ideas from the *Geography*. Keith Lilley summarizes research demonstrating that "Ptolemy was neither 'lost' to the West in the Middle Ages nor was the Renaissance characterized by a switch from 'medieval' to 'modern' modes of cartographic and geographic representation."[8] The *Geography* was not itself available in the Latin West until 1407, but evidence shows its broad influence in medieval geographical writing and in maps before then. As mentioned, I will be concentrating on Ptolemy's ideas about chorography. Jesse Simon writes that the term *chorography* "may have disappeared completely [in the West] sometime after the ninth century. The disappearance of the word, however, should not imply the disappearance of the practice," and other historians have seen influences of chorographical practices in medieval writing and *mappaemundi*.[9] When the *Geography* became available, it was immediately popular; in England, it was studied at Oxford and Cambridge in the fifteenth century. The more familiar *Almagest*—the *Mathematical Systematic Treatise*, as Ptolemy called it until it was translated into Arabic under the title of *The Greatest*—preceded the *Geography* in Ptolemy's oeuvre.[10] As is well documented, knowledge of Ptolemy's *Almagest* was transmitted earlier in the West via Martianus Capella and was known within and outside the universities from the twelfth century on with Gerard of Cremona's translation of the Arabic. In English writings, Ranulf Higden in John Trevisa's translation of the *Polychronicon* refers to the *Almagest*, Chaucer refers to Ptolemy and his *Almagest* six times, Gower notes it, and Stephen Scrope's *Dicts and Sayings of Philosophers* (later printed by Caxton) draws on it.[11] Both Ptolemaic works are on the arrangement of the world, but the *Almagest* is an astronomical work on the cosmos that lays out the order of the spheres, stars, and planets down to earth at the immovable center, while the *Geography* addresses only the earth, its latitudes, and its regions and individual countries.

Ptolemy begins his *Geography* with categorical distinctions among geography, chorography, and topography. He is interested in geography as a method of representing "the entire known part of the world," but his understandings of *chōrographia* and *topographia*, while less known today, are of equal significance for insights into spatial thinking and representation in the late Middle Ages.[12] A *khôros* is literally a room, although its closest modern equivalent in usage would be *land*, as in "a piece of land" or "my land" (referring to open land or owned land). A *khôros* is a space not in an unplaced sense; it is always somewhere, emplaced within a location.[13] Since the Enlightenment, the three fields of study—geography, chorography, topography—came to be associated with disciplines based only on the size of the area studied: geography as a discipline that studies the whole earth, chorography a region, and topography the landscape within a region. These appear to be later Enlightenment distinctions, which drew on early modern readings of the *Geography* that emphasize disciplinary particularities and also privilege geographical isometric projections of the heavens and of the earth's coordinates on the flat surface of a map.[14] The late Middle Ages, on the other hand, was not interested in disciplinary and quantitative distinctions, nor did it concern itself in any clear fashion with the *Geography*'s presentation of graphic projection. Instead, it engaged more with the qualities of each of the different but overlapping ways of seeing and representing areas of any size that geography, chorography, and topography encourage.

Ptolemy in the *Geography* distinguishes between the kind of map one might produce if one is attempting to depict the whole of the *oikoumene*, "the entire world," and a chorographical map of a region.[15] Even though the two practices share some methodologies, Ptolemy notices a problem with the global cartography of his day—namely, that some mapmakers "distort both the measures and the shape of the countries," of landforms, of the sea, and of the whole *oikoumene* to fit all of the known world within the space of a page. This tendency he attributes in part to the greater amount of information about some parts of the earth (e.g., Europe) versus other areas (e.g., Asia and Africa). A map of the world, he notices, may enlarge Europe because more is known about it and therefore more material needs to be fitted into the space, whereas less is known about Asia and Africa, so their spaces are made smaller. Ptolemy is concerned with correct proportionality, and maps of smaller regions can more easily avoid problems of distortion, he says, because the mapmaker is more liberated to vary the parameters of the map, making a map of a smaller area if there is a lot of information about that region or larger if there is little to record and the features are farther apart. The continent of Europe can be subdivided into secondary areas, for example, each with a map, while lesser-known places, such as the whole of Africa, might require only one or two maps. Ptolemy is concerned that each map should be self-consistent in terms of scale and drawn using mathematics.[16] Simon posits that Ptolemy's response to other cartographers of his day is likely registering the fact that chorography already "represented a point of divergence in the geographic tradition. Instead of populating an immutable shape with known contents, the contents may have started to dictate the shape of the land."[17]

Medieval maps of all sized areas are not particularly concerned with constant scale. The important point that is retained in terms of local maps is that the size of the map and the size of the land area presented are not bound to firmly established preexisting conventions; the map size and the land size can vary. This makes local maps different from maps of the heavens, *mappaemundi*, and maps of Britain because the chosen land area and the map itself are more adaptable to immediate circumstances. Moreover, the items on a map of a smaller region do not derive from prior authorities as much as global and national maps. Macrobian zonal maps show the same frozen, temperate, and torrid climes, and so-called *mappaemundi* (mainly the northern hemisphere) are built out of shifting but generally received knowledge that was quite established by the late Middle Ages. The organization of a map of the *oikoumene* is usually oriented East in the Latin tradition with Jerusalem at the center and the Mediterranean prominent. The sources of a *mappamundi* are also established: Homer, Strabo, and other ancient works; the Bible and biblical commentaries; crusade, pilgrimage, and other travel narratives; and so on. World maps may incorporate changing knowledge but not to the extent of local maps. As Ptolemy says, maps of the world need to depict "the more noteworthy things."[18] A local map is, conversely, more heuristic, less subject to repetition and a long history of authorities, all of which mean more possibilities in terms of map sizes and the items that could be included.

Another way that Ptolemy describes the goal of chorography versus that of geography is that the former involves creating "an impression of a part, as when one makes an image of just an ear or an eye," whereas the latter might depict "the whole body" or "a general view, analogous to making a portrait of a whole head."[19] He does not really follow up on the part-to-whole ratio of choro- versus geo-graphy, but a number of implications of his observation are possible if one meditates on how medieval audiences probably considered local maps and local areas in relation to larger areas. A smaller area on a map or on the ground might have been thought of as emblematic of a larger area, or, vice versa, a whole world or larger area could potentially have been thought of as condensed into a smaller one. Another possibility is that a smaller area could be judged by its conformity to, or deviance from, a larger geographical scheme—for instance, a region of the earth appearing limited in its resources in contrast to the bounty of the whole planet. Or perhaps the relation of part to whole could enable what geographers describe as "scale jumping," in which a social or other phenomenon jumps from a small sphere of influence to a much larger one or vice versa. The power of scale jumping is that small events can affect large ones, or a small or large phenomenon is exposed as limited because of its scale.[20] The site of the Crucifixion would seem to be the most significant in terms of scale jumping in the Christian West during the Middle Ages in that its locationally specific events came to stand for a global geography that will last until Judgment. *Mappaemundi* such as the Ebstorf map have the Crucifixion in the center and Jesus's head, hands, and feet extending out to the top, sides, and bottom of the whole earth.

Less speculation is needed when considering what Ptolemy has to say about the nature of the landforms and manmade objects on a local map, the characteristics of which again have implications for how a viewer interacts with a local space and a local map. He explains that *chōrographia* "sets out the individual localities, each one independently and by itself, registering practically everything down to the least thing therein (for example, harbors, towns, districts, branches of principal rivers, and so on)" in contrast to the presentation of the entire "known world as a single and continuous entity."[21] The individuated features on maps—rivers and streams, cities and towns, major roads and pathways, large and small islands, and so on—can encourage an immersive experience where the viewer is, as it were, motivated to enter the space of the map and move incrementally, sometimes along a road or other way and sometimes in another direction, from one feature to another. Daniel Connolly and others have explored how Matthew Paris's pilgrimage itineraries and map of Britain encourage an "imagined" and also haptic and multisensual journey for the audience of monks who could not actually travel but could virtually do so. They could trace their way from town to town and shrine to holy site on each segment of Paris's itinerary maps, considering and, in a way, experiencing the travel and each location.[22] Ptolemy says chorography can "present together even the most minute features."[23] To experience local space is often to interact with its individual items visually, aurally, and tactilely, as well as imaginatively.

A map of a local space often reacts to the location more directly than a map of a larger area. Now, as mentioned, in some cases, a mapmaker might produce a map from documents alone. The evidence of the local maps that survive does not always tell us whether a mapmaker was from an area or had ever visited an area, let alone surveyed an area in some way. When Skelton and Harvey say the local areas were "entirely . . . known personally to their authors," they may have meant areas that the mapmakers knew *of* from documentary sources but had not encountered personally.[24] Nevertheless, we know of some mapmakers who were familiar with a local area in person, and a local space and map could therefore be compared back and forth in the mapmaker's mind, unlike a map of the whole realm or the known world. Perhaps with a global map a person might compare a familiar place or region with a *mappamundi*'s presentation of that particular part of the earth, but this seems a very limited way to consider this kind of map; furthermore, *mappaemundi* do not facilitate extensive comparison because their individual areas and items are often not very detailed. A local map and its local area allow for more comparison and contrast possibly on the part of not only the mapmaker but also perhaps its audiences, who might contemplate a map and an area they know.

The local map, therefore, implies a different hermeneutics of both a map and a space in which veracity could come into play. Ptolemy says that the chorographical map "attends everywhere to likeness, and not so much to proportional placements," but my findings are that the opposite is true with the local maps.[25] The veracity of a local map lies in the fact that items on it are in the same relative position as the objects on the ground rather than accurate representations of land areas, villages, trees, crossroads, and so on, which would be unusual (though not unheard of) among medieval illuminations. Some local maps show that there has been a certain amount of measurement of the ground and/or of a page without too much emphasis on accurate or proportional scale but nevertheless implying a rough match in terms of position between objects on the ground and on a map. This is not true of all local maps, but the hermeneutics of some can include realism in the sense of correspondence between a map's placement of landforms and actual landforms in addition to, or in place of, conformity to documentary knowledge. Authority is put aside for veracity. This is in part because the knowledge a local map presents is not as subject to hierarchies of authority as the knowledge on maps of the heavens, a *mappamundi*, or a map of Britain. John Moffitt contrasts medieval global maps with more local ones on this point when he writes, "On the one hand, there are the timeless and immutably fixed 'topographical' features of the planet, which are distantly viewed from above. In contrast, there are, simultaneously, all of those non-flattened, 'normally' viewed chorographic details that provide pictorial information about the particular characteristics of various regions."[26]

Finally, Ptolemy states that "regional cartography deals above all with the qualities rather than the quantities of the things that it sets down."[27] Ptolemy's point is again to contrast chorographical cartography with the world map, in which the former has the potential to individuate each location. Where a *mappamundi* might depict all cities, for instance, using the same building forms in a variety of numbers and combinations, local maps have the ability to individuate the depiction of cities or towns. Michael Curry and Jesse Simon characterize Ptolemy's chorography as a way of thinking about the physical world in terms of "relationship[s] between events and the places and times at which they have occurred. . . . The surface of the earth was not seen as a surface of infinite variation." Instead, "the world itself—terrestrial and celestial—acted as what one today might think of as a kind of information storage device, one that operated via what amounted to a set of signs or symbols."[28]

The characteristics of the objects on a local map have several implications, ones that draw attention to important degrees of distinction among chorographical maps. One implication of a focus on the "qualities" of individual items on a map is the more "immersive" nature of the map I mentioned earlier in that these qualities encourage viewers of some chorographic maps to engage more intensely with singular objects, such as being drawn to a particular village or following along the route of an individual road or river. But this is not true of all local maps, and indeed we see two general tendencies in the maps that survive. Some local maps do not individuate the characteristics of arable areas, trees, ditches, drains, and buildings but instead employ less detailed and more symbolic or generic signs for these things. The result is a map that creates an impression of consistency across the items within an area. The viewer's experience is therefore less absorbed in terms of the individual features; he or she can look at the whole map area "in a single glance," seeing with a certain sense of mastery the general patterns and relations among the objects.[29] The conclusion the local maps will reveal is that these differences between what might be called an individuating map and a systematic one are of degree rather than kind. All maps are to a certain extent systematic in terms of the symbols they use, but their categorizing aspect is more apparent and significant in some maps than others. The important point is that the difference affects the viewers' experiences in significant ways.

The differences between individuating maps and more systematic ones are significant because careful consideration helps further to see how *chorographia* and local maps can reveal the characteristics of local space itself. Sir Isaac Newton, the historians of cartography Harley and Woodward, and the philosopher Maurice Merleau-Ponty all draw a contrast between what they call "heterogeneous" and "homogeneous" spaces. The challenge in perceiving the characteristics of local space in the late medieval era is compounded by our living in a post-Cartesian, post-Newtonian world and at a time in which geography tends to follow established cartographical conventions. Newton's notion of "absolute" space as distinguished from "relative" space contributes significantly to our understanding of space today. Newton writes in the *Principia* that space is "absolute" in that it exists "without reference to anything external" and is "homogeneous and immovable" (*Spatium absolutum natura sua absq[ue]; relatione ad externum quodvis semper manet similare et immobile*). The opposite is merely "relative space . . . determined by our senses from the situation of the space with respect to bodies" (*relativum . . . quae a sensibus nostris per situm suum ad corpora definitur*).[30] We continue to follow Newton today in thinking at times of space as the same everywhere, independent of ourselves, and not "relative."

In the field of cartography, modern conventions of mapmaking lead to prejudice against premodern representations of local (and larger) spaces, a distinction that has also been described using the terms "homogeneity" and "heterogeneity." Harley and Woodward's conclusion to their *History of Cartography* identifies two kinds of maps existing side by side in the late Middle

Ages. One type appears to point forward in time to postmedieval cartographical practices in which space is "homogeneous," where "[e]ach point on the map is, in theory at least, accorded identical importance." The other type of map strikes the modern viewer as antique and "inaccurate." The latter is what they call the "heterogeneous" map, in which "particular areas were given significantly different weight and map space on a single map, to the extent that any notion of a uniformly scaled image is absent. . . . Here, certain parts of the map may be endowed with particular meaning and importance."[31] "Heterogeneity" in terms of variable scale and the artistic rendering of objects on maps is more prevalent in the Middle Ages (although we will have occasion to compare pairs of maps that exhibit both styles). It is as valid a method of representing space as the prevailing homogenizing cartographic projections with which modern viewers are most familiar, but it remains difficult to appreciate the older styles of maps on their own terms because they do not reflect current practices and seem objectively incorrect.

Philosophy, particularly phenomenology, is helpful in considering different kinds of spaces because it asks fundamental questions about them and so challenges us to put aside preconceptions about how space can be organized. Merleau-Ponty uses the terminology of "homogeneous" space as opposed to heterogeneous "oriented" space. He seeks a return to appreciating and understanding the latter manner of perception instead of the former, which he describes as a second-order way of seeing that is contingent on the first. He says, "one must . . . reject as an abstraction any analysis of bodily space which takes account only of figures and points." (He describes objects in geometry, but one could substitute modern maps.) He continues: "The truth is that homogeneous space can convey the meaning of orientated space only because it is from the latter that it has received that meaning. . . . The multiplicity of points or 'heres' can in the nature of things be constituted only by a chain of experiences in which on each occasion one and no more of them is presented as an object, and which is itself built up in the heart of this space. And finally, far from my body's being for me no more than a fragment of space, there would be no space at all for me if I had no body."[32]

Later, Merleau-Ponty reiterates that "every conceivable being is related either directly or indirectly to the perceived world, and since the perceived world is grasped only in terms of direction, we cannot dissociate being from orientated being. . . . [W]e cast anchor in some 'setting' which is offered to us." Further on he adds that "the system of experience is not arrayed before me as if I were God, it is lived by me from a certain point of view; I am not the spectator, I am involved."[33] Jeff Malpas and other philosophers follow Merleau-Ponty in paying attention to relative, inconsistent, and embodied space.[34] It is the challenge of the current chapter to recover the characteristics of premodern local space, and phenomenology is useful because it asks questions that urge a reconsideration of how people experience space. In terms of the distinctions elaborated by Newton, modern cartography, and phenomenology, all three contrasts between homogeneity and heterogeneity—two post-Enlightenment and one postmodern—signal a need to return to underlying ideas. They especially suggest that it is important to be aware of distinctions between local, experiential presentations of space and abstracting ones.

Local Edges

In contrast to space in the Newtonian and modern scientific sense, a chorographical space has edges. The philosopher Edward Casey proposes three aspects of platial edges: they "delimit," "generate," and "surprise." Edges delimit in that "[a] place comes to its own edge. If it did not come to *some* edge, it would not count as a place at all. Places require edges," and "each edge serves to contain, to define, to arrest movement of the eye if not the foot." Edges also less intuitively "generate" in the sense that an "edge is at the same time the start of another place." Edges "are in effect places of origin." Casey notes that, of course, "it is often the same edge that serves in the two roles." They have a "Janusian character." Finally, edges can also "surprise" in the sense that they "play an occlusive role: the outer edge . . . prevents us from seeing what is to come on the other side. . . . When these things or happenings arise, they do so unexpectedly to some significant degree."[35]

I quote from Casey's analysis of spatial edges to think about how a local space is defined and determined (*de-finire*, to end; *de-terminare*, to end). The English local maps have a variety of edges. Some appear *on* maps: walls, rivers, or cartographical symbols that do not appear on the ground, such as property lines or abstract lines drawn by the mapmaker. Another edge is that *of* the map, that is, the margin of the page or pages. The local maps show some variety in that a page is sometimes a single leaf and sometimes a bifolium, and the size and material of the page can vary as well. A further significant factor is the space on the page between a wall, a river, or a boundary mark and the edge of the page, a gap that becomes more complex in the absence of a defining border on the page; several of the local maps lack a wall or other structure that outlines the space within the page, and they also lack lines or other symbols from the mapmaker. The trees, fields, roads, streams, property lines, and so on presented on the map simply run out to the edge of the page or end in an empty space as they approach the sides. In the absence of physical or other boundaries drawn on a map, the interrelations among the "internal" features on a map gain in significance because the relationships rather than an edge contribute to a space's definition. Pressure comes to rest on the ways the objects interact with each other and form a coherent space.

Consider a pair of maps of Canterbury Cathedral's grounds, the earliest surviving diagrams of their kind in England from about 1160 (see figures 1 and 2). They are chorographical in that they depict a smaller defined space. The evidence explains that they register buildings and other structures because of innovations in waterworks. The images of the buildings, pipelines, plants, and so on, however, do not correspond with the objects they depict in a particularly realistic fashion; the objects on the maps instead operate as signs of their functions and connections. Also, map pairs are worthy of note; two pairs of maps of the same areas (four maps total for two areas) survive from the late medieval period in England: the Canterbury Cathedral pair and another set that presents an area in Inclesmoor, Yorkshire. Because each pair shows the same area but each map is different from the other (even though contemporary), the pairs are particularly revealing of the hermeneutics of local space because they enable comparisons between the designs and the understandings of the same area's features. The pair of Canterbury maps offers the remarkable opportunity to compare and contrast different presentations of the same area, one that includes boundaries on the page and one that does not.

The two images, one a two-page spread and the other a single surviving page, currently occupy the last folios of the trilingual manuscript known as the Eadwine Psalter. The well-known illumination of the monk—Eadwine of Christ Church, Canterbury—is included nearby; evidence suggests he oversaw the book's creation.[36] The maps appear to have been drawn in the manuscript after it was bound. The first image, on folios 284v–285r, is probably now upside down, having been inverted in the seventeenth century (I have rotated it back). The pages have been trimmed slightly on the top and bottom.[37] The second image, on folio 286r, is half of what was originally a two-page spread; the remaining page was a verso page (the crease and sewing holes are visible on the right) with the other half of the image, the recto, now gone. The two complete images would have formed a pair, with both set off from other manuscript pages in their original form; they are blank on their opposite sides.

The images are what we would today call *plans*, but a distinction between a map and a plan is not one that can be read back into this time in history. Take the two-page illumination on folios 284v–285r first. The local space is a bounded area, has a largely self-consistent internal scale, and shows a coherent network of items. The edges of the space are almost completely designated while its contents are carefully laid out not in a particularly accurate

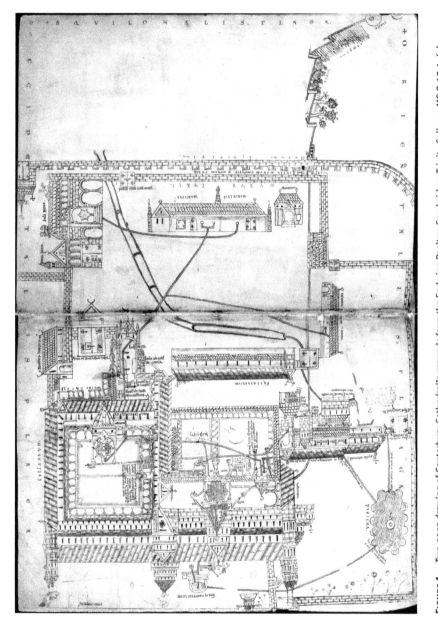

FIGURE 1. Two-page drawing of Canterbury Cathedral's grounds from the Eadwine Psalter. Cambridge, Trinity College, MS R.17.1, fols. 284v–285r. Reproduced courtesy of the Master and Fellows of Trinity College, Cambridge.

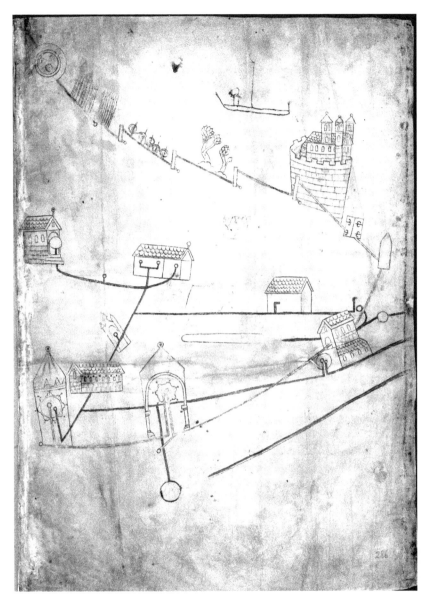

FIGURE 2. Single-page drawing of Canterbury Cathedral's grounds from the Eadwine Psalter. Cambridge, Trinity College, MS R.17.1, fol. 286r. Reproduced courtesy of the Master and Fellows of Trinity College, Cambridge.

fashion (if we compare the map with existing archaeological remains) but in a way that makes spatial sense on the page. Large, carefully placed letters border the map on three sides: "ORIENTALIS PLAGA," "OCCIDENTALIS PLAGA," and "AQUILONALIS PLAGA" with a fourth likely trimmed off. Within the page, the structures and words do not suggest any particular orientation to the map. The walls around the Canterbury Cathedral grounds instead frame in the other structures. Inside this perimeter, additional walls and the buildings parallel the outer walls. The buildings are depicted in elevation, so the whole map combines a bird's-eye view with ground-level elevations of the buildings from different points of view. Each of the buildings is proportional relative to the other structures, although again not to the actual buildings. The words within both illuminations mainly have to do with plumbing, and historical records tell us that the point of the maps was to record a recent prior's renovations to the water supply, rainwater drainage, and waste drains, which are rendered in the illumination in different colors. The only exception to the symmetry of the frames around the edges of the pages along with the interior buildings occurs in the upper left corner of the map, which has an extension of jagged lines to show the source of the cathedral's water supply. Analyses of both maps have struggled to match the objects in the two drawings with historical records and surviving structures, more often than not failing to identify even the waterworks with what we know was built or with functional plumbing.[38] Pipes go in directions that they did not go, catch tanks are insufficient for any adequate supply, and buildings are different from how they existed at the time, meaning that even though the drawings appear to memorialize the prior's innovations, they sacrifice veracity for visual coherence. The walls, pipelines, buildings, and other features instead work together on the page to give the sense of an internally consistent system.

While it is a presumption to extrapolate from one mapmaker's graphic methodology to a historical hermeneutic impression of local area, the map so far suggests that a local space is a self-contained unit, bounded and "systematic," as Robert Willis characterized it in one of the first analyses of the Canterbury plans in 1868.[39] Turning now to the second image, the remaining half of the diagram on folio 286r reinforces these impressions. The map on this folio is more clearly oriented to a static viewer as you see it because, even though there are no labels and it combines overhead with ground-level views, all the structures and other features, such as the crops, vineyard, and orchard trees in the upper left, are oriented in one direction. Unlike the edges of the two-page image, the edges of the area on the folio are not confined within symmetrically arranged walls or other structures. The three built structures on the left side of the diagram and from which, or to which, pipes attach nevertheless offer the beginning of an internal systematicity on the page. The structure at the top is a "turris" or catch-pit for initial water supply, which features a perforated drain; in the middle is an "aula nova" or guest house; and at the bottom is a washing place. The three structures are in a line and are laid out differently from how they appear on the two-page map; there, the "turris" is outside the city wall in the top left, the "aula" is inside the walls in the lower left, and the washing place is inside the cloister. The items on the single page are also rendered differently from those on the surviving two-page map in that they are significantly less proportional in size to one another. The logic behind this presentation appears to be related to the significance of the structures to the plumbing. Thus, the washing place and water tower lower down are exaggerated in size. The pipes then make their way, via the settling tanks at the top and various outlets and inlets within structures, to congregate at and near the water tower in the center of the right-hand side of the page. We can imagine that in the complete two-page layout, the pipes would perhaps splay out again, providing a loose, bow-tie-shaped symmetry to the image as a whole.[40]

A much later map from the fourteenth or early fifteenth century also shows waterworks and innovations but on a different site. The map is of a London charterhouse, and it is interesting in several respects, but I will treat it only briefly. The map is in many ways remarkably similar to the two Canterbury maps in that it shows the origins of a watercourse outside the boundaries of the charterhouse and a number of the features of the waterways inside the cloister. Some buildings are shown from an overhead perspective while others are shown from a side or raised view. William St John Hope, who reproduces and discusses the documents pertaining to the waterworks and the map, describes how the plan was likely updated several times to reflect innovations in the water supply and plumbing, thus revealing a close, if not direct, relationship among the innovations in the ground, legal records, and the plan, a striking example of a map attempting to reflect actual features on (and in) the ground with some veracity. Here too, however, the map does not always accurately represent those features.[41]

The two earlier Canterbury plans together indicate some of the initial characteristics of a local area. It is a space that can be outlined by manmade structures (and, we can assume, natural ones), such as the walls of the cathedral. In addition, or alternatively, an internal regularity can order the area. The structures on the maps are rendered quite differently on each plan, but the second, one-page map makes clear that both images are internally consistent in terms of scale and presentation. The supply, drainage, and waste systems are unrealistic and nonrepresentational, but each of the systems flows through its various structural devices in a somewhat comprehensible manner as a closed scheme. It remains to be seen whether the Canterbury plans can be said to be representative of renditions of local spaces more generally, but the evidence so far indicates at a minimum that local space is organized.

Comparison between the Canterbury plans and other maps demonstrates that a physical edge can frame in a local space, as in the walls around Canterbury on the two-page spread; several other local maps that survive are framed in this way by manmade or natural (or quasi-natural) features, or a combination of different features: roads, rivers, streams, and so on. However, as in the one-page map of Canterbury, it may not be an actual structure on the ground or a physical feature of any kind that tells where a local area ends and that provides definition of a space. In that case, internal systematicity can produce different degrees of spatial coherence (just as it does today with a subway map, for instance), but spatial integrity can become tenuous if the items on a map are too diversely represented, if there is little apparent arrangement of objects, and if the ratio between the objects on the page and open, blank space becomes too large. Consider figure 3, a map of Cliffe, Kent, from about 1400 that shows two distinct and labeled areas below the rippling waters of the Thames at the top of the page.

The top, almost square area is a larger flood zone (below the arc of the sea wall), and beneath it is a smaller rectangular marsh. A drain divides them with a labeled rectangular gate on the left and an encircled cross (likely a sluice) on the right. Two thinner, long rectangles of land parallel the central area on both sides; the one on the left has words in it, and the one on the right has a labeled building with a peaked roof. This map registers a problem with the local area's boundaries because documents suggest that it records a sudden Thames flood of the area, which had made an incursion on owned land.[42] It is therefore ironic yet logical that the two rectangles of local space that are best defined on the map are the ones under threat.

The Thames's waves, the two main rectangular areas, and the other features of the Cliffe map are like the two-page Canterbury plan with its walls bordering the area; the physical features edge the map. The Cliffe map's rectangular lands do not, however, enclose all of the page as on the two-page Canterbury map, making the Cliffe map more like the one-page Canterbury plan of the waterworks. As on that single-page Canterbury diagram, the map of Cliffe includes substantial blank space, in this case opening out at the bottom of the parchment. On the one-page Canterbury plan, the internal consistency of features helped determine a local space not bounded by manmade or natural features. The same is true on the Cliffe map, except there is less coherence. The map's symbols are consistent in that they are all a single brown color

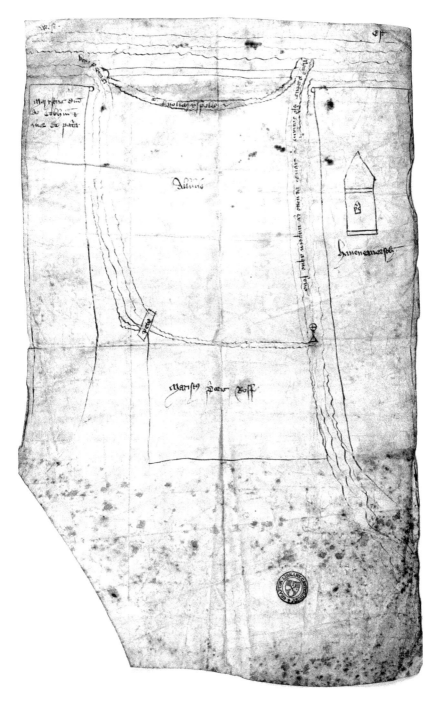

FIGURE 3. Map of Cliffe, Kent. Records of the Dean and Chapter of Canterbury Cathedral, Charta Antiqua C.295. Reproduced courtesy of the Dean and Chapter of Canterbury.

and made with the same hand in the same ink; there is little to no shading or other filling in of spaces. But blank areas on the map are prominent, and the bare, schematic nature of the map means integrity is de-emphasized. At the bottom of the map, the lines become less careful, trailing off in increasingly rough squiggles, and also become lighter. The building on the right, which might offer some definition to a side of the map, is objectively problematic; it is not identifiable, or it may be Cliffe church, which was "actually over a mile away in a different direction."[43] The openness of the page space and the schematic nature of the liminal zones provide a sense of ample spatiality.

The edges of the Cliffe map suggest other contiguous areas of an unidentified kind that are continuous with the properties of the current space. This is a different kind of edge. Casey clarifies that places have a "capacity to reach out into their environs in subtle ways while also conveying a sense of their own character and content as places. Such edges are open-textured and porous. Not only do they define or enclose a place; they extend that place into what lies around it—they take it into the circumambient space."[44] Instead of one surprise, therefore, there may be a decreasing series of distinctions or simply an end to defining characteristics and internally coordinated items until a chorographical area loses its coherence enough to then start up as another. The other locality may have similarities, of course, with the place one has just left, perhaps uncanny similarities, but it will also be different, with objects and experiences that match and mismatch what has come before. A paradox exists: a local edge may surprise gradually.

Local Contents

Turning now from the edges of local spaces to look directly at the internal features that accompany them, the pair of Canterbury maps and the single Cliffe plan suggest that a space's contents can cohere because they are organized into systems that can be standardized to a certain degree or more open. The bordering features (where they exist) and the internal systematicity moreover lead to different effects: they encourage different kinds of attention in which, in some cases, the eye follows the objects within a space, moving from one item to another, intent on how they interact; or, in other cases, one notices the way internal features unfold toward other spaces, including nearby ones. Additional maps enable other kinds of attention to the contents of maps, and the conclusion they suggest is that the mapmakers of local areas deploy a range of hermeneutics in attempting to understand local space and to present their areas to audiences in recognizable ways. At one end of the spectrum is a way of apprehending space as heterogeneous. Objects on maps are more individuated, each item a "miniature" in the sense that Roland Barthes describes in which the "miniature does not derive from the dimension but from a kind of precision which the thing itself observes in delimiting itself, stopping, finishing."[45] This mode of presentation does not primarily derive from an attempt at objectivity but from the way in which a local area allows for paying attention to discrete and diverse items. In contrast, objects on maps might be depicted using more of a kind of shorthand. Objects on these maps become types or genres of items in that they are represented by the same signs: roads are depicted with similar lines, waterways by another kind of mark or marks, buildings are largely alike, and cultivated land is indicated by the same tufts of grass or symbols for crops. The maps exhibit a certain homogeneity. Here, symmetry or consistency among a map's contents is prominent. This second style of local map spreads attention across the whole scene because each genre of item and all of the types of items taken together have a similar weight. Modern map viewers are familiar with this second kind of cartographical sign system since it is dominant in present-day maps, but this kind of cartography was just appearing in the late Middle Ages. In fact, the heterogeneous map and the homogeneous map coexist at this time, and, as mentioned, they involve differences of degree. The modern viewer needs to put aside modern prejudices and a progressive model of cartographic history and instead take the two kinds as representing two equally valid hermeneutic tendencies.

Taking the more homogeneous map first, some local maps at this time have schematized sign systems for their contents: trees, rivers, houses, hillocks, and so on. The systematics indicate a shared apprehension between the makers and their audiences that local things are not infinitely individual but have comparable appearances and qualities. Items become signs, and those signs are transferrable across a local area and thus across a page. The cartographic uniformity of the maps is easiest to perceive in plans that have geometrically arranged lines, and the clearest example of geometric systematicity is a map of an area within Exeter city from about 1420 that displays only a set of tenter frames for drying cloth and little else. The diagram shows a small open space within the city, probably a square, and within it five long rectangular boxes parallel to each other. Each oblong box is divided into roughly equal portions of four or five individual squares. The plan is overtly systematic in that it has simplified the information it provides to the tenter frames alone, each of which is labeled with the name of its current owner.[46] A similarly rigorous spatial symmetry appears on a quite different map from about 1442, one of a larger area of Durham. Because of its scale, the Durham map is more complex than the Exeter plan showing just the cloth frames. The Durham page is dominated by geometrically rendered lines. The upper third of the map shows parallel horizontal lines evenly spaced to create rows to indicate tenement farmlands. The lands systematically diminish in length, so the rows get shorter and shorter as they go down the page. The whole forms a right-angled triangle with horizontal stripes. The majority of the remaining two-thirds of the map below the tenements indicates an area that is open but that has also been pinstriped, this time with parallel, evenly spaced vertical lines (even though this area was not divided into plots of individual ownership). A diagonal line—the line in dispute at the time of the mapmaking—divides the horizontal from the vertical strips. A ruler was used to create the lines, and little else is marked on the folio.[47]

A third map shows an even larger area and is yet more complicated, but it nevertheless demonstrates a hermeneutics of space that perceives, understands, and presents a local area as systematic. The map of Sherwood Forest (figure 4) from the late fourteenth or early fifteenth century addresses a less cultivated area than the Exeter and Durham maps of an urban space and farm tenements. Forests, we have come to realize, were not just wild and uncultivated places, and the Sherwood Forest map has to grapple with a large amount of information, including recently built structures. On a map of approximately 31 inches by 23 inches, it displays over three hundred names and many graphic features: woods, barrows, hills, ways, villages, parks, fens, closes, cliffs, gates, castles, and so on.

To handle the number of natural and manmade features, the cartographer has systematized his material. The map separates the whole area from what might surround it with a (now faded) red double-lined box, with circles at each of the corners. This makes the map initially similar to the two-page Canterbury plan, with its walls around the edges of the page, but here the lines do not indicate any objective feature on the ground. The double lines that edge the Sherwood map also divide the resulting large square into quarters, and each of those quarters is further divided into a pair of right-angled triangles. The circles in the four corners of the map contain the names of the cardinal points, and all of the map's lettering and features are oriented east, so east is at the top, north at the left, and so on, as in a typical *mappamundi*. We do not apprehend anything about the circumstances of the map, so it is not possible to know whether the mapmaker was drawing on the conventions of the world maps, but the circles and orientation suggest that the influence was there, an important point since the impact of *mappaemundi* on local maps is a matter of some dispute.[48]

Our interest lies in the mapmaker's desire to subdivide a large area into smaller parcels in order to make sense of the space. The lines that make the

LOCAL SPACE, EDGES, AND CONTENTS   41

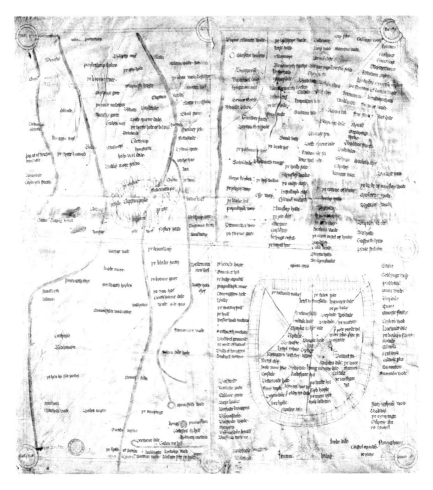

FIGURE 4.   Map of Sherwood Forest, Nottinghamshire. Archives of the Duke of Rutland, Belvoir Castle, Leicestershire, British Library, map 125.

quadrants and the triangular spaces, like the demarcation of the whole area, do not correspond to natural or legal boundaries. The thicker lines showing rivers (also red) that come down the map challenge this schematic layout since they cross from one quarter of the map to another, but even here, they nearly parallel the double edging lines, and they almost all fit into the left quadrants. The majority of the remaining visual elements, though large in number, are also schematically distributed in that they tend to sit within each of four boxes and the eight triangles; most easily seen are the oval palisades of Clipstone Park in the upper left and the larger palisaded enclosure of Bestwood Park in the lower right. The many names on the map, with a few exceptions, tend to fit within the lines of the artificially divided areas, and the names also tend to be laid out in columns and rows, further indicating the mapmaker's systematicity rather than attention to accuracy.

Comparison of another pair of maps emphasizes that the ways of understanding and rendering the contents of local areas were not settled into a systematicity that solely produced homogeneous spaces. A pair of Inclesmoor maps again makes possible the comparison of different forms of contemporary mapmaking relating to the same area (see figures 5 and 6). Both Inclesmoor maps exist in relation to one set of legal records and one purpose, but their depictions of what is easily recognizable as the same area point to divergent cartographic tendencies. One map is more of the kind we have been discussing, in which schematization dominates in terms of the symmetries it employs and its use of the same sign to represent items of the same type. The other map engages more with the miniature's precision in that each detail is rendered in order to make individual things distinct and self-contained, if not also self-enclosed. Here, at the same time in history, are two understandings of local space that appear to diverge, one toward the homogeneous end of a range and the other toward the heterogeneous end. Again, this is a difference in degree rather than kind, and we will find that even heterogeneity can give way to systematicity.

The two maps are remarkable not only because they allow for comparison but also because they accompany a legal dispute that provides specific information about their genesis, including that they are based in part on direct observation. Public Record Office, Duchy of Lancaster, 42/12 and Maps and Plans C 1/56 show approximately 240 square miles of Inclesmoor in West Riding, Yorkshire, southeast of Leeds. The maps were occasioned by Henry IV's royal *inspeximus* of 1410, in which two stewards were directed (in 1405, 1406, and 1407/1408) to research archives and accumulate their findings concerning a contestation over land use from over a hundred years earlier. These named men, Richard Gascoigne and Robert Waterton, have been grouped "among England's first historical geographers."[49] In 1300 the abbot of St. Mary's, York; Henry de Lacy, Earl of Lincoln; the Abbot of Airmyn; and a certain William of Hook challenged one another to rights to arable land and the economically significant turbaries, or peat beds, on the moor that were used as fuel for industries such as brick making, pottery, and dying.[50] A hundred years later, the stewards Gascoigne and Waterton, in addition to looking at the archives for the rulings on these challenges, were commanded "alez personelment veer le soil de Inklesmore en Snayth."[51] They were to look at the area in person, presumably to compare what was on the ground with what was in the records and to either confirm or update the documentary evidence. The intent seems to be that legal, historical, and sensory accuracy coincide.[52]

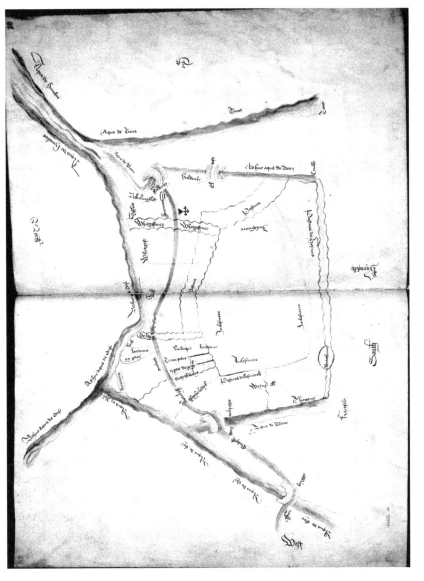

FIGURE 5. Small map of Inclesmoor, West Riding, Yorkshire. The National Archives (TNA): Public Record Office (PRO), Duchy of Lancaster, 42/12, fols. 29v–30r. Reproduced by permission of the National Archives of the UK.

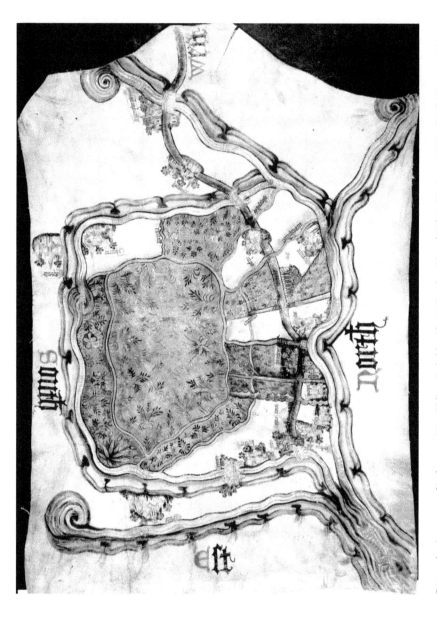

FIGURE 6. Large map of Inclesmoor, West Riding, Yorkshire. The National Archives (TNA): Public Record Office (PRO), Maps and Plans C 1/56. Reproduced by permission of the National Archives of the UK.

LOCAL SPACE, EDGES, AND CONTENTS 45

Produced at the same moment in history, with the same legal motivation and from the same observation of the land, the maps nevertheless render what is recognizably the same area very differently. Together they make more apparent than just single instances of maps that graphics such as these represent attempts to understand a local space, and they show their makers intentionally creating particular intellectual, emotional, and aesthetic effects. Taking DL 42/12 first (figure 5 originally measures 15 × 21 inches and is referred to as the small map), the most outstanding feature is that its contents tend toward what I have been calling homogeneity. Its perspective is one in which the viewer looks into the space from any angle; it presents a local world that can be surveyed as a whole and one in which each class of elements has a similar weight. It seems to draw on ready-made graphic signs, such as appear in the Exeter and Durham maps and which also are seen in manuscripts of other kinds (for example, in historical tables or in contemporary poems that use square brackets to draw parallels between similar written items on a page or to mark rhyming lines). Cartographic layouts on pages such as cartularies or lists of rentals street by street also increasingly use symmetrical graphic forms.[53] We find it hard not to feel at work the science of the mathematician, the surveyor, or the modern geographer who could just as well map another area with similar graphic symbols.

The small map is carefully laid out on the pages, centered with surrounding space, which is labeled only with the cardinal points and one or two words. The names there and in the interior are not oriented in any one direction, nor are they oriented to the nearest edge of the paper (unlike the two-page Canterbury map); the mapmaker seems purposely to vary the orientation of the names so as not to privilege any particular orientation to the map as a whole.[54] The map can be rotated in any way to facilitate looking into the space. The systematicity of the names also emphasizes homogeneity. All of the names on the page have an initial capital, and the names fall into groups depending on the map item they identify. Broad, blue, shaded rivers have names that are closely parallel to their waters to facilitate their easy identification. Smaller watercourses are labeled with writing that in contrast goes over their undulating lines (for example, "Foxholedyk" is written over a small watercourse within the large rivers at the center upper left, to the right of a large bridge). Boundaries are named parallel to or within pairs of straight or wavy lines: in the northwest, within thin vertical strips of land, "Landemer" is repeated and "The Kay"; in the northeast, within scalloped lines, "Whitgiftmer." The names of towns tend to be superimposed over the brown, thick and long, roughly horizontal stripe of Moregate road. Three of the four humped bridges, meanwhile, cleverly split the names, so the names are like the bridges themselves in that they cross the waterways (on the left are "Feri-brigge" and "Tour-nebrigge," and on the right "Es-toft"). The other names, such as those of the villages along the banks of the rivers, parallel the waterways. There is also a system of lines for the roads.

The other map, MPC 1/56, (figure 6 originally measures 25 × 34 inches and is called the large map) creates a more heterogeneous effect with its contents. The land spaces within the island that the rivers create are distinct from each other. Careful shading provides a sense of the landforms, including on the edges of unpainted areas. The towns are particularly differentiated and are miniaturized. Each town has clearly discernible rounded turrets, conical or rectangular roofs that are either thatched or tiled, bell towers, and pinnaces from the top of steeples; and each town has a different combination of structures. As on the small map, the names on the map are not oriented in any one cardinal direction but rather are worked into or next to a chorographical item. However, the words are not all written in the same script of roughly the same size but are instead distinct from each other, like the buildings' roofs. The map belongs more to a tradition of deluxe manuscript arts; the mapmaker's sensibility is that of the experienced illuminator.

The two images are different both in terms of the understandings of Inclesmoor each map displays and in terms of their desired effects. The more schematic small map encourages the eye to follow the lines of the rivers and roads, and its open space and the graphics emphasize relations among things, even scalar relations among different kinds of topographical features and areas. The more detailed large map leads the eye to distinct local areas and from there draws attention closer and closer to the miniature. Nevertheless, their similarities are less obvious but actually offer even more compelling insights into how the features of a local area were understood. Consider the layout. While the legal interest may have been in arable land and peat, and even though the maps render their rivers quite differently, both mapmakers use the major geographical feature of water to structure their images. The waterways provide a sign to which other items on the maps—boundaries, buildings, flora, roads, words—relate spatially. If we start with the small map with north at the top and work from the north down, notice that the waters form the inland islands containing Inclesmoor: the large ones Ouse and Humber, Trent in the east, and Aire in the west, then within those, the Torne (west) and Doon (east and continuing along the bottom). The map also pays attention to smaller watercourses; in the upper left, for example, within the larger rivers are the wavy lines of Foxholedike and Braythwayt. All the waters have the same scalloped lines—brown lines outline the edges of large rivers, and blue or brown lines by themselves or in pairs show smaller streams and other waters such as the pond, Braithmed, at the lower left. The larger rivers' waters are carefully shaded to provide depth and substance: heavy (blue) strokes at the beginnings or endings of the Ouse, Humber, Trent, and Aire and darker shading along one shore and one edge of the Torne and Doon graduating to lighter blues. The beginnings or endings of the rivers do not extend to the edge of the page but instead typically start with some space around them as they fan out to indicate even larger waters.

Attention to the waters on the maps makes clear that the maps are in fact similar in that they are self-consistent, although to different degrees. That is, despite the fact that the large map individuates more, it is also to a certain extent systematic, classificational. Like the smaller schematic map, the large map also uses the water to structure its space, and renders its cerulean rivers and lesser waters, like the smaller version, with lighter and darker shades of blue and stripes to indicate their substance. Except for the Humber in the upper right (which heads off the page), the rivers whirl out of themselves from the corners of the vellum. A similar helix produces "Braythmer" at the center bottom. The rivers are stylized and wavy, again as on the smaller map, and the streams are similarly rendered with undulating lines. There is a clear hierarchy of three kinds of watercourses even though the riverbanks are all drawn in the same braided manner. Sprays of foliage are similarly repeated across each of the interior wild spaces of the moor. The same style of plants decorates the fields, and similar trees ornament the villages. The names on the large map appear at first glance to be rendered in an individual fashion but are in fact hierarchical and categorized: large red and black for the cardinal points; gold initial and red for the moors, the marshes, and the major rivers, (e.g., "Humbre," "Doen," "Ouce," and "Trent" in the upper right); red initial and black for meres, towns, and so on.

Another similarity between the two maps is that the aquatic elements are not natural; they are human-nature hybrids. Rivers accrete and unravel, but their defining distinction is they move, so it is tempting to consider them in terms of hydraulics, tidalectics, flow, or fluid dynamics. In this situation, it may be especially attractive to follow William Howarth in his writings on wetlands, which he describes as "multivalent": "water-land." He offers that in "their wildness," they "dispossess readers of old codes and lead toward new syntax, where phrases may begin to reassemble."[55] But the waters on the Inclesmoor maps have been shaped. The rivers on the large map are almost subsumed under the emphasis placed on their riverbanks with stacked sides and puckered ditches or sluices for irrigation. In the east on the large map, the artist indicates water or silt flowing from the earthworks beside the Trent. In fact, nearly all the areas on the Inclesmoor maps are historically the result of land reclamation, the diverting or wholesale creation of waterways, and the earth is also shaped into edged lots and is crossed with roads. In this way, these maps are similar to other local maps that draw attention to water not only as a means to lay out their images but also as a means to show the mechanical arts of drainage, water supply, fertilization, power for mills, and fishing.

The comparison of these two maps of Inclesmoor, along with assessment of other local maps, shows that each apprehends and renders local space in distinct ways. Each has a particular way of understanding the things before it, whether those items are in the world in front of the mapmaker's eyes, in the historical documentary sources, or, in the Inclesmoor case, in a combination of observation and text. Local contents also can be systematic and to a degree symmetrical. Although one map individuates items more than another, even down to a miniscule scale, the mapmaker and the viewer still discern items in patterned ways by using common signs in a finite number of combinations. Human activity and shaping of the natural landscape come to the fore. The real difference between a homogenous and a heterogeneous space lies in the approach to the contents that each kind of map invites. The homogeneous map allows the viewer to stand back and take the whole space into view, to assess with ease the relations among different objects pertaining to the space, comparing and contrasting items. The contents of the heterogeneous map encourage the viewer to immerse himself or herself within the space with "a kind of precision which the thing itself observes in delimiting itself, stopping, finishing," as Barthes describes. Once closer, the viewer follows the internal logic of the space, comparing dissimilar items that happen to be contiguous or continuous in the case of roads and rivers. But the attention does not end there, because patterns gradually reveal themselves. A systematicity emerges like that of the homogeneous map, but it is a patterning that arises more organically from the individual items on the page. The differences are of degree rather than of kind. The more homogeneous approach to space uses abbreviation, abridgement, the précis. The heterogeneous perspective involves singular objects, discerning relations among objects, and tracing along routes, but it nevertheless reveals patterns.

Local Times

The Inclesmoor maps furthermore suggest that the edges of a local space and the categorization of its contents can define it, yet it can also be temporally diverse. In general, the local maps display present features, or recent items, alongside past ones. On the one hand, the items on the maps are phenomena that the mapmakers have just surveyed ("personelment" as the Inclesmoor mapmakers were commanded) and their audiences have known or wanted to know, and they are what the viewer sees as a presentation of a current space. So, for example, some local maps depict what has recently happened: the Canterbury maps appear to record innovations in the water and drain systems, the Cliffe map responds to a recent flood, and the Sherwood Forest map shows new structures within the area. On the other hand, maps such as the Inclesmoor ones show items that are present, but part of their purpose is to provide contrast with the past. The viewer is supposed to look at the maps' depictions of pasture, turbaries, "les boundes," and so on, and compare them with the documents from a hundred years earlier to see how the features match or do not match.[56] The viewer is considering a present local area and a past one at the same time, shifting perspective and imaginatively moving the pieces of the map in order to think of how the present is different, how it is an accretion of intervening alterations in land. In this sense, the Inclesmoor maps are not that different from many others. The space of the local in each case opens out to the dimension of time so that a present thing is explicitly marked or implicitly registered as adjacent to a former space, superimposed on it, or otherwise set in relation to it. The map is, as Curry states, an "information storage device," except there are two or more times of information.[57]

Some maps follow this temporal logic and explicitly display present and past information. Further comparisons and contrasts with maps of the whole of England or Britain, and with maps of the *oikoumene* (as in *mappaemundi*), are revealing on this point. That the present and the past can exist on the same visual plane on local maps makes them similar to maps of larger areas, such as the Hereford, Gough, and Matthew Paris, which may be thought of as visual encyclopedias that display information from a number of different times. The large-scale maps are expansive in terms of time, including as they do Homeric incidents, the matter of Rome, biblical and other phenomena, alongside, in some estimations, contemporary events.[58] Again we might think of the mapmakers and their audiences comparing and contrasting the present with the past and vice versa. Differences, however, remain in terms of the temporality and spatiality depicted in chorographical versus geographical maps. The time scale is generally simpler or shorter on local maps. Local maps also are not prophetic or apocalyptic like *mappaemundi*; Earthly Paradise and Gog and Magog do not await in the East, nor does Jesus embrace the world or bless it, promising a second coming. The phenomena in the spaces are also different in that *mappaemundi* show places based on biblical, classical, and otherwise mythical history, whereas local maps tend to present features that are managed and owned or otherwise primarily subject to human actions.

It is not possible to generalize across all local maps, because some ask the viewer to compare events from longer ago in a process that appears similar to the viewer's experience of large-scale maps. A map of the Isle of Thanet, Kent, from the late fourteenth or early fifteenth century shows a complex of interactions between past and present (see figure 7). The map is curious in that, unlike several other local maps, it does not seem to be made explicitly for judicial purposes. The map is revealing for the sense it gives of how writing and image can interact because, unlike many other local maps, it has text that surrounds it on the same page, passages of writing that interact with the illumination. It is also interesting because it is one of the first images in history to be referred to as a "mappa." Measuring about 21 × 15 inches, the chart appears in Trinity Hall, Cambridge, MS 1, which is Thomas of Elmham's *Historia Monasterii S. Augustini Cantuariensis*, a manuscript that also contains a diagram of the high altar of St. Augustine's Abbey, Canterbury.[59] Thomas completed the chronicle about 1414 when he left Canterbury and when he also wrote the *Liber Metricus De Henrico V*. It is believed that he produced the map as well as the text. Thomas would later become chaplain to Henry V.[60]

The map's buildings and words are oriented to the east. Above it, the left-hand column of the *Historia*'s writing abuts the northeastern top of the island and the sea, and the right-hand column wraps around a southeastern part of the island and ends between the island and an additional block segment of sea in the top right. On the right, close examination reveals that the letters are written over the border of the Isle of Thanet and the square piece of sea. This implies that the map was drawn before the main text's writing, possibly suggesting a certain priority given to the image. The narrow outlined box that sits on the waters at the top of the map and juts out to the left and right in two pieces separated by the two columns of main text contains the map title: "Mappa Thaneti . . . Insule." *Mappa* is a surprisingly difficult word to understand in the late fourteenth and early fifteenth centuries. The text of Cambridge MS 1 subsequently refers to the image as a *"figura"* ("*In ista enim præcedente figura . . .*").[61] *Mappa* is rarely, if ever, used as a single word in this form in contemporary sources, instead appearing almost exclusively as part of the compound *mappemounde/mapemonde/mapamounde/mappamound*, notably in English in Gower's *Confessio Amantis*, Chaucer's "To Rosemounde," and Robert Henryson's *Orpheus and Eurydice*.[62] Earlier Latin and French meanings of *mappa* refer more generally to a cloth, even a napkin, as well as to material on which things may be written or drawn. When Osbern Boken-

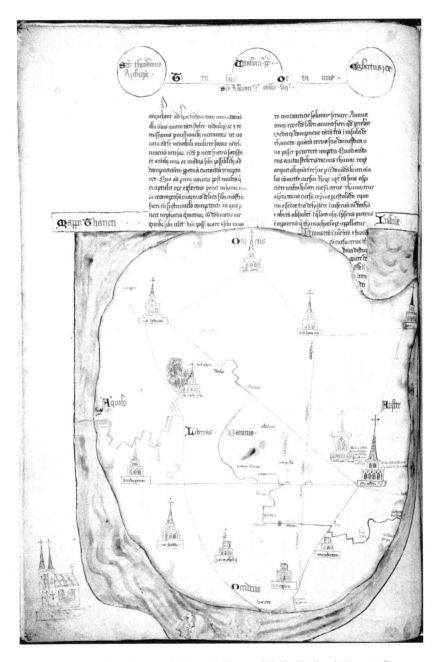

FIGURE 7. "Mappa Thaneti . . . Insule," Kent, in Thomas of Elmham's *Historia Monasterii S. Augustini Cantuariensis*. Trinity Hall, Cambridge, MS 1, fol. 42v. Reproduced by permission of The Master and Fellows of Trinity Hall, Cambridge.

ham uses a version of the stem to title his fifteenth-century chronicle *Mappula Angliae*, he simply means a short treatise. In fact, the compound *mappamundi* could mean a written description without any drawn image or, if it referred to a map, a chart of any sized area.[63] Consequently, "Mappa Thaneti Insule" could signify a range of things, but the fact that it precedes the image and, to a lesser extent, that the image is oriented toward the east clearly demonstrates awareness on the part of the creator that he was making a map, perhaps within a cartographic tradition that included world maps.

The drawing does not appear to meet judicial needs, unlike some other local maps such as the Cliffe map or Inclesmoor pair. F. Hull suggests that, because of objective accuracies, the map displays places that Thomas probably knew.[64] It is somewhat schematic with its simplicity of line and sign systems, but it also individuates, signaling further interest in actual locales and their distinctive qualities. Thomas uses a red line to outline the island, and he employs the same red to draw roads, which are sometimes named, with straight or neatly curved lines. He employs green ink as a form of shading around the inside edge of the island, and he often paints faint hooked tails into the island as if to indicate coastal hills and/or inlets. He uses the same green boundary outline to make a detailed, markedly crooked line that cuts across the island from the center left to the lower right. The other features depicted are the built structures. The churches are individuated; for example, in the extreme lower left (on the mainland, not on the island) stands the church of Reculver with round twin towers, and elsewhere are Romanesque and steepled buildings. Other details include beacons with their ladders (one on the left below center and another in brown just to the right of center), a windmill below the left beacon, wayside crosses toward the lower left, a dark pit near the middle, and groves of trees (a large one near "Wodecherche" in the upper left). The relative locations of places on the isle, the characteristics of buildings, and the names on the map suggest this is, as its editor suggests, an "extremely accurate piece of work," in the sense that it has objective veracity to contemporary geographical features even though the landmass of the Isle of Thanet is not accurately shaped.[65] Like the Inclesmoor maps especially but also the other local maps, the Thanet diagram makes a kind of realist truth claim.

It is therefore surprising that the point of a map that might be called a "mimetic topographical map"[66] is not primarily to record some boundary or other recent property dispute but to recall an event from seven hundred years earlier; the manuscript contains an outstanding example of "mappa" and text interacting with each other on this temporal point. The image offers the reader further explanation, and, following the map, the text then explains the graphic by linking the places on the map to events in the written account. The text preceding the illumination tells the legend from the late seventh century of (Abbess Domne Eafe), niece of King Eorcenbert of Kent, who received wergild for the murder of her two brothers in the form of all the land her pet deer could delineate in a single run. The manuscript leads up to the picture, saying right before it in the wrapping text, "Here I intend to describe the true site and form of this island, with the dominions marked out by the run of the deer" (Huius enim insulae situm et formam, cum cursu cervae inter dominia distinguente, describere hic intendo).[67] The deer run is the jagged line below the center from the left in the north ("Aquilo," the North Wind) across the island to the south. It divides the land into western and eastern halves. A deer is visible at the left below "Aquilo," seemingly the start of the run, and below it is a label *"cursus cerue,"* "the run of the deer." Following the map on the facing page, the columnar writing explains at some length the line of the deer's run, the lines of the roadways, and other lines, all of which reinforce the sense of wanting to provide the viewer with information that ties the map to the text (and vice versa) and to the historical legend. Thomas encourages the viewer and reader to look at the map and the text as interacting with each other, to look from the story to the map and back. For example, he begins the explanation immediately following the map with the fact that the deer's run is "three feet wide, without interruption, and is all well preserved," and "The area of the ground, which the pet deer ran around, enclosed in length and breadth forty-eight ploughlands" (*tres pedes in latitudine contineret, absque interruptione aliqua integraliter conservanda. . . . Spatium vero terrae, quam cerva domestica circuivit, in longam et latum xlviii. aratrorum tenebat*).[68] Thomas seems to be encouraging the viewer to compare the legend with the map *and* with the actual locations on the island. Thomas of Elmham's *Historia Monasterii S. Augustini Cantuariensis* and its map suggest that written and visual identifications of chorographical features mutually reinforce each other and interact with objective reality.

By way of conclusion, I concede that it may appear anachronistic or potentially naive to see a form of realism in the local maps, including the Thanet one, and their accompanying texts. Scholarship has become wary of such pronouncements and tends to avoid older arguments that claim realism in any uncomplicated way, such as John Manly's observations about Geoffrey Chaucer's "development" from formal sophistication to "methods of composition based upon close observation of life."[69] The scientific writings I discuss in subsequent chapters also have a complicated relationship with empiricism and empirical phenomena. Most of the maps are accompanied by historical records of some kind, and it may be argued that chronicles and other nonfictional writings provide the closest analogy to the maps. In this regard, the historian Antonia Gransden has pointed out that realism is increasingly valued in chronicles beginning in the twelfth century. Gransden also cites the historian T. D. Kendrick's observation of a "nascent and rather casual medieval topography" in history writing.[70] The chorographical depictions of local places that we see on the maps nevertheless suggest that even their realist depictions of local area conform to conventions. Realism is, after all, a mode of description, in part a set of conventions, and the maps, although largely individual and idiosyncratic, also reveal systematicity concerning edges and contents. This is not, however, to simplify a relationship between realism and systematicity; the two seem to work in tension, even opposition, with each other. What we can conclude about local space from the evidence presented is that the edges of a local space tend to belong to one of three types, each with distinct effects: actual features such as walls or rivers, or, on maps, lines of a cartographical sort can enclose and delimit; maps can lack drawn borders but still be systematic enough to achieve a sense of integrity; or they can lose coherence so that the edge of one space imperceptibly transitions to become another. We can also conclude that although maps might be heterogeneous in terms of their depiction of local space and contents, they are all to a certain degree homogeneous or systematic. The effects of these two kinds of maps are different: the perspective and experience are, as it were, outside-in for the homogeneous and internally circuitous for the heterogeneous one. We will see next how Middle English literature reflects, refracts, and differs from these spatial hermeneutics when the literature also addresses local spaces.

CHAPTER 2

Local Literature
Vernacular Local Space and John Lydgate's
Siege of Thebes

The previous chapter explored the hermeneutics of local space, the ways that late medieval mapmakers and their audiences perceived and understood local areas. Local spaces, I have argued, were the most common and significant kinds of space: bounded, defined, and emplaced. The maps brought to light that, first, local edges are complex. Most obviously, a local space's edges enclose an area and provide borders that divide one locale from another, but they also join one place to another. The edges can, moreover, exist in the physical world as features on the ground—walls, ditches, watercourses, or roads—real borders that are natural, manmade, or nature-human hybrid structures. Edges can also be up to the mapmaker, subject to a decision to end a local area by drawing lines or edging with blank space on a page without anything on the land to establish the border. Second, the maps are revealing because late medieval conventions of local mapmaking were not yet established. As opposed to *mappaemundi*, local maps were often based on less canonical (or at least qualitatively very different) authority; more of the sources can be immediate and situation specific than with maps of the cosmos or the whole earth. Some of the local maps derive their ideas about a space from direct observation and a comparison of what is on the ground with a historical source, and so they indicate awareness of observational veracity. This *commensurable* process (from *com* and *mensurare*, to measure together) can confirm a historical authority over the present situation of what is on the ground, or it can favor the present situation so that the past is superseded or otherwise compared with the present. In either case, it seems that mapmakers could be responsive to largely unconscious understandings of local area that were shared between cartographer and audience. Third, we also learned that a local area's contents fall along a range from individuated depictions of heterogeneous contents to generic kinds of items that are more homogeneous, each type of space encouraging different responses from the viewer. In the heterogeneous case, the trees, lands, buildings, streams, and other features are depicted in such a way that they signal an understanding of contents as distinct with attention paid to their particular characteristics. The observer's eye moves in closely to explore an individual feature while also tracing or following a route among things on a map. In the second kind of map, a more symbolic depiction of items presents a whole area from the outside as a sign system. Either the more heterogeneous or the more homogeneous way of apprehending a space could provide cohesion if an edge was present, but in the absence of a border, a local space's integrity is more dependent on the systematicity of its internal features.

The following discussion takes up these ideas: the effects of edges, relationships between historical sources and present-day realities, and the varied organizations of local spaces with their different effects. I use insights gained from Ptolemy and cartography to examine the Middle English semantics of local space and John Lydgate's *Siege of Thebes* of 1421–1422 along with the (later) iconic portrait of the Canterbury pilgrims in the *Siege* in British Library, Royal 18 D 2. The first section of the chapter assesses the Middle English vocabulary used to name local areas, a broad but definable semantic field of words that describe these spaces. These semantics reveal particular ways of thinking about local area in thirteenth-, fourteenth-, and fifteenth-century culture just as the maps emerge out of intersections between philosophy and practical knowledge, between the world of the mapmaker, the observed world, and history. The second section investigates Lydgate's *Siege of Thebes* and the portrait of the Canterbury pilgrims in the Royal manuscript. The *Siege* and the illumination show in different ways additional complexities of the local—namely, how Lydgate and the illuminator employ a local area to explore specific tensions that local pressures create, especially when the local space becomes incoherent.

The semantics of local space and Lydgate's *Siege* and the illumination share many of the cartographic characteristics of local space but also diverge from these presentations. Together, the semantics and Lydgate's *Siege* along with the illumination emphasize that local space is tied to the identity of an individual person and group of people more explicitly than in the maps. Recall that for Aristotle and the tradition that followed him, place and being are inextricably bound together: "things that are are somewhere."[1] Where Aristotle, Aquinas, and others after them were concerned with individual entities of a human and nonhuman sort, Lydgate's poem and the illumination draw attention to group identity, *communitas*. One difference between, on the one hand, the maps and, on the other hand, the vocabulary and Lydgate is that humans do not commonly appear on the maps, whereas spatial edges and cohesion have implications for human sociality in the semantics, the *Siege*, and the illumination. Moreover, Lydgate's poem is of course a fictional work, so the question of veracity is not applicable in the same way as it was with the maps in which the placement of watercourses, buildings, trees, roads, and forests on a map could correspond with the placement of items on the ground. Aquinas added to Aristotle's definition of space and being by saying that a goat-stag and a sphinx could not be anywhere because they are fictions. But Lydgate's poem, along with other writings, implicitly reverses this idea and posits that it is possible in fiction for unreal beings to be in a place. The illumination, although a later work and therefore separate from Lydgate's own production, shows the community of characters in a way that seems to respond to the *Siege* text and also to its own demands for pictorial presentation. It therefore intersects both with the poem and in some ways with the cartographic (that is, visual) presentations of a local space.

Despite its fictionality, Lydgate's *Siege* attempts a certain kind of spatial realism when it creates the illusion that its narrator—Lydgate apparently *in propria persona*—joins up with an actual group of Canterbury pilgrims, including Chaucer, on a real pilgrimage. Furthermore, as Stephanie Trigg writes in analyzing the paperback cover of the *Riverside Chaucer* and its reproduction of the Royal 18 D 2 image, the scene presented is a "combination of realism and wish-fulfillment," the latter on the part of Lydgate and also the individual reader of the image, who can also join with the Canterbury pilgrims, including Lydgate and Chaucer. In her interpretation, the image confirms and reinforces a sense of a "Chaucerian community" that extends beyond Lydgate's time to ours.[2] In contrast, in my reading, Lydgate's attempt to expand the social coherence of the pilgrimage group to himself and to potential readers fails. Lydgate's communality is, after all, based on a group that was never coherent in his Chaucerian model. Moreover, when it comes to Royal 18 D 2 and its interpretation of Lydgate's text and the visual presentation of *communitas*, the local space in the image also fails to offer full social cohesion. The insights from chorographic cartography on edges and contents draw attention to the fact that the illumination exhibits an incoherent space that works against the veracity of a collective identity of the pilgrims.

By way of an initial example, Chaucer's Prioress's Tale shows what can be at stake in terms of a local space's characteristics, edges, and contents in literature. Attention to space in the tale shows why the topic is important and not just a backdrop or setting for action and character.[3] The Prioress's Tale investigates a local space with particularly terrible consequences. Chaucer focuses the audience's attention so that the narrative comes to involve local space centrally, the tale's tensions revolving around a contrast between enclosed spaces and expansive ones. The tale depicts a local Jewish spatiality as nominally confined within a particular area within a city and the opposite of the surrounding Christian area at the same time that it is paradoxically rendered as open to Christian areas and able to move out into those neighborhoods. It may be said that the tale's prejudice is spatially organized in that the local space is systematic, its coherence depending on unresolved but narratively productive boundaries between enclosed and open areas.

The Prioress's Prologue lays out the first space, and it is not local. Indeed, it may be said to be the opposite of local in its employment of universals from scripture, the Office of the Virgin, and Dante. The Prioress-narrator opens with an invocation of God's name horizontally "in this large world ysprad" so that everyone—from "men of dignitee" to suckling children—praises him. Next comes a different orientation—namely, a vertical lineage of the Holy Spirit "ravyshedest doun fro the Deitee," which also descends temporally from Moses's burning bush to the Virgin and thereafter to the Prioress. The remaining two stanzas of the Prologue employ different but equally expansive metaphors: first, of the Virgin bringing every human being light in darkness, and second, of the Prioress herself as a little child before the Virgin and scarcely able to "expresse" "any word."[4] The effect, therefore, when the reader turns to the tale is of a drastic narrowing from these vast scales as if the increasingly constricted spaces of the tale are much less significant, mere exempla, and therefore limited, narrow, deficient.

The opening of the Prioress's Tale is categorically more restricted in both space and time than the Prologue. It begins with a spatial telescoping effect, common in Chaucer's introductions to tales, in which the focus narrows from a large geographical area to a smaller and smaller location. It is also more constricted in time than the vast temporal scales of the Prologue given the past tense of the tale's opening—"Ther was in Asye"—which is the opposite of the Prologue's eternal present:

> Ther was in Asye, in a greet citee,
> Amonges Cristene folk a Jewerye,
> Sustened by a lord of that contree

> For foule usure and lucre of vileynye,
> Hateful to Crist and to his compaignye;
> And thurgh the strete men myghte ride or wende,
> For it was free and open at eyther ende.
>
> A litel scole of Cristen folk ther stood
> Doun at the ferther ende.[5]

The nonspecific, continental placement of the tale in Asia pushes the tale "out there" in a geographical sense and differentiates it from Europe. The opening stanza's next gesture is spatially to narrow down from the continental scope but also to entangle and conflate geography with population; the sequence is Asia—city—Christians—Jewry. The Jewish area thereafter initially appears to be only one section of road since it is called "the strete." The next spatial detail is that "Doun at the ferther ende" of the open street is the "litel scole of Cristen folk." These are nesting boxes of scale, constricting and bringing readers to the main action while also concentrating the effect of the narrative to come.

The tale thereafter provides some relief from the spatial claustrophobia of the opening when readers learn that "thurgh the strete men myghte ride or wende, / For it was free and open at eyther ende." The street, the smallest unit of space, at this point near the tale's beginning suddenly appears open in a geographical sense and also in a social sense in that any people, Christian or Jew it would seem, can ride or make their way down it. The beginning of the tale therefore introduces thematic concerns in spatial terms in that, at the same time as it draws focus in to bear on the school, it has already set up a contrast between the Jews and the Christians in the city. That is, the tale's thematic tensions are spatially established, especially because the description is explicit in condemning the Jewish space as symptomatic of the city's lord, who "sustains" it "For foule usure and lucre of vileynye." As Louise Fradenburg writes, "Immediately, the tale attempts to fix location, to map a terrain with an inside and an outside, with entrances and exits. The goal of this cartography is in part the location, even the detection, of the Jewry; from its embedded position, the tale will uncover it, bring it to light, make its inwardness available to view."[6] The openness of the street, a space that has been conflated with the whole "Jewerye," may be welcome amid all the spatial concentricism, but this very openness is also already a threat to the Christians around it.

The climactic moment of the Prioress's Tale, when the dead child breaks into song, takes up the tale's spatial tensions and renders them in terms that exploit the flexibility of the word *place*—its ability to signify a small location and a larger area—in a way that explicitly contrasts a constricted space with a more expansive one. It is as though at the key moment the telescopic and concentrated narrowing at the opening of the tale is everted through a sudden efflorescence. There is a symmetry at the climax of this compact tale in that the temporary relief from enclosure in the form of the open street at the tale's beginning, which has now in the course of the narrative become a threat, is repeated at least twice at the tale's climax. One moment contrasts the hired murderer's "privee place" within an alley with a larger area.[7] The opened, extensive area appears once the clergeon has disappeared when the child's mother explores a large space, including when she discerns that her son was last seen in the Jewish area, and she "gooth, as she were half out of hir mynde, / To every place where she hath supposed / By liklihede hir litel child to fynde."[8] That is, the mother's frantic and unwilled search opens up the whole city, including the Jewry. The second contrast initially builds on the sense of a potentially more dispersed Jewry and contrasts it with the boy's body being in a tiny space, one of the most compact in all of Chaucer's poetry:

> She frayneth [asks] and she preyeth pitously
> To every Jew that dwelte in thilke place,
> To telle hire if hir child wente oght [at all] forby.
> They seyde "nay"; but Jhesu of his grace
> Yaf in hir thoght inwith a litel space
> That in that place after hir sone she cryde,
> Where he was casten in a pit bisyde.[9]

The repetition of "in," "inwith," and "in" again, and the extra internal rhyme of "place" within the rhyme royal ("place," "space," "place"), as well as some awkwardness of syntax, compress these lines in a way that increases the sense of the tightness of the space itself.[10] The words entangle with each other: "in thilke place," "in hir thoght inwith a litel space," "in that place." The space is bound up, condensed.

The subsequent action continues the constriction versus opening structure of the tale. After narrowing in on the tight pit where the widow's son has been hidden, the space opens up again. The child, throat slit and sitting upright, sings the *Alma Redemptoris Mater* "[s]o loude that al the place gan to rynge." Small and constricted—the Jews' place—becomes filled with the clear and expansive Christian voice. The "strete" of the opening of the tale reappears at this point in the narrative in that "[t]he Cristene folk that thurgh the strete wente" hear the song and seek the town official, who transfers the boy's body to a nearby abbey after processing through the streets. It is thereafter emblematic that the Jews responsible for the boy's death are "bound" before being drawn and hung.[11] The mother's searching the whole city, the boy's voice echoing out over all space, and the Christians' search and subsequent procession throughout the area are structural reflections of the initial spatial openness of the narrative. Where at the beginning of the tale the Jewish area is open to the rest of the city, the freedom of movement turns back on itself and becomes a threat fulfilled. In the final reversal, both the murderer's attempt to hide the body within the Jewish neighborhood and the Jews' attempt to conceal that location are suddenly unfolded outward so that the child's presence spreads back out over the whole city. That is, the tale's main action concludes with the bordered Jewish area transformed at least three times by the mother's passionate search, the boy's voice, and the Christian people so that its accessibility, and indeed the openness of the whole city, is now improved.

There are similarities between the concerns of the local maps and the interests of the Prioress's Prologue and Tale. The former suggested that edges are significant and can define a space in different ways. One is that local edges enclose, separating one space from another, and the borders also join one space to another in either abrupt or gradual ways. Boundaries are significant in the Prioress's Tale in that the edges of the Jewish area within the city are the ends of the street, which seem to both define a "Jewerye" and suggest ease of travel into, through, and out of the area. These potentials, the tale suggests, are problematic, threatening. Also, the maps' attention to contents within an area showed two tendencies, one toward individuation and the other toward categorization of items. The tale individuates most when it comes to the murder, concentrating spatial elements and the reader's attention on the singularity of that site, which is repeatedly named: "a privee place," "a pit," "a wardrobe," "a litel space," and a "pit" again.[12] But the tale also puts the places into categories by distinguishing between Christian and Jewish areas, with the latter at some points suggested as being a smaller area (maintained, as mentioned, for "foule usure and lucre of vileynye") within the larger city. Last, we might ask about the veracity of the space since the maps suggest some matching between an area on the ground (or the presentation of a local space in records) and the representation of the space on the page of a map. The brutality of the crime and the repeated details about the pit create powerful effects that suggest some sense of realism. However, the quantity of spatial individuation and specific or chorographical detail is not that great in the Prioress's compact tale, drawing attention away from any potential realism and toward the symmetries and asymmetries between religions, peoples, and spaces. All of these spaces in the tale are framed by the Prologue and the tale's ending, the latter of which returns to the most expansive senses of time and space, to universals. At the end of the tale, the Prioress desires to "meete" the clergeon's tomb "Ther he is now," and she notoriously evokes Hugh of Lincoln, "slayn also / With cursed Jewes," desiring that Hugh pray for the Prioress's audience, "us," to increase the mercy we might receive from God through Mary.[13] The space is largely "systematic" in the homogeneous sense, and it is this sense of systematicity that operates in creating the tale's anti-Jewishness, but the tale's framing ultimately potentially extends that prejudice to all readers. It is as though the whole action in Asia is supposed to be read as narrowly, even obdurately, constricted on this earth as opposed to the eternity and capaciousness of a Christian universe.

The Semantics of Estral Space

Where Chaucer's Prioress's Tale explores implications of space, including, to a certain extent, the word *place*, other Middle English vocabulary reveals the chorographical characteristics of local area. The first idea of note is that *local* as an adjective was not used in a geographical sense in Middle English. *Local* appears primarily in medical contexts to refer to places on or in the body. Works such as the early fifteenth-century translation of Guy de Chauliac's *Grande Chirurgie* employ "locale" as an adjective to describe "perticuler" ailments, processes, treatments, and remedies.[14] In a more psychological context, Caxton describes how "the comune understondyng" is better suited to an "ymaginacion local" as part of his justification for translating the *Lyf of the noble and Crysten prynce, Charles the Grete* in 1485.[15] I use the word *local* throughout this discussion, but *local* was not transferred in English to a larger spatial meaning until the sixteenth century, notwithstanding its Latin root in *locus* (clearly used in the classical era to refer to a place) and despite the Latin *locus* and *localis*, and French *local* existing in the Middle Ages and referring to geographical space.[16]

The adverb *localliche* appears in Middle English, although it was also uncommon, but here the meaning is more closely a spatial one. It matches almost exactly the scholastic, scientific sense of chorography but in a vernacular (or at least translated) context. One of the few places *localliche* survives is in the anonymous English version of Guillaume de Deguileville's *Pilgrimage of the Life of Man* from about 1450. In the poem, Sapience counsels Aristotle about the Eucharist, and the conversation turns to the size of a space. Aristotle wonders how large things such as "al the sovereyn good" can be contained in a small thing like bread. Sapience answers with three analogies: a heart containing a sizable quality like the good, a mind containing thoughts of Greece and Athens, and shards of a mirror reflecting back the whole of a person as much as a whole mirror. The philosopher asks her further about her description of the bread, whether "vnderstonde ye that localliche, virtualliche [potentially], or oother wise" and, although she answers that she does not understand the bread "localliche," she says that she employs her examples to show that "in diverse wises in the litel places these thinges ben put."[17] "Localliche," a translation of the source French *"locaument,"* here shifts the focus of the word *local* from a physiological realm to a psychological and indeed a spatial one in the sense that *localliche* means a part standing for the whole.[18] The sacramental iconography is clear, but it also echoes Ptolemy's initial definition of a chorographical space as "an impression of a part . . . rather than the whole body."[19]

Because the words *local* and *localliche* are only partially revealing, it may be tempting to turn to words in Middle English such as *space, place, there-aboute,* or to similes that describe discrete spatial areas; but the fact is that these words and the similes were applied to areas of any size. *Place* as a noun or verb could refer to areas and objects of unspecified extent, or *place* could refer to a particular location, or it might again be used in a medical context to refer, for instance, to how much volume an organ takes up. In the sense of physical space separate from human bodies, *place* in the Middle English corpus refers to an area as small as an individual room or as large as a shrine, a public square or marketplace, a battlefield, a town, a whole city, a whole realm, or even the unspecified and very broad abstraction of "time and place." Middle English *place* and *space* can also indicate a physical area in the abstract sense of a spatial physical extension, and the words have a common temporal sense of a period of time or an occasion.[20] But all this is not meant to imply that medieval writers neglected to consider the word *place* and its various meanings. They noticed the valences of the meaning of *place*, some of them pertaining to the characteristics, edges, and contents that are typical of a local area, and they occasionally use the word to exploit a certain productive tension between a smaller place and a larger space or other area as Chaucer did in the Prioress's Tale.

The fungibility of *place* and *space* is perhaps inherent in the words, a flexibility that continues today, but such pliability also appears to reflect scientific discussions of the concepts of place and space that were in flux in the late Middle Ages. Aristotle's definition of place or space is as "the limit of the surrounding body, at which it is in contact with that which it is surrounded" and "the first unchangeable limit of that which it surrounds."[21] Medieval scientific thought often followed Aristotle in that "Latin scholastics . . . conceived the place of a body as the innermost, immobile surface of the containing body in direct contact with the contained body. The containing surface was held to be exactly equal to the body it contained and distinct and separable from it."[22]

Ancient and medieval philosophers, including Roger Bacon, Robert Grosseteste, Thomas Aquinas, and Giles of Rome, however, found problems in Aristotle's definition of place since he insisted that place be immobile. For instance, they gave the example of a ship and its surrounding water, both of them moving relative to a shore. They also pointed to the fact that the outer circle of the heavens is in motion and so should have a place, but since it is the outer boundary, there can be nothing beyond it that encloses it. Aristotle's theory of space ultimately seems to require that a body's limit or the surface of contact between an object and its container also be somewhere, meaning that his idea could lead to infinite regress. Various solutions were proposed; for instance, in mid-thirteenth-century Oxford, the scholastics supplemented the body-container idea with a further qualification that the containing surface had to be locatable in terms of fixed points in the cosmos.[23] That is to say, the Greek *topos* idea of place as container gave way to other ideas about where something is, Latin *ubi*. Some earlier medieval commentators on Aristotle discussed the location of things in the sense of a position fixed in the cosmos. Thomas Aquinas questioned the meaning of place when he examined the problem of whether a place is a quality that inheres in a body or is something external and added to it; he held that place in the *ubi* sense is external to a body, but others suggested a place was a property of extension related to a body in itself. Albertus Magnus proposed that *where* may be expressed by the relationship between the body and the containing place, thus returning it closer to *topos*. Others argued that the place *where* something might be was a kind of imagined void, which anticipates the sense of space found in Descartes and Newton. William of Ockham denied that *where* was a reality.[24] The *where* of something is, in addition, important to Aristotle's idea of natural place, which will be addressed in a later chapter.

In Middle English, the words *place* and *space* are fairly synonymous, but the distinctions between the *ubi* idea of positional place and *topos* as a container-contained relationship are at play in Middle English literature, which occasionally in quite explicit fashion acknowledges the scientific ideas of *topos* in relation to *ubi*. Chaucer exploits the rhymes of *place* and *space*, and the scholastic differences between *topos* and *ubi*, and he does so also to indicate character and to develop themes. In the Knight's Tale, for example, Theseus's lists are built according to precise measurements, their contents compactly designed: "And shortly to concluden, swich a place / Was noon in erthe, as in so litel space."[25] The arena is spatially efficient in an Aristotelian sense of place in that one object (the lists's "place") fits into the earthly area of the setting (the "litel space"), which is also of course where Palamon and Arcite first fought over Emelye. The thematic point being made is one in which the characteristics of a local physical area subtly foreshadow or otherwise indicate the heart of the narrative. In the Knight's Tale, the arena's physical suitability is testimony to Theseus's organizational acumen, which readers see later in the tale in his political ability to arrange the marriage of Palamon and Emelye, an arrangement that not coincidentally occurs in the context of what Fradenburg calls the "boundless credence" of the prime mover speech—that is, a speech that begins with an evocation of medieval physics.[26]

A similar rhyming acknowledgment of scholastic descriptions of *place* and *space* occurs in the Reeve's Tale, set in the outskirts of Cambridge. Symkyn taunts the students Aleyn and John that they can construe a larger space out of a smaller enclosure. He says,

> Myn hous is streit, but ye han lerned art;
> Ye konne by argumentes make a place
> A myle brood of twenty foot of space.
> Lat se now if this place may suffise,
> Or make it rowm [roomy] with speche, as is youre gise.[27]

William Woods finds the closest echo for these lines in the writings of the fourteenth-century philosopher Albert of Saxony, who described it as within God's power to enclose any sized object or physical area within a millet seed. Woods analyzes Symkyn as mocking the clerks' knowledge, a taunting that in the end exposes the Miller's own pretensions, spatial and otherwise: "[T]he central emptiness must be that of Symkyn himself, driven to expand into outer or inner space, because he is unable to accept the nature of his own small place." Symkyn's cleverness and Chaucer's multivalent insinuations are effective only if an audience recognizes the scientific allusions and ideas.[28] Chaucer seems to be appealing to the audience's awareness of the scientific senses of place and space, relying on this knowledge to cast a skeptical eye on spatial inflation for political, scholastic, and personal ends. *Place*, therefore, has an important semantic field that Chaucer explores and that contributes to the structure of some tales. However, he, like other Middle English writers, also uses the word in a variety of situations that are not spatially significant (to do with an unspecified length of time, for instance) or that could mean a space of any size. Other words to denote defined areas existed of course, such as the specialized language of the administrative units of a *shire*, *hundred*, or *wapen-take*, but many words are as nonspecific as *place*. Otherwise, adjectives are added to a noun to denote the size or characteristics of a certain area, such as *litel*, *particuler*, *private*, or *roughe*, *grene*, and so on.

Estre is a more common noun in Middle English to denote a defined area in the absence of the noun *locality* (which also does not exist at the time).

Estre derives from the French verb *estre* or *ester*, meaning "to stand," and its closest antecedent in French appears to be a substantive meaning "a stay," which then appears to have developed to mean "a place." Middle English uses *estre* quite frequently to denote places of different sizes, but the semantics of *estre* tend to cluster around more narrowly chorographical places.²⁹ *Estre* has both the sense of a defined piece of land and the connotation of a condition in John Gower's *Confessio Amantis* when the emperor Constantine addresses Saints Peter and Paul, who have appeared to him in a dream, and asks them "of youre name or of youre estre."³⁰ In Robert Mannyng's *Chronicle of England*, estres are most frequently cities; Exeter, Portchester, Gloucester, Winchester, Circester, and Rochester are each called an "estre."³¹ Chaucer's narrator in his translation, *Romaunt of the Rose*, says he has been in all the "estres" of the garden, suggesting a certain degree of individuation to the locations within the larger area.³² Estres can even be interior, such as the "estres of the grisly place . . . the great temple of Mars in Trace" in the Knight's Tale or, as we saw, the "estres" of the more modest bedroom in the Reeve's Tale.³³

The scientific background I have outlined reveals that the central ideas about space were unstable at this point in history. Middle English literature embodies the uncertainty, and Chaucer exploits the constricted sense of *estre*. If *estre* is tied up with situation in a more abstract sense, and the fundamental underpinnings of space are uncertain at the time, then we might expect writers to use imagery of space or place to explore ambiguities of being. As mentioned, this is where literature differs from the local maps because literary works are much more invested in the implications of space not for just any objects but specifically for human identity, activity, and sociality. Lydgate's *Siege of Thebes* and the illumination explore them all but especially question space in relation to social cohesion.

Estral Complexities in John Lydgate's *Siege of Thebes*

The *Siege of Thebes* and the well-known illumination of the opening of the poem that accompanies it in British Library, Royal 18 D 2, provide an opportunity to explore space in relation to collective being. The Prologue and the opening of the first part of the *Siege*, and the illumination in the manuscript (figure 8), illustrate the characteristics of estres, the variously edged nature of a local space, the characteristics of the contents of a local area and how they relate to each other, and the multiple temporalities of estral space, all of which we saw in the local maps. The poem and the illumination are, however, more than illustrative of these ideas. The illumination, for instance, engages in some ways with a presently realistic depiction and in other ways militates against it. This may be because the Royal manuscript and its illumination are from the second half of the fifteenth century, some forty years after the poem's composition and after Lydgate's death, and so is an interpretation on the part of the illuminator. Nevertheless, the Prologue, tale opening, and illumination graphically display the complexities of estral space in separate but overlapping ways.

The Siege of Thebes opens as a story of interpellation, and this is where most interpretations of the poem and the Royal image have tended to focus. Analyses of the *Siege* and the illumination see the Prologue, tale, and image working together to incorporate Lydgate into a lineage of Chaucerian *auctoritee* and

FIGURE 8. Illumination of John Lydgate's *Siege of Thebes*. British Library, London, MS Royal 18 D 2, fol. 148. © The British Library Board.

into Chaucer's fictional world of the *Canterbury Tales*. The *Siege* further invites communities of readers to join intimately in a Chaucerian and Lydgatean world of pilgrimage and tale-telling. Trigg, for example, analyzes the temporal extension of Lydgate's imaginative inclusion into the pilgrimage to current readers of Lydgate and Chaucer when she discusses the illumination's appearance on the cover of the *Riverside* paperback edition.[34] These and earlier interpretations of the image, 15.5 × 11 inches in the manuscript, argue that the monk in a gray caul in the near center at the rear of the traveling men is Lydgate, and the man to his right in white with a mantle is either the Host or Chaucer himself, which is interpreted as an interesting ambivalence, even conflation, in its own right on the part of the illuminator. Nearly all the pilgrims are speaking, gesturing, or otherwise interacting with each other, suggesting a comfortable *communitas*, and at the center, Lydgate the monk leans in to listen eagerly to the Host/Chaucer. It is as though the pilgrims are an enlivened picture of Lydgate's Prologue because of the vivid coloring of the clothing and horses; the orientation of the people and the animals; the familiar hand gestures and horse poses, clearly indicating a journey to the viewer's left, to the West and London; and the familiarity among the pilgrims.

The argument Trigg and others make is that these avid conversations and the gestures enable even modern viewers of the illumination to identify with Lydgate. Indeed, the Prologue, which precedes the image in the manuscript, encourages this committed engagement with the persons of the return journey from Canterbury when Lydgate characterizes Chaucer as the "Chief registrer" of the pilgrimage.[35] This descriptor, often overlooked in analyses of Lydgate's characterization of Chaucer perhaps because of other lengthy and more florid descriptions of the *auctor*, is significant because "registrer" indicates Lydgate's characterization of Chaucer as a truthful recorder. In his *Fall of Princes*, for example, Lydgate describes Diligence as "registreer to suppowaile [assist] trouthe,"[36] and elsewhere in Middle English a "registrer" is used in legal contexts and as an appellation of praise for a historian. Lydgate is admiring Chaucer because he is accurate, and readers are similarly drawn to the image because it promises a chance to witness the pilgrims as though in person. They are invited to project themselves into the scene to become intimate with Chaucer and hear his words directly.

These kinds of interpretations, though compelling, nevertheless have a tendency to overlook the local matters in the poem and the Royal illumination. An emphasis on the characters neglects the estral chorographies that display diverse areas. These regions often work in tension with each other in the *Siege*, thereby disturbing the sense of community inside the image as well as between the image and audiences for Chaucer's and Lydgate's worlds. The tensions also bring to light differences rather than parallels between the poem and the image so much so that it is tempting to see the illuminator as deliberately working against the communal coherence that Lydgate intends. Taken together, the poem and the image might reinforce a sense of community even beyond the temporality of a Chaucer-Lydgate-audience complex, but they also have the potential to displace the contents of the area and the characters within the narrative and visual frames, and therefore to disrupt the viewer's interactions with the space. How the local is defined and the effect of its edges, contents, and times in the *Siege* and its illumination are the subject of what follows. Instead of *communitas* across time, attention to the local space suggests a more disrupted, less harmonious sense of group identity.

What is significantly local in the *Siege of Thebes* and its illumination? To begin with the Prologue, although the prosody slides from one topic to another in a sixty-five-line incomplete sentence, the opening is spatially structured: a large scene telescopes step by step into a smaller and smaller one until we end at a very specific location set within a delimited estre. Like other telescoping beginnings, such as the Prioress's Tale, this one is also allegorical, in this case standing for literary genealogies of authorial, vernacular, and national kinds, as the scholarship on the poem has pointed out. Like its most direct model, the General Prologue of the *Canterbury Tales*, Lydgate's opening dependent clause describes at some length the astral-temporal time of spring, which is also the time, we are informed, when the *Canterbury Tales* were told "complet."[37] The Prologue goes on to eulogize Chaucer before continuing to skate, after another twenty lines and without pause, back to the *Tales'* opening setting to name the Tabard at Southwark. From that specific place, Lydgate's next marked interpretation of the *Tales* transitions to a later time when the pilgrims (only male pilgrims are named in the Prologue to the *Siege*) have apparently reached Canterbury, and that is where Lydgate says he meets up with the company. The constriction of space thereafter, as it were, accelerates when, by chance, Lydgate says he has stayed at the same inn as the pilgrims in Canterbury. Here the Host appears and pressures him to join with them and to "be bound to a newe lawe / Att goyng oute of Canterbury toune," namely to tell another tale. The Prologue states that the first tale will be told the next morning before the pilgrims, now including Lydgate, have their first main meal at "Osspryng," some ten miles west on the road from Canterbury. The Prologue goes on increasingly to define the physical location, for that morning once they have exited the walls of the town, "whan we weren from Canterbury paste / Noght the space of a bowe draught," the Host demands the tale from Lydgate.[38]

Time and place, incidentally, are similarly drawn together later in the poem. In part one, for example, Lydgate's narrator, like Chaucer, employs the rhymes of "space" and "place" to describe in some detail the building of Thebes and soon thereafter goes on to contrast that building project with the time he has remaining to tell his tale, which will be long but within spatiotemporal limits: "my tale which that ye shal here / Upon oure waie wil lasten a longe while, / The space as I suppose of seven myle."[39] The opening of the Prologue and this moment later in the poem therefore function to telescope attention to the environs of the town of Canterbury as an estre, a local space that contains a limited amount of physical distance/time within which to tell a tale: between just outside Canterbury and the town of Ospringe. The prospect of the rest of the return journey to London thereafter appears to offer other spaces that can accommodate more tales in additional episodic locations.

Where the Prologue might be said to begin at an edge just outside the gates of Canterbury and point toward a terminal end point at Ospringe, the illumination defines a smaller estre. The Royal illumination comes after the Prologue and before the poem proper in the Royal manuscript, framed, as it were, by the red letters of "Explicit Prologus" that precede the picture and "Prima Pars" that follow it. Most interpretations understandably focus exclusively on the illumination as illustrative only of the Prologue, construing that this part of the poem and image are complementary. Analyses concentrate on the pilgrims who emerge out of the frame on the right and are now disappearing off to the left, heading toward London. The company of men provides a sense of boundedness to the space even though we cannot see the end point on the left of this stage of the pilgrims' journey or, incidentally, the far right walls of Canterbury town. A viewer can see how the illuminator suggests this neat sense of an estre and of the human community. On the right, behind the last visible pilgrim, the road leads back to the open portcullised gate of the walled town of Canterbury with its rounded fortifications and arrow loops, brown-roofed buildings, and outstanding larger structures. The pilgrims are clearly tied in the space to that town while they move off to the left.

The estre outside Canterbury in both the Prologue and the illumination appears to afford a precise and satisfying sense of place, an effect that critical interpretations understandably follow. But fault lines in this interpretation soon appear if one looks again because the local area in the poem and the illumination is a more complicated space than the neat correspondences suggest. One of the tensions might be due to being outside the town while also moving the ten miles to Ospringe, which is out of sight, a slight motile strain that is the first hint of deeper conflicts. To begin, if we recall the homogeneous signs and heterogeneous contents of a local area, then Lydgate's allotted space-time between a bow-shot from Canterbury and Ospringe is concise, perceivably uniform, and therefore classifiable as homogeneous. It is a ten-mile unit to which Lydgate will fit his tale despite the fact that the whole poem of the *Siege of Thebes* ends not with Ospringe (the poem does not return to the frame setting) but instead with a vision of a more abstract ideal future where no war or conflict exists. Maps like the more detailed Inclesmoor one clearly show that a local space was understood in terms of homogeneous area but that it also could diversely weight content within a space and thus create a more individuated and dispersed sense of attention. In the itemized Inclesmoor map, attention was drawn to particular sites and details in an uneven fashion. The eye moved closer and closer to individual features, each of which assumed distinct characteristics. In the case of the *Siege*, these effects are evident, and the image amplifies them. The reader and the viewer notice, for example, the gray friar's "thred bar hood" in the Prologue and, if one looks very closely, the fraying along the shoulder and hood in the illumination.[40] Or, for example, each of the rocks on the road that winds its way out of the portcullised Westgate of Canterbury is individuated. We are back in the realm of the miniature with all its attempts to draw the eye in to marvel at the distinctiveness of each item in a local area.

So, on the one hand, there is a sense of uniformity and neatness of a ten-mile extent, and on the other, attention is pulled unevenly toward distinct items. This second irregular sense continues in the illumination because the edges, the contents, and the very definition of the estre do not work in harmony. First, criticism of the image that focuses only on the pilgrims is incomplete. Royal 18 D 2 contains three principal areas: the pilgrims in the foreground linked to Canterbury on the right; indigo and pale azure hills, trees, and buildings in the far background; and the set of large tan buildings on the left. Unlike the Prologue with its clear demarcation of one pilgrimage portion and one narrative length, the illumination shows three individuated estres within the one larger space of the image. The illumination is therefore similar to the map of Sherwood Forest, where the mapmaker schematically divided the whole area into separate self-contained parts, which was in part a solution to a cartographic problem of a large area. But even this formal strategy to organize space does not work so well in the illumination.

Each of the three areas within the illumination appears at first glance to be self-contained. Potentially the most diverse main portion is the one with the pilgrims and the town on the right with, as I pointed out, the road linking them together. This estral space may be divided further into two parts—the pilgrims, horses, and road bordered by grassy areas in the center foreground, and the walled town in the right rear. Each of those spaces has edges that contain self-consistent elements even if they do not correspond to reality. They are fictional in relation to the characters of Lydgate and Chaucer in the *Siege*, and

Canterbury looks as though it rises up a hill toward the rear of the image, but the actual town is in a river valley. The illuminator has rendered the roofs, walls, and buildings from a slightly elevated position so that they appear to be sloping downward toward the viewer to make the town features visible. The second main area of the illumination is far in the background, a dominantly blue area that contains a larger estre of leafy trees, a town or two, and tall hill peaks in the back beneath a soft sky. Its blue shadings suggest a Netherlandish influence. Like the town, that estre is a largely contained space, and where the maps created veracity sometimes by means of self-consistency, the town and blue background space are self-consistent even if they do not correspond to reality.

The third main edged and potentially self-contained part of the image is the large tan building or cluster of buildings immediately behind the pilgrims on the left. These buildings present the greatest challenge to the coherence of the illumination as a whole. Contradictions abound. The buildings are rendered in a distinctly different manner from that of the town and the blue background. They appear nearer to the viewer than the town, and they present side-on, ground-level elevations with the buildings' walls flat against the spectator while its roofs and sidewalls recede in perspective, unlike any of the town's structures. On the left, the buildings continue off the side of the image but seem as though they will stop. The space containing this imposing group of buildings is not coherent, and even though its edges are defined, they too do not function in a clear manner. Behind and to the right of the structures, the land seems to drop away and not be connected to the land beside the left walls of Canterbury; but the right-hand edges of the tan building or buildings are nevertheless associated with Canterbury because in the right foreground is a strange, darkly colored jagged fence that reaches from the right side of the buildings to the tower adjacent to Canterbury's Westgate. Furthermore, the area in front of the estre containing the tan buildings is joined to the pilgrims. There is a worn patch of grass that extends from in front of the door of the nearest of the tan buildings to the pilgrims, a path that is of the same nature as the one the pilgrims are on. The grass behind the pilgrims extends to the front of the buildings on the left and the dark fence on the right. If a viewer looks closely at the doorway in this very clear illumination, he or she can make out a half (or stable) door and behind it a tiny figure. The building cluster therefore initially appears to form a distinct area from the pilgrims or pilgrims-Canterbury space—it is rendered differently and appears separated by the large depression in the ground on its right—but it also is demonstrably part of the pilgrims' area in the foreground and therefore also connected to the town farther back.

If there is some form of separation between the pilgrims and the tan buildings, it is, like the edges of some of our maps, an edge that is gradual, revealing in transitional steps the similarities and differences between the group of pilgrims in the foreground and the buildings in the left rear. Even if we overlook this spatial tension and the inconsistencies, the building group on the left does not correspond with any single historical situation. The imposing tan buildings could be Canterbury Cathedral, but the cathedral resided, and still resides, within the city walls. It is quite clearly present in the illumination in the walls of the town on the right, standing up as a large gray and blue gothic tower; it is even correctly placed in the town in relation to Westgate. The tan buildings on the left nevertheless have a building that seems to look like the cathedral: in the rear of the structures, a taller blue-roofed structure with crenellated terraces and elongated windows rises up behind lower-peaked but still impressive structures in the foreground. But if we want to identify the cluster with the cathedral, the large building appears to have two transepts where the actual cathedral has only one. So what are the large tan buildings outside the walls? No such building, religious or secular, existed in medieval Kent near Canterbury. Ospringe had the Maison Dieu, but that was a small hospital; Boughton under Blean had a parish church, but it was not nearly as large as the pictured structure; and Harbledown had a hospital, but it too was not sizable. Rochester Castle might be a candidate, but the image looks nothing like its buildings, which were some twenty-six miles away from Canterbury and well beyond Ospringe.[41]

Art history does not cast any further light on the spaces of the image. Art historian Jean Givens has helpfully distinguished among realistic, naturalistic, and descriptive modes of illumination in the late Middle Ages, wherein a realistic style refers to real-world things, naturalism may or may not refer to external items but "register[s] the overall irregularity and variety inherent in living creatures," and the "descriptive" mode is not necessarily lifelike but "visually communicate[s] information concerning the external and, sometimes, internal physical structure of real-world objects and phenomena."[42] It may be tempting to focus on the realism or even the naturalism of the illumination, but if anything, the image includes all three modes, even all three modes within each estral part. The pilgrims, road, and rocks in front would be realist except they are fictional figures, the town on the right is ostensibly naturalistic but may just as well be descriptive, and the buildings on the left may be similarly naturalistic and descriptive, while the blue hills in the background seem descriptive but also lifelike.

One final detail also pulls against the coherence of a local area that seems to contain and not contain the pilgrims and Canterbury along with the tan buildings, and it might even be said to strain against the closed masculine community of pilgrims in the foreground: the figure behind the half door. Close inspection of that tiny part of the illumination reveals a woman in a (secular rather than ecclesiastical) white caul with forehead clearly visible and a red bodice. She has placed her left hand on the half door, and she looks in the opposite direction from the pilgrims, not toward Ospringe and the road where Lydgate's telling of the tragic history of Thebes will transpire, but instead back toward Canterbury. It is tempting to read her as a tiny exception to the boisterous community of male pilgrims and those in the audience of this dramatic scene who wish to join with them, a supplement that looks away from the direction of the pilgrims and their tale-telling voyage, a kind of resistant reader who remains behind and thinks about something else.

The edges and contents of the estres in the illumination—indeed the very definitions of an estre—are strained and work in tension with the attempted unification of the *Siege* with the "complet" *Canterbury Tales*. The temporal diversity of the image is complicated beyond any neat coherence of Lydgate joining up with Chaucer's *Tales* and the audience finding satisfaction in vicariously joining the pilgrimage or at least witnessing its intimacy and communality. Recall that a local area in literary and graphic presentations can depict the present and one or more moments from the past. The Prologue and the image initially seem to work well together in this regard, reinforcing each other to create social and literary harmony. The Prologue attempts to bring past and present together, albeit in metonymic and somewhat uncomfortable fashion as it slides from the time of year, to the fictional time of the *Canterbury Tales*, to the time when Lydgate enters that fiction and meets up with the pilgrims. The illumination might also be attempting a similar harmony by presenting three (or at least two) estres. In a temporal reading, the present in the image is the pilgrims within bow-shot of Canterbury, which is also the time when Lydgate must tell his tale. The past in the illumination is most immediately Canterbury town, the place the pilgrims have left and, as we know, the never-reached end point of the *Tales* themselves. The tan buildings could therefore be a different temporal past entirely. They might be Bury St. Edmunds, Lydgate's own monastery that he has left to visit Canterbury, although the Benedictine abbey had only one transept and the secular woman also militates against such an interpretation.

Another possible interpretation of a temporal kind lies in the poem proper. While the Royal illumination has been read in relation to the Prologue, it actually comes between the Prologue and the main narrative of the *Siege*. A part of the image could refer to the tale to come. The tan buildings could be a rendering of Thebes, to be sure an anachronistic placing of the city in the "present" of the poem with Lydgate and the other pilgrims, but an anachronistic device that is not unusual for medieval illuminations. The opening of the first part of the poem details at some length the construction of the city, "Bylt and begonne of olde antiquité."[43] One story Lydgate recounts in this opening is of Amphion building it out of music in the middle of the Greek's land. As we have noticed, including a scene from the past in a present estral location is a technique that was not unusual for local spaces in general. The Isle of Thanet map contained both contemporary buildings and natural features, *and* the tale of Domne Eafe and her deer. But here in Lydgate's case, the illumination's relationship to the poem is complicated. There are no Theban walls "made of lym and stoon" as Lydgate describes them in the opening lines.[44]

The image may ultimately be too ambiguous to interpret with certainty. What we can observe, however, is revealing: that this estre is not a simple, or simply unified, space where each part works easily either with each other or even internally within individual localities. In the Prologue, the temporal slipperiness is unnerving, and in the image the separate areas also upset closure because they do not go together or cohere internally. The illuminator, it would seem, intentionally or unintentionally echoes the spatiotemporal uneasiness of the poem. Both may foreground the present vibrant reality of the pilgrims with Lydgate and possibly his *auctor* among them, but both also subtly strain against that easy coherence and community. Lydgate's Prologue and the image in the Royal manuscript not only embody but also complicate the kinds of presentations of local space that we have seen in contemporary semantics, literature, and maps. After telescoping in its opening lines, the *Siege* defines the estral via the space Lydgate is allotted to tell his tale, a quite precise distance to Ospringe, but we are aware that the voyage back has many more potential segments in which someone else might tell a tale. That openness is extended and troubled in the illumination's clashing spaces, whose boundaries overlap with each other and separate.

A local area in these samples of Middle English literature, maps, and other phenomena is a fairly defined space whose edges are boundaries of connection and difference. An estre may contain objects with a sense of neatness, but its contents are usually also complex, often hovering between, on the one hand, a sense of specificity and, on the other, self-consistent uniformity. Looking forward in history, one can see that uniformity will tend to take over so that objects are thought of, or at least depicted, generically according to a system; this is especially true of cartography. It is a misconception, however, to impose that sense of consistency, or spatial hermeneutics, on the late medieval past. The local space, one that appeals even in a sensual manner to characters in a poem and an audience, is not a simple area without contradiction and conflict.

CHAPTER 3

Horizonal Space
Measuring Local Area with Astrolabes, Quadrants, and *Topographia*

In *The Perception of the Visual World* (1950), the psychologist James J. Gibson sought to describe the underlying physical and mental principles of how people see spatial area. Scientists in the Second World War, including Gibson himself, had attempted to improve the ability of pilots to land airplanes, but Gibson found that their experiments were largely ineffective because they were simulations of flying performed in controlled laboratories. Gibson realized the underlying problem lay with aerial research, so he turned his attention to examining objects as they appear at ground level. His conclusions led him to begin new lines of inquiry that resulted in his theory of how people perceive things visually by means of the fundamental properties of edges and surfaces (before introducing the further problem of depth). Gibson called his new approach a "ground theory" rather than an "air theory" because the important factor for pilots was not the air through which they flew but instead the face of the earth and the horizon along with other edges. His research fundamentally changed the direction of perceptual psychology.[1]

When one considers how science, literature, and the mechanical arts perceive space in the late medieval era, the striking fact is that the overwhelming quantity of observations are from the ground. It was possible for someone to imagine vision from the point of view of someone or something high up—classical myths, biblical dream visions, and other sources contain imagery from a high elevation, and these visualizations continue in the Middle Ages—but the bulk of sights are from ground level. This dominant mode of spatial hermeneutics may be called *horizonal*.[2] By *horizonal*, I mean two things, first that the area perceived is ordinarily horizontal, largely parallel to the earth in a zone or band near the ground or sea. Viewers in historical, literary, and scientific writings, and in the visual arts most frequently look out from a place on earth (occasionally slightly elevated) rather than down from a position that is far above the ground. The second aspect of a *horizonal* understanding of space in science, mechanical arts, and literature is that a viewed area is usually bounded by the horizon, defined in John Trevisa's late fourteenth-century translation of Bartholomaeus Anglicus's *De Proprietatibus Rerum* as "þe cercle to þe whiche þe siȝt strecchiþ and endiþ."[3]

This preponderance of horizonal perceptions is largely overlooked in scholarly study, perhaps because the spaces and views of them seem unremarkable, plain, or without significance. While the area near the earth and bound by the horizon appears somehow neutral or meaningless, it is not. There are, of course, variations in kinds of horizonal spaces, such as maritime areas and terrestrial ones, mountainous regions and riverine ones, but that is not so much the point as to consider the possibilities of what horizonal areas "afford" a creature, to borrow Gibson's best-known term from his later 1979 work, *The Ecological Approach to Visual Perception*. An "affordance" in Gibson's theory is what an environment "*offers* the animal, what it *provides* or *furnishes*." An environment affords an animal certain possibilities for interactions involving motion, eating, sociality, and so on. Gibson reasoned further that affordances exist in the relationship between an animal and the environment; space is not fundamentally reducible to "the scales and standard units used in physics" but instead depends on a "complementarity" between a creature and an environment, so it is different for different animals.[4] The evidence I present suggests two initial deductions that may seem evident but are important beginning points for analysis of horizonal space in the late Middle Ages. First, interest lies in distinct objects in locations around a viewer's position that are seen in relation to a typical ground level. Perception is in this sense *objective*, keyed to individual items. Second, each of the objects is *complementary*; each is apprehended in relation to the position of the viewer. This "complementarity" can be turned around so that the viewer is another object within the landscape and can be seen in relation to other objects.

The sources I examine in this chapter are necessarily diverse—mechanical devices, visionary literature, historical records, and more—but the main evidence of what horizonal space affords viewers in the late Middle Ages is drawn from astrolabes, quadrants, and related treatises on geometry. These devices and writings show the ways one can measure the heights, lengths, and areas of objects within a view on earth. In the previous chapters I looked at one geographical area—local space—via maps, Lydgate's *Siege of Thebes*, and the later manuscript illumination of the *Siege*, all of which define local space while occasionally also troubling a sense of spatial cohesion. The current chapter registers the sense of the local, but here the focus is more directly on horizontal space itself and only implicitly takes into account the scale of the areas involved. In the next chapter that is the counterpart of this one, I look at two narratives about travel for their literary insights into horizontal spaces. Other ways of thinking about horizontal space, and other kinds of texts and devices are also likely revealing about the fundamental qualities of space in the period, and it is hoped that this discussion will lead to further investigations into the nature of spaces and the meanings they afford.

One particular advantage of astrolabes, quadrants, and treatises on geometry, however, is that they address horizontal space directly. Also, their parts and their qualities, which at first glance seem in some respects mundane and straightforward, turn out to be advantageous for thinking about space because they address located objects of an everyday kind. Furthermore, these devices and writings belong to the mechanical arts and therefore lie at the intersection of theoretical and practical understandings of space. Finally, they implicate the viewer with the world in revealing ways. Gibson's *Ecological Approach* goes on to extend the senses of objectivity and complementarity so that a study of a space's affordances can disclose "the 'values' and 'meanings' of things in the environment."[5] Astrolabes, quadrants, and the treatises on them may not explicitly address what I have called a hermeneutics of space—they do not typically rise to a level of consciousness about the connotations or implications of space—but they nevertheless implicitly explore the meaning of objects; they do not contain overt theorizations of space but reveal what Jeff Malpas calls "complex spatial and topographic frameworks."[6] They provide a sense of the significance of objects within an area, the ways that people perceived objects, recognition of the viewer's self, impressions of the ability to move in a space, and more.

A smaller body of evidence I investigate in this chapter for other features of the horizontal is the rhetorical trope of *topographia*. *Topographia* links objects to their immediate locations with a particular set of cause-effect relationships. The classical heritage of *topographia* that the Middle Ages inherits states that objects and activities within an area—hills, plants, animals, farming techniques, and so on—arise in fitting or unfitting fashion out of a particular place and time. In a sense, then, it is possible to read *topographia* as going beyond an Aristotelian idea of place as a container-contained relationship.

Recall Aristotle's definition of where something is as "the limit of the surrounding body, at which it is in contact with that which it is surrounded" and "the first unchangeable limit of that which it surrounds." An object's place is the intersecting surface between the item and the space around it, a surface that is indistinguishable from the object and the space.[7] *Topographia* extends the Aristotelian physics of place to include causal and even ethical relationships between a place and an item within it. It also diverges from Gibson's affordances in that the space not only "provides or furnishes" certain items and qualities. There exists a kind of geographical determinism at the time, a theory that sees biological and cultural phenomena as the consequence of regional environmental features. The point in topographical writings is sometimes made that both contents and locales are better or more natural if they suit each other, although this idea is subject to some debate.

This chapter's focus on the horizontal and the mechanical arts is offered in part as a corrective to previous scholarship on astrolabes and cosmography. Studies have tended to focus on the astronomical functions of astrolabes and theories of the universe, and they have obscured the arguably more customary terrestrial uses and implications of astrolabes and the more common horizonal perspective. Readers are likely aware of the many valuable histories of medieval astrology and astronomy, scientific implements, and literary uses of astral phenomena.[8] My argument urges that it is equally valuable to study the more everyday areas that occupied late medieval culture and to take note of scientific and systematic thinking about them. Nevertheless, it is unreasonable to deny that perspectives that differ from the horizontal existed. This chapter therefore addresses the exceptional cases of spatial perception—namely, written and graphic depictions of viewpoints that are from above the earth rather than bound to it. What is the significance of these views? Do they interact or not with horizontal outlooks? What are their implications for spatial hermeneutics in the late medieval era?

Abstraction from the Earth

According to the French geocritic Bertrand Westphal and the historian of cartography David Woodward, the late Middle Ages were a time of change in the fundamental qualities of spatial perception. Westphal, following Paul Zumthor and others, writes that "a vertical relationship with Heaven ended up by erasing itself for the benefit of a horizontal projection" and, at the same time, this horizonality also began to "open" up space "beyond the horizon, beyond the perceptible limits of an area." Western Europe "opened . . . onto the idea of undefined space, which is virtually infinite" rather than one that was in some sense closed in, and "the territory that is the equivalent of *here* started to vibrate with curiosity and interest, and then to gape wide open at its margins."[9] Woodward suggests something similar. He acknowledges that there have been exaggerated claims about a transition in geographical thinking from the Middle Ages to the early modern period, but he posits that "the overwhelming conclusion is still that a rapid and radical change in the European world view took place during the fifteenth century . . . a transition . . . in the way people viewed the world, from the circumscribed cage of the known inhabited world to the notion of the finite whole earth."[10] Woodward draws his evidence from maps of the earth that emerge in the fifteenth century that begin to utilize Ptolemaic mathematics to lay out a geometric vision of the earth. His evidence appears initially problematic in terms of its significance and its teleological narrative that arrives too conveniently at modern forms of mapmaking, but if taken within the context of similar forms of maps from the late Middle Ages, as Woodward does, one cannot deny that something different is occurring in terms of thinking about geographical space in the late Middle Ages.

Westphal's propositions are fascinating for what they say about one aspect of the horizontal, namely the outside edge of the viewer's perspective. In this argument, the horizon comes to stand not as an enclosing device but rather as a realm of interest and possibility. And Woodward's observations address the other aspect of the horizontal; instead of the preponderance of viewpoints being located near the ground, one could view the world and its spaces from an overhead perspective. The latter is indeed a "vertical relationship with Heaven" in which one can identify in some way with a divine perspective and look down on the earth even in a judgmental fashion. As opposed to the horizonal view, the overhead viewpoint renders a different shape, a space whose nature is quite different, and a place whose contents are organized in particular ways. The evidence for exceptions to the ground-level horizontal perspective includes the local maps discussed in chapter 1. Recall that the local contents of the maps could be singular with distinct characteristics, the space broken up into individuated sites that attract the attention of the viewer one by one, or the maps could contain genres or types of objects in arrangements so that the local space became more flattened and equable, encouraging a view of its area as a system of interacting spatial information. These two ways of understanding local space were tendencies with degrees between "heterogeneity" and "homogeneity" as Sir Isaac Newton, Maurice Merleau-Ponty, and Harley and Woodward characterized the two fundamental ways of thinking about space. Either way, the overhead perspective appears to enable thinking about a space as homogeneous more than a horizontal viewpoint. The farther one removes oneself from the ground, the more one perceives it as disembodied, in a word, *abstract*.

The word *abstract*, from *ab-straho*, means to draw away from something, typically a place or person; in medieval Latin, *abstractus* was transferred to philosophy and grammar to distinguish an abstract object from a concrete thing. A *tractus* is a drawn-out or stretched space, giving us a *tract* of land. In a sense, then, *to abstract* means to move away from the ground and particularly from a specific locale on earth. In Ptolemaic terms, it is to withdraw from a topos or choros to the geos, a worldview. In Middle English, *abstracte* came to have philosophical and grammatical meanings associated with the word, and it also meant an abbreviation or summary of a longer piece of writing (a *tractatus*, *tractus*, or "tracte"), but it principally meant to remove one's self from earth, which is generalized to mean to remove one's self from worldly affairs.[11] The sense of being physically raised or raising one's self above the earth, however, remains even within philosophical tendencies, the terrestrial sense of existence enduring in texts that describe, for example, biblical scenes or various kinds of *contemptus mundi*, but the earth in these views is different from a horizontal sense of it.

The first difference from a horizontal perspective is that an overhead view renders a geographical space that does not really have a horizon in the usual sense of the word. There is no straight or slightly curved line that can surround a viewer. The horizon is transformed into a circular edge on a world that can be taken in in one glance and is separate from the viewer, which is a very different sort of entity and perspective. Philosophy and literature offer many examples, of course, of people and characters being lifted away from earth so that it becomes a circle. Macrobius, Dante, Boccaccio, Chaucer, Julian of Norwich, and others depict such a perspective from the heavens. Probably the most well-known and authoritative example in vernacular literature is when Beatrice beckons Dante to look back down from the heavens to earth:

> Col viso ritornai per tutte quante
> le sette spere, e vidi questo globo
> tal, ch'io sorrisi del suo vil sembiante;
> e quel consiglio per migliore approbo
> che l'ha per meno; e chi ad altro pensa
> chiamar si puote veramente probo.

(With my sight I returned through all and each of the seven spheres, and saw this globe such that I smiled at its paltry semblance; and that counsel I approve as best which holds it for least, and he whose thought is turned elsewhere may be called truly upright.)[12] The "thoughts" that Dante draws attention to at the end of the passage may of course be turned to God, and indeed one can achieve a godlike, omniscient perspective. Chaucer, for example, translates Boethius as saying that "God, whan he hath byholden from the hye tour of his purveaunce [foresight], he knoweth what is covenable [suitable] to every wight [person]."[13] The most commonly mentioned Chaucerian example of a human being in a position similar to that of God occurs in the conclusion of *Troilus and Criseyde* when Troilus has died, rises up to the eighth sphere, and looks down on the *oikoumene*, "This litel spot of erthe that with the se / Embraced is."[14] In still another genre of writing, the perspective close to the Divine offered to Julian of Norwich in the long text of her *Shewings* is even more distant from the earth. In the famous image, when God reveals to her a "ghostly sight," she regards not just the earth but all creation as "a little thing the quantity of an haselnot, lying in the palme of my hand as me semide, and it was as round as any balle."[15] The divinely abstract image develops into a three-dimensional object in Julian's vision.

The direction of movement in these abstracting gestures can vary. Julian's *Shewings*, for example, might begin this scene with a hazelnut, which is actually earthly (or "homely" in her language),[16] but any terrestrial reality that might be tied to a grounded space or entity tends to dissolve into eternal reality in the *Shewings*. In Julian and other texts that ab-stract, the viewer moves away from the earth so much so that the earth shrinks to a tiny entity and the viewer leaves the earth behind entirely. John Metham's adaptation of the Pyramus and Thisbe story has a similar approach to space when it eulogizes the zodiacal signs as "nobyl deyfyid sygnys" because they "abstracte / From erthly mancionnis to the asuryd fyrmamente."[17] On the other hand, Chaucer's movements in relation to an overhead perspective are varied. They can be like Troilus at the end of *Troilus and Criseyde* or like "Geffrey" in the *House of Fame*, in which the characters move away from the earth so that it becomes a "litel spot," a "prikke."[18] Taken as a whole, however, Chaucer's writings tend to describe the opposite motion. The narrative, often an opening of a narrative, transitions from a large perspective over a vast area to narrow in on smaller and smaller spaces, even to the compact site of the lists in the Knight's Tale or the tiny constrictions of the pit into which the little clergeon is cast in the Prioress's Tale.

Cartography offers an equivalent, though comparatively static, set of graphic works that ab-stract from a horizonal perspective. *Mappaemundi* and especially Macrobian maps of climatic zones—maps of one side of the earth with climate zones of frigid, temperate, and equatorial bands—afford an overhead perspective that sees the earth, or large parts of it, as homogeneous.

This is not so much the case with detailed *mappaemundi*, such as the Hereford map, that have Jerusalem as their center and Earthly Paradise as a kind of spatial, moral, and temporal terminus as though one is supposed to "read" up the map from an earthbound standpoint. Some *mappaemundi* can be schematic to an extreme, outlining only the three known parts of the Northern Hemisphere—Asia, Europe, and Africa—in the T-O form and not much more. In the English tradition, the famous image of the archer shooting at the earth in three manuscripts of John Gower's *Vox clamantis* combines the T-O with the three elements of earth, air, and water in each of the three parts, an image that is moralized in the manuscripts.[19] Macrobian maps are more abstract than even very schematic *mappaemundi*. The orientation of these common maps of the climes is invariably north, the zones are symmetrically laid out on the round earth, and the earth as a whole is measured not by its distance from Earthly Paradise or the Godhead, neither of which typically appears on or around the hemisphere. It is of no importance to an understanding of these maps whether the viewer is looking at the "side" of the earth that contains Europe, the Near East, and Ethiopia, as Africa was often called, or at a side of the earth that shows the Far East.

A lack of orientation on maps also speaks to the degree of abstraction involved in perceiving space from an overhead perspective. Maps that are oriented seem to encourage more immersive experiences, such as in the Hereford map and its presentation of materials leading up through Jerusalem to the Earthly Paradise. Many Western maps are oriented east like this, Arabic maps tend to focus on the Indian Ocean and are therefore oriented south, and some maps of the *oikoumene* are oriented north.[20] But what of maps that have no orientation and can be, or indeed necessitate being, turned to see their features clearly? Of this class of maps are the portolan charts of coastlines and seas, typically displaying the Mediterranean and the Black Sea, and sometimes the coast of Spain, France, and southern Britain. Some of the significant portolan charts of coastlines and seas—the Carte Pisane (ca. 1275–1300), Giovanni da Carignano map (ca. 1300), and the Pietro Vesconte atlases (1311–1327)—display, with great variety, flags, mountains, structures, and so forth that are all oriented in one principal direction. These maps, however, at the same time list the names of ports and coastal towns in all directions and largely perpendicular to the coasts so that the charts have to be turned to be read. They are both oriented and not oriented to the extent that Tony Campbell concludes that "there is no way of telling which, if any, of the four main directions they were primarily intended to be viewed from."[21] Space here is abstracted in the further sense that it is not reliant on a stable viewer or page angle, and its internal space can make sense from any angle.

Of a similar nature is the use of constant scale. Geographers describe objective scale as the kinds of ratios we are used to today in which the sizes of objects on a map correspond at a different scale to the sizes of objects in an area of the world. It is difficult to prove the use of objective scale in the Middle Ages. Constant scale, on the other hand, does not necessarily involve a correspondence between the page of a map and a topographical area. Constant scale on a map instead occurs when diverse kinds or genres of objects are drawn as approximately the same size (and often in the same manner). Different mountains or mountain ranges are similar in size, buildings are roughly the same size despite the fact that some are in reality larger than others, and trees and other phenomena are similar in size. Many *mappaemundi* and some of the local maps in chapter 1 tend toward this kind of sign system (as do modern maps of the whole earth, which necessarily distort physical area because of the transformation of a curved planetary surface to the flatness of a page). The earliest verified use of mathematical principles to create constant scale is usually attributed to the Albertinischer plan of Vienna and Bratislava of 1421–1422.[22] Here again, a process of organization takes over from the tangible presence of diverse objects. The contents are abstracted to become signs, and the viewer's experience of space is of the entities on maps—oceans, continents, islands, rivers, fauna, and so on—as forming a system.

The appearance of geometric lines on maps also pushes them toward creating an impression of space that is all of a whole and as a system that could be applied to any geographical area of any scale. Some of the earliest lines that are arguably of this kind are rhumb lines, the sunburst shapes of straight lines that begin to appear about 1300.[23] Rhumb lines typically originate at a series of points that is arranged in a circle on a map. The portolan charts are the most famous for these; they appear on the Carte Pisane, the Giovanni da Carignano map, the Pietro Vesconte atlases, and other maps. Portolans typically have sixteen originating points, with thirty-two symmetrical lines extending out from each one. The originating points and the lines do not correspond with landmass features, and it is unclear whether they coincide with compass directions (as they are sometimes thought to do). They are perhaps an example of pure abstraction in that they do not key to geographical phenomena of any sort.

A further clear example of map lines that seem to render area in an abstracted manner is graticular, or grid, maps with meridian (longitude) and parallel (latitude) lines that begin to appear in the fourteenth century in Italy. This is a new technology that is often pointed to as signaling a change in the direction of cartography and a shift in the hermeneutics of space itself. This idea is asserted even though the graticule is not part of a clear trajectory of development from ancient to modern mapmaking. Graticules first survive on the ancient maps of Strabo (and probably Hipparchus before him). Soon thereafter, Marinus of Tyre and Claudius Ptolemy offered their own methods for rendering the Northern Hemisphere on the two-dimensional and rectangular area of a map.[24] The few examples of maps with grid lines that begin to appear in the mid-fourteenth century have obscure origins, probably not deriving from rhumb lines but perhaps from Arab or Chinese sources instead.[25] The rarity of these maps is probably not due to a loss of knowledge about mapmaking practices in the Middle Ages as some modern studies imply. Medieval cartography has different interests; medieval *mappaemundi*, for example, seem to have served different purposes—namely, as visual encyclopedias more akin to palimpsests of overlaid biblical, historical, mythical, and other information.

The graticule maps nevertheless are systematic and seem to demonstrate a way of thinking about the earth that sees its phenomena not so much according to their individuated characteristics but in terms of placement and relative location. The earliest surviving example of a clearly graticule map is of Palestine from the thirteenth century, Paulino Veneto and Marino Sanudo producing another version of it in the fourteenth century, which was likely illustrated by Pietro Vesconte. This later illustrated version survives in nine copies. The map shows two arrays of parallel lines set perpendicularly to each other to create graticules of one league square. This means that the map has constant scale *and* is an attempt at objective scale (or at least the relative placement of objects on a map). A further remarkable quality is that the Palestine map is accompanied by a list of towns identified according to the graticule squares. That is, its graticle has a key.[26] The map occurs within a narrative that also enhances a sense of space subordinated to the graticule. One version, for example, Bibliothèque Nationale de France manuscript Latin 4939, contains a sequence of maps beginning on folio 9r with a *mappamundi* subsequently labeled with the names of principal places and brief descriptions of each of the places in the Northern Hemisphere. Next on folio 9v is a map of Syria and Egypt, and a list of places with descriptions of those locations. Then the map of Palestine with graticule lines spreads across folios 10v–11r with its list of places there. There is in the manuscript, as it were, a zoom effect from the globe to the region to the area in a stepped fashion.[27]

It nevertheless remains difficult to discern whether *mappaemundi*, Macrobian zonal diagrams, portolan charts, or graticular maps express a fully abstract-homogeneous sense of physical space or rather some variant on a heterogeneous one. Perhaps it is fairer to say that they combine the two tendencies, or that it is difficult to generalize across the different kinds of maps. Constant scale, the lack of orientation, and the rhumb lines of the portolans may suggest homogeneity that finds itself in the modern form of latitudinal and longitudinal lines, while the fact that the maps tend to individuate the items they depict, can be selective in the areas they portray, and often show no interest in either constant or objective scale pulls toward seeing the areas as heterogeneous. One almost always finds diversity on the page owing to explicit economic, political, and theological interests in the construction of the charts. Given that all maps are in some respect ideological, it is arguable that no map can be completely homogeneous. We remain aware, for instance, that most maps today privilege the place of the map's origin by distorting land size, framing the earth to give the place of origin prominence, and so on.

In addition, that *mappaemundi*, portolans, and the even rarer graticular maps are memorable and have received substantial attention in literary and cartographic analyses might be a consequence of the very medieval theological or other hierarchies for which they are offered as evidence. Or they might have received attention because they appear to lead to modern mapmaking practices that typically employ constant scale and a sense of overhead objectivity. The same tendencies may account for the attention paid to literary examples of overhead perspectives in Macrobius, Dante, Chaucer, and others. These moments in literature are also rare, especially in contrast to horizonal views. It should also be recognized that a narrator or character's ab-straction in the literature invariably occurs under special circumstances as in dreams or other visions or the afterlife. Moreover, even in these extraordinary situations, the narrator can still be interested in the specifics typically included in a horizonal view: a particular place, a person who remains behind, a topography with individuated characteristics, a hill, a stream, a garden.

Horizonal Space and Things: Astrolabes, Quadrants, and Practical Geometries

Attention to the more common earthly functions of astrolabes and quadrants, to the practical geometries, and to *topographia* (another kind of "device" in the rhetorical tradition) reveals the features of the more commonly encountered horizonal space, an area near earth and within the bounds of the horizonal edge. This evidence suggests that the horizonal space is made up of "object-relations" in the sense that space is experienced through objects in relation to the body. Vice versa, the eyes, head, feet, and physical height of the body are considered in reference to external things such as towers, pits, bodies of water, and areas of land so that the self is considered almost as another object: "The subject is the object."[28] In rhetoric, the relationship between place and object can become causal, in which the individual topos of a location and of a trope in writing can give rise to objects and sets of practices that are thereafter perceived to be fitting or unfitting for that particular location.

Astrolabes calculate when and where one is located in relation to the heavens and on the earth. They enable relatively large-scale measurements showing, for instance, the time of day, the time when a star or planet will be in a particular position in the sky, when Easter has been and will be in any given year, and one's latitude on earth. Their astronomical functions are some of the most complicated and intricate, and that may be one reason why they have received emphasis in the various tracts that explain how to make and use an astrolabe as well as in modern analysis of the devices. Chaucer's *Treatise on the Astrolabe*, for instance, is principally composed of sections that address the horological and astronomical functions. The other functions on an astrolabe and its compact derivative, the quadrant, not only measure the large-scale features of the universe and the sky but concern themselves with spatial measurement of a more local and horizontal kind. The quadrant is a wedge-shaped device that integrates some of the astrolabe's functions in one quarter of the whole astrolabe's circle, hence its name.[29] The horizontal objects that astrolabes and quadrants measure are buildings, hollows in the earth, wooden poles, and people. We are back with the more intimate contents of a horizonal area.

The first observation to be made is that the astrolabe and the quadrant are object oriented, not necessarily in an ontological sense but more modestly in that they work by sighting objects. They are measuring devices that produce numbers (hours, dates, times within seasons, latitudes, and so on), but their measurements are ultimately of individual things. The complicated astronomical calculations are to do with objects such as the sun, the pole star, planets, and other lighted phenomena. One side of the astrolabe and the quadrant contains, among other instrumentalities, two devices: numbers from zero to ninety running around the outside edge, or *limbus*, and a shadow square, or *umbra recta*, which is a smaller square on the plate marked with twelve degrees (see figure 9). In contrast to the complications of astronomical calculations, the uses of the *limbus* and the *umbra recta* are relatively simple: they can be used for altimetry and planimetry to measure the heights, depths, and lengths of objects, the latter enabling the calculation of the area of a piece of land.

To gauge the height of an object, the user measures the distance he or she is from the object (by pacing or using a measurer's rod, cord, or chain), lines up the entity through the raised sights, and notes the number that the plumb line, the silk thread (or *perpendiculum*), crosses on the *limbus* or the *umbra recta*.

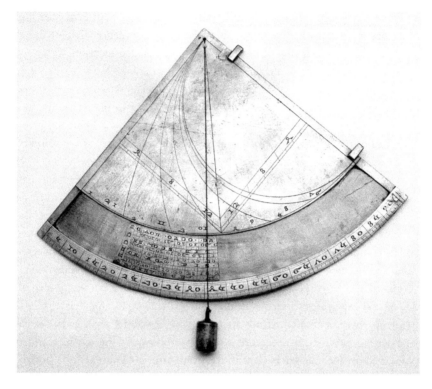

FIGURE 9. Quadrant with sliding latitude plate, shadow square, and plumb bob with a pearl bead. Quadrans Vetus, Horary Quadrant, French?, c. 1300. Inv. 52020. © Museum of the History of Science, University of Oxford.

If the operator can position himself or herself at the right distance from the object so that the *perpendiculum* crosses forty-five degrees on the outside rim's protractor-like *limbus*, or twelve on the shadow square's *umbra recta*, then the distance from the object is the height of the object. If the measurer is at a greater or lesser distance from the object, then a simple calculation using the tangent of the angle, multiplied by the distance from it, gives the height because the object is at a right angle from the viewer. If one does not know trigonometry, one can use the *limbus* of ninety degrees or the shadow square with a simple ratio, as will be shown. The point is that these aspects of the instrument are available to users who vary in their proficiency in measuring, none of them complex.

Chaucer's *Treatise on the Astrolabe* is a practical geometry, and his and many other *practicae geometriae* describe astrolabes, quadrants, and geometrical problems; a majority of these common treatises are illustrated with diagrams of the instruments.[30] The devices and the practical geometries belong to the category of mechanical arts. They are "philosophically empirical" in the sense that they are geometrical, or abstract, yet also related to observable phenomena.[31] One of the earliest of such writings is probably by Hugh of St. Victor, who contributed to elevating the status of the mechanical arts to being disciplines worthy of study and worthwhile practices in themselves. Hugh describes his *Practica geometriae*, composed in the early twelfth century, as a summary of geometrical knowledge for students.[32] Elsewhere in his *Didascalicon*, also an introductory text to summarize knowledge, Hugh further distinguishes between the mobile objects that astronomy addresses and the stationary objects that geometry treats.[33] In other words, he was interested in both the horizonal and the vertical axes of investigation.

The kinds of objects that one can measure with an astrolabe or quadrant are those that appear within sight. The species of objects are named in practical geometries, and, while it is somewhat difficult to discern whether they are hypothetical or part of a textual tradition rather than evidence of actual objects and "practical" usage, they seem to indicate at least the *kinds* of items that can be measured. Hugh's *Practica geometriae* distinguishes among geometry's branches in a traditional manner: planimetry, cosmimetry, and altimetry. Planimetry applies to the length and breadth of planes, cosmimetry measures spherical objects, and altimetry calculates heights as well as what is above and below an object. Altimetry is useful, Hugh says, for measuring a tree's tallness and the sea's depth.[34] Hugh later uses geometry to calculate the measurements of Noah's ark in *De arca Noe morali*.[35] The *Practica geometriae* discusses using the astrolabe although its measurements principally employ only the limbus and shadow square within the astrolabe as well as a measurer's rod (also called a *perch* in Middle English), an instrument of standard length made out of wood or iron.[36] The geometrical measurements discussed in the *Practica geometriae* are largely traditional and abstract, but they also provide some sense of the everyday areas to which they may be applied. It describes measuring the altimetry of tall objects in general if they are set on a plain, or if the space between the observer and the tall object is impassable "because of an object such as a river or a gorge" (*fluvii vel vallis*),[37] or if the object is in a valley while the observer is on a hill and vice versa. Its explanation of planimetry begins by noting that it involves the measurement of a "planar area within the visual horizon" (*plani inter orizontem*).[38]

Geometrical works that involve the astrolabe, quadrant, and other devices and that follow the treatise attributed to Hugh are many. They include a tract that begins and is called the *Artis cuiuslibet consummatio* from the late twelfth century (translated into French in the thirteenth century), and the *Quadrans vetus* or *Tractatus quadrantis* by Johannes or Robertus Anglicus at the University of Montpellier, from the thirteenth century. The *Artis cuiuslibet consummatio* treatise survives in some fifteen manuscripts, and it describes how to measure areas, heights, and volumes, as well as discusses fractions. Several chapters address the number of houses (*domuum*) that can fit in an area, including the number of houses in a city, and how to measure the height of a tower (*turris*). In another chapter, it describes how to infer the height of an object by shooting an arrow with a string attached at its top, which will create a physical triangle with the hypotenuse that, together with the distance from the object, can be used to calculate the height.[39] The *Quadrans vetus* survives in over one hundred manuscripts and was adopted early into the elementary curriculum at the University of Paris.[40] The tract contains a brief guide to making and using the various parts of a quadrant for solar measurements, altimetry, planimetry, depths, and volumes. Like the *Artis cuiuslibet consummatio*, the *Quadrans vetus* wants its succinct ideas to apply to the height, width, and depth of any object, so it generally guides the reader to measure "things," "holes," "wells," or "vessels" (*rei, putea, vasa*), and occasionally it offers an object such as a tower (*turris*) as an example.[41] The *Quadrans vetus* describes several methods of measurement for the height, length, and depth of objects.

Bodleian Library MS Ashmole 1522 is a fourteenth-century miscellany of mathematical, trigonometrical, and cosmological treatises, including one on the astrolabe. Folios 70–78 contain the *Quadrans vetus* and include a diagram of a quadrant as well as five other illuminations. Figure 10 shows different ways of taking the altimetric measurements of a tower. The left-hand illumination on the page labels the tower's height a-b: a at the top and b within the base. The viewer sights along the top of the quadrant to a, and if the angle is forty-five degrees on the *limbus* or 12 on the shadow square, as it is in the diagram, then the distance from b, the foot of the tower, is the same as the height: a-b is the same as b-c. The diagram shows that one has to take into account the height of the observer. The *Quadrans* explains a second way of measuring the height by comparing triangles, and the left illumination also shows this. One can compare the small triangle d-c-e with the larger triangle b-c-a via a simple ratio of the foot of the small triangle, d-c, to the larger side of the triangle of the object itself, b-c. As Chaucer's *Treatise* explains, one can sight the top of a tower and note that the rule (equivalent to the plumb line on a quadrant) crosses the shadow square at, for example, the number four (out of twelve possible on the *umbra recta*). If the distance between the base of the tower and the viewer is twenty feet, then the ratio of four to twelve is three, so the tower is three times twenty feet high.[42] Thus, one does not need to know trigonometry and tangents to measure the height of the tower. The right-hand image on the illumination shows the slightly more complicated

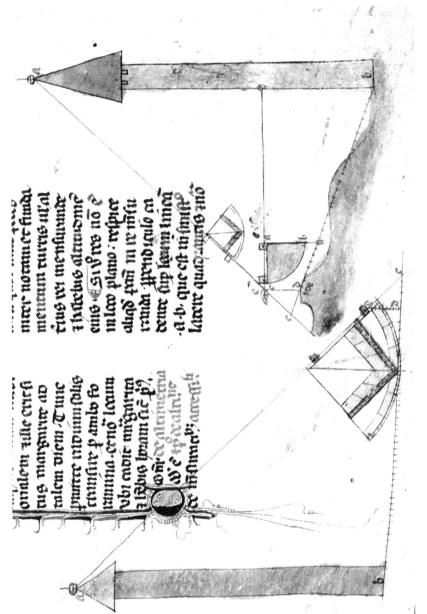

FIGURE 10. Measurements of towers using quadrants. *Quadrans vetus*, Bodleian Library, Oxford, MS Ashmole 1522, fol. 75r.

calculation of the height of a tower that stands in a valley where the observer is higher than the base of the tower, which is again labeled *b*. The calculation involves inverting the quadrant to measure the depth and ultimately adding the height to the depth.

Another work that is similar to the *Artis cuiuslibet consummatio* and the *Quadrans vetus* is by Dominicus de Clavasio, written in Paris in 1346, and surviving in some fifteen manuscripts, including one copied in England in 1373.[43] Dominic describes using the eye, the quadrant, and the astrolabe to measure the distance between two mountain summits, the height of a mountain from its summit to its foot, the length of a valley, the distance between the eye and a point in a valley, the height of a tower, the depth of a well, and the distance between two points in general with the eye being at a measurable height or on a level with the observed object. The work goes on to describe, with diagrams in many of the manuscripts, how to calculate the area of geometrical figures such as triangles and circles, as well as the surface areas of spheres, cylinders, and cones. It then provides methods for measuring the volumes of cubes, spheres, wine casks, and other shapes. The bulk of the treatise and its language remain mathematical, indicating an audience who is able to use the instruments of the day with some sophistication. The treatise, like the *Practica geometriae* and the *Quadrans vetus*, is nevertheless also a "practical geometry," written as a series of instructions or guides (*accipe, vide, pone, divide,* and so on), with the author inserting himself in elaborations of the instructions by saying *tunc ymaginor*.[44]

Horizonal Space and Viewer Relations

The diagrams showing the use of mechanical devices and the treatises on them have suggested the kinds of objects that can be measured and the ground, bodies of water, and valleys that can intervene between a self and a thing—that is, the varying kinds of topography that objects can occupy in a local space. Hugh's *Practica geometriae* and the other tracts also describe the calculations required if one cannot measure the distance from the base of an object because of "an obstacle, such as a river or a gorge" (*fluvii vel vallis alicuius objectu intervallum*). In this case, the viewer can take one measurement of the height of the tower on an astrolabe, then move back to a second position, measure the height from there, then calculate the ratios to again take the height of the tower.[45] This kind of necessary movement of the observer is one example of the substantial information the practical geometries provide about a viewer's own body in relation to objects and locations. The diagrams and treatises depict the relation between a measurer's own person and an object in terms of position and scale as well as a kind of reciprocity or interchangeability between self and object. For example, they instruct the operator of an astrolabe or quadrant to lay his or her eye on an instrument in order to sight along it. Figure 10 includes an eye looking through the lower quadrant's sights. Even when measuring the height of the sun, the light cast through an astrolabe's sights and across its measures is viewed by the operator. The treatises do not focus only on sight but involve the whole body since they begin with the simplest forms of measurement in which a person paces out or otherwise measures the distance between the person and a tower, pit, or another thing.

In the case of altimetry, treatises on astrolabes, quadrants, and other devices also take into account measuring the distance of one's eyes from one's feet on the ground. The *Artis cuiuslibet consummatio* notes how the viewer can lie down, or "if you measured standing, add your height" (Si stand mensuraveris staturam tuam adde),[46] so one's own height up to the eyes is a significant and useful unit of measure that can enable other measures. The operator's eye height means that one thinks about not only the measurable object in front but also the ground behind, an area that, as we have seen, forms another triangle that can be used in calculations. Hugh's *Practica geometriae* details how some operators of another instrument, a right-angled triangle, can use it to measure heights, and that users may put these devices on stands or hold them because "many find it injurious to bend down often and get their eye on the ground" (quod injuriosum videatur mensori totiens prosterni et oculum humi defigere).[47] The *Quadrans vetus* describes a variation on altimetric measurement, using the parts of the quadrant and also calculating height without it, in which the viewer places a mirror flat on the ground so that one can see the top of a tower. The ratio of the operator's eye height and distance from the mirror, with the known distance from the tower, enables a calculation of the height of the tower. The same principles of altimetry are applied to planimetry, using a mirror that is now placed vertically on the ground, and in this case the viewer lies on the ground and sights the end of the plane.[48] British Library manuscript Lansdowne 762 contains short texts in English to calculate an area of land as well as an altimetric measurement of the height of a steeple not using a quadrant but a set of simple comparative calculations using a rod. This late fifteenth-century compilation also contains prophetic verses, an excerpt from John Mirk, and the Langlandian "God Spede the Plough." The person measures a known distance from the steeple, lines the rod up with the eye while looking at the steeple, and the treatise says "thann let hym leye hym downe alonge upp righte beyonde the staffe from the steple warde his feet juste to the staffe." Presumably the ratio of the distance from the measurer's eyes to the base of the rod, and the distance from the staff to the base of the steeple, can be applied to the height of the staff to arrive at the steeple's height.[49] In all these treatises, the eyes and the body become a part of the instruments of measurement, and their position is significant for calculation.

What influence did the astrolabes, quadrants, and treatises of practical geometries have on medieval culture and a broader sense of objects, topographies, and relations between human bodies and the areas around them? Exactly how practical were the devices and tracts, or were they considered only in the classroom? What does the evidence of schoolroom or broader application imply for the more general sense of horizonality and the hermeneutics of local space? The stakes of the discussion appear high in one respect because the answer to these related questions about the practical or exclusively academic use of the instruments and tracts has been offered as a "rupture narrative" between medieval ways of thinking about space and early modern ones.[50] No less a figure than Michel de Certeau's is involved here, whose influence on medieval spatial studies of various kinds is in some respects peculiar because he insists on a sharp break in spatial understanding in the fifteenth century that divides medieval philosophy from the "birth of modern scientific discourse."[51]

It may be possible that knowledge of how to use the instruments and people who actually read the geometrical writings largely existed within the walls of the universities because of Latin and the difficulty of knowing trigonometric operations.[52] J. A. W. Bennett in *Chaucer at Oxford and at Cambridge* describes Bishop William Reed's connections to Chaucer and his *Treatise on the Astrolabe*. Among other substantial gifts to Merton College, the bishop bequeathed a book containing a description of quadrants to Merton in 1375 (Merton College, Oxford, MS A.3.9), and Bennett also describes how Merton was a place to study the astrolabe, in part because it owned them.[53] It is furthermore difficult to find evidence that an astrolabe or quadrant was used in the field to survey lands or actually measure towers or other objects for building, warfare, or other uses. There also seems little relationship at this point in history between, on the one hand, the instruments and the geometrical texts and, on the other hand, maps. It has been argued that geometrical practices and maps did not inform or enhance each other even though today we easily think of sketching a plan of a piece of land to understand an area and to present it to others.[54] It also seems strange to us today that the instruments and treatises would be produced but not used, yet perhaps that was what happened.

On the other hand, evidence suggests that knowledge of how to use the instruments extended beyond the universities to other schoolrooms and elsewhere. Medieval lands were measured by means of perches and other devices. Skelton and Harvey note that we do not hear of the astrolabe, quadrant, or any other device being used for this planimetric purpose, but the devices would enable measuring right angles along with other degrees and therefore seem at least potentially useful.[55] Also, astrolabes were known outside the university. Chaucer's *Treatise*, which survives in thirty-two complete and incomplete manuscripts, is addressed to his son who was ten years old. Four manuscripts, including Bodleian Library, MS Bodley 619 (a preferred textual version), title Chaucer's work *Brede and Milke for Children*.[56] In England, a quadrant was made for Richard II, and another probably for his half brother, John Holland, Duke of Exeter, but ownership of quadrants was not limited to royalty or the aristocracy more generally.[57] In addition, the uses of the devices I have discussed can be simple; trigonometry is not always necessary because of the shadow square and limbus measure of degrees. Furthermore, the intention of what are after all called "practical" geometries was certainly that the texts should be applied. The thirteenth-century *Practica geometriae* of Fibonacci, or Leonardo of Pisa, circulated outside of universities, and works of practical geometry did not remain in Latin but were translated into French as early as the twelfth century and into English by at least the fifteenth century.[58] Hugh of St. Victor seems to be drawing on actual measuring practices when he describes a viewer moving around in a space to obtain measurements and the aches and pains he or she might have because of having to get down to the ground so often. His stated intention in the *Practica geometriae* is "to teach practical geometry to our students, not as something new, but rather as a collation of older, scattered material," and where "theoretical geometry uses sheer intellectual reflection to study spaces and intervals of rational dimensions . . . practical geometry uses instruments and gets its results by working proportionally from one figure to another" (Practicam geometrie nostris tradere conatus sum, non quasi novum cudens opus sed vetera colligens dissipata. . . . Theorica siquidem est que spacia et intervalla dimensionum rationabilium sola rationis speculatione vestigat, practica vero est que quibusdam instrumentis agitur et ex aliis alia proportionaliter coinciendo diiudicat).[59] One modern editor of Hugh's treatise says that while the Latin of the texts points to an academic audience, the content "suggests surveyors and technicians."[60] The *Artis cuiuslibet consummatio* begins with a critique of "modern Latins" (*moderni latini*) who neglect practice for only theory in the quadrivium. "The perfection of any art," it begins, "depends on two aspects: theory and practice" (Artis cuiuslibet consummatio in duobus consistit, in theorice et practice, ipsius integra perceptione).[61]

However, it does not appear strictly necessary to determine whether astrolabes, quadrants, and geometrical treatises were used or applied. That they are being produced and reproduced in some numbers, along with local maps and other instruments of land measurement, suggests that people inside and outside the university shared a sense of horizontal space as a set of relations between objects and viewers, relations that are measurable in a variety of ways: scientifically, legally, and practically. Each method involves the viewer looking out to an object on the ground or within a zone near the ground but always within the horizon. The viewer, in addition, does not see abstract space within the horizontal area but instead looks at particular objects within the space, attention moving to one or another item according to current demand. Despite the scientific nature of the measurement, the space is largely heterogeneous. Also, the viewer's own physique, his or her eyes and body, is part of that space; it is in a sense the origin of the measurement but also integrated in the scene. In Gibson's terms, the ground-level horizontal perspective affords a self-aware, self-situated objectivity.

Topographia and Elevation

As discussed in chapter 1, Ptolemy's *Geographia* opens by distinguishing *geographia* from *chōrographia* and *topographia*. In that chapter, I noted that *chorographia* becomes a device for understanding and rendering the individual items of an area. Ptolemy goes on to define *topographia*, like chorography, as distinct from the geography of the world cartographer. In his formulation, topography is "landscape drawing" and, also like chorography, requires artistic proficiency, while the world geographer has to be proficient in mathematics.[62] Topography in the Enlightenment came to be associated with elevated views of landscapes (an aesthetic term used in painting), and today topography entails a set of features often presented on maps; but an even earlier sense of topography, as in "place writing," exists most clearly in the rhetorical arts.

Indeed, Quintilian associates *topographia* with representation itself in his definition of the concept. In his *Institutio oratoria*, Quintilian says representation "is something more than mere clearness" (*repraesentatio quam perspicuitas*) in which facts are "displayed in their living truth to the eyes of the mind" (*oculis mentis ostendi*). He defines *topographia* as "clear and vivid descriptions of places" (*Locorum quoque dilucida et significans descriptio*). It is a subspecies of the larger trope of *descriptio*; more than a representational technique, it is an indispensable part of persuasion. Quintilian refers to his explanation of Ciceronian "ocular demonstration" (*sub oculos subjectio*), which becomes necessary whenever people want to do more than just state that something occurred; instead, they "proceed to show how it was done, and do so not merely on broad general lines, but in full detail" (*cum res non gesta indicatur, sed ut sit gesta osten-*

ditur, nec universa, sed per partes). *Descriptio* can present what has happened, what is happening, and what might happen or might have happened.[63]

In the Middle Ages, Matthew of Vendôme took up ideas such as these from the classical rhetorical tradition in his *Ars versificatoria*; he also addresses the utility of topographical description, but he extends the definition of *topographia* in crucial ways. He notes that the subject matter of verse "is drawn from place when, because of fitness of place, it is inferred that something had been done or not done" (*Argumentum est a loco quando per opportunitatem loci aliquid factum fuisse vel non fuisse conjecturatur*). That is, on occasion, places signal occurrences and come close to being causal agents. Some descriptions of place, he continues, are effective, while others may detract or distract from the main argument of a poem. He offers an example from Cicero, who, in a courtroom, described the natural delights of Sicily to effectively imply a contrast between its features and the unnaturalness of the adultery that occurred there in the legal case he is describing. Matthew then continues to offer a lengthy poem of his own as illustration of *topographia* in which an unnamed locale is blessed with sheltering trees, refreshing waters, and many individual birds. According to Ernst Robert Curtius, the poem is a possible source for Chaucer's own depiction of the garden and birds in the *Romance of the Rose*.[64] The *Ars versificatoria*'s birds all sing appropriate songs. Matthew also says that a description of a place is always tied to time, and both are "attributes of action" (*attributa*).[65]

At almost exactly the same time as Matthew in the twelfth century, Gerald of Wales employs the word *topographia* in his *Topographia Hibernica*.[66] This work, more than Gerald's descriptions of Wales in the *Itinerarium Cambriae* and *Descriptio Cambriae*, hinges on the fitness of matching human culture to other geographical information, an outlook that he may have first explored in an earlier scientific work, now lost, titled *Cosmographia*.[67] Gerald refers to his task in writing the *Topographia Hibernica* as "to rouse the reader's attention, by setting before him some new things, either not before noticed or very briefly noticed; exhibiting to him the topography of Ireland in this little work of mine, as in a clear mirror, so that its features may be open to the inspection of all the world" (*Aggrediar tamen utcunque novis quibusdam, et quae vel nullis hactenus edita sunt, vel perpaucis enucleata, lectoris animum excitare; expressamque Hiberniae topographiam hoc opusculo quasi speculo quodam dilucido repraesentare, et cunctis in commune palam facere*).[68] While his motivation for writing about Ireland comes, he says, from the fact that a description of it has been neglected, his explicit outline of the work is indicative of the rhetorical tradition that underpins the book. Despite that, as Jeffrey Cohen says, his book "overflows with unsystematic detail,"[69] it is structured in three *divisiones*: the first situates Ireland in relation to *majore Britannia* before describing its geology and natural phenomena, the second part details its prodigies, and the third gives the history and characteristics of its inhabitants.[70] Indeed, Gerald exploits the divisions between locations and the phenomena, extraordinary characteristics, and people within his topographical scheme when he shows how Ireland's inhabitants are perversely unlike their good land. As Kathy Lavezzo points out, Gerald "radically separates the Irish from their physical settings, thus rendering their barbarism as disturbing as possible,"[71] with a concomitant sense of a marginal and aberrant monastic culture. Ireland's discord between what were traditionally fitting elements of culture and topography would have struck readers as even more willfully ignorant on the part of the inhabitants.

Before concluding, I would like to consider a third possible point of view between an overhead and a terrestrial view in the late Middle Ages. Romances, chronicle descriptions, visionary writings, and other literature occasionally contain figures who ascend towers or other high points and look over areas of land. These views appear reasonably frequently in illuminations, which sometimes contain a viewing figure who looks out from an elevated position, or they simply present a scene from a height. A character, narrator, or the audience is raised up with a view that consequently stretches farther in distance. These vistas are not to be confused with Enlightenment aesthetic preoccupations with the sublime or landscape painting. Medieval sources for raised viewpoints are many, and some may again be found in ancient or religious texts (Jesus in an elevated position, for instance, in the Temptation or at the Resurrection), but sometimes the visualizations are contemporary and apparently experiential. Consider the case of Richard II during the Revolt of 1381.

On Corpus Christi Day in 1381—Thursday, June 14—the main body of the rebels from Kent and Essex entered London, some through Aldgate on its eastern side, where Chaucer was leasing accommodation. It was the beginning of two days at the climax of the revolt. Their numbers were estimated at fifty or sixty thousand, though they soon grew to as many as a hundred thousand. They killed people, and they sacked and burned buildings, including the Savoy and the Hospital of St John in Clerkenwell, both outside the western city walls. When they found they could not immediately speak with King Richard II, they besieged the Tower of London on its east side from the adjacent area of St. Katherine's hill, hospital, and church. The *Anonimalle Chronicle*, often described as an eyewitness account, narrates three related occasions on that day when Richard, fourteen years old and in his minority, looks out from high viewpoints from the tower. The first is on the morning of June 14, when the king is described as surveying the burning buildings from a turret. It seems he evaluated what lay before him because the chronicle follows the description of him with the narrative that he summoned the lords there with him in the tower to receive counsel. After the meeting, he had it publicly declared that the rebels should leave and meet him the next day at Mile End. The chronicle says that announcement was intended to end the siege and to let the lords and others in the castle escape, but the plan "came to nothing" (*mes ceo fuist pur nyent*).[72]

Later that same day, "thoughtful and sad," Richard tried to appease the rebels with a grant of reprieve for their crimes, and this time "he climbed onto a little turret near St. Katherine's where a large number of the commons were lying" (pensive et trist . . . amount sur une petit toure devers seint Kateryne ou furount gisauntz graunde nombre des comunes). He said to them that he would have a bill that guaranteed their freedom and other demands drafted, he had a clerk write it up while they looked on, and he signed it. He then had two knights go around the rebels gathered there and read the bill aloud to them. The *Anonimalle Chronicle* gives a text of the bill, and the account adds that the knights "who read it stood on an old chair above the others so that all could hear" (*et cestuy qe list la bille estea en une auncien chare amont les autres, issint qe toutz purroient oier*).[73] But the king's plan did not work and seemed only further to affront the rebels because they called for the heads of lawyers and others who could write documents such as the one read to them. The rebels then set more fires. After this, for a third time, "the king himself ascended a high garret of the Tower in order to watch the fires. Then he descended again and sent for the lords to have their counsel, but they did not know how they should counsel him and were surprisingly abashed" (le roy mesmes alast a une haute garett de la Toure pur veer le feu et puis descendist a vale et mandast pur les seignurs davoir lour conseil; mes ils ne savoient coment ils purrount conseiler, et fueront si abaiez qe merveille fuist).[74]

Several causes might explain Richard's behavior and the silence of his counsel at this point in the afternoon of June 14. They were all secure but trapped in the tower, surprised by the numbers of the rebels and the violence of their protests as well as the participation of several Londoners; the lords around the king seem genuinely unable to offer him advice or were perhaps unwilling to risk themselves for their young lord.[75] What is of interest for our purposes, however, is not so much the psychology and politics of the group but the *Anonimalle Chronicle*'s depiction of Richard's reaction when he gazes from his lookout points. He is depicted each time as singular, perhaps not literally alone, but the chronicle narrows attention to the king himself as if he were by himself. In these moments, Richard gazes out from the tower, and his attention is obviously drawn to the fires; but where we might expect his higher standpoint to lead him to muse on the troubles throughout and surrounding London, or for the narrative of the chronicle to extend his consideration to what was occurring beyond his sight across England to the fraught difficulties of the realm, especially in the semicircle to the east where the rebels were most active, he instead considers what is in his physical view and no more. We also do not read of his thoughts ascending to heaven and to a contemplation of fate or the Divine's role in the course of human events, or in a plea to God, as the religious figures in London had done the day before.[76] The *Anonimalle Chronicle* is not hesitant to enter the minds of the people involved in its history, as it does in saying the council was afraid, but here the narrator is restrained and simply notices the king's physical behavior. In the last moment in the tower, the king turns away from the scene and back to his counsel, who are themselves speechless. Richard's situation exhibits two related features of late medieval visual perception that are typical of the period. The king in this account of the Revolt of 1381 is elevated from the earth, but he remains close to the ground; he is not separated from the terrain, suggesting that his interests lie in the local and the horizontal.

In conclusion, it may be arguable that the systematic forms of maps and aerial flights of characters in literature express an interest in geographical space as an abstract or abstractable entity. The horizon might in some cases appear to "vibrate with curiosity and interest," as Westphal says, or it may become a small circle within which one can perceive "the finite whole earth," as Woodward claims.[77] Whether on the ground or raised in the air, however, space was not frequently thought of as an expanse available to perception from any point of view. It rarely signals some sort of frontier that was of interest in itself because it continued, as was the case much later in history, with the American West and expanding Manifest Destiny. In Middle English and Middle French, a *frounter* or *frunter* is a border region and also a stronghold that guards that edge.[78] The horizon *contains*, and the vast majority of interest lies in the area to which "þe siȝt strecchiþ and endiþ."[79] The horizon could, of course, advance when a person or other thing moved toward it, it changed its qualities as a person moved around, and it moved farther away if someone raised himself or herself up off the ground, as Richard does in his turrets. But the interest tends to lie in what the horizontal perspective affords—namely, a space that may on rare occasions be thought of as homogeneous and able to be subjected to abstraction, but more commonly afforded objects with individuated variations that included an embodied self.

CHAPTER 4

Horizonal and Abstracted Spaces
The Book of Margery Kempe and *The Book of Sir John Mandeville*

This chapter takes up the previous chapter's findings about late medieval scientific understandings of space in order to examine *The Book of Margery Kempe* and *The Book of Sir John Mandeville*. Chapter 3 began with the idea that the horizonal outlook is the common mode of spatial hermeneutics, a way of apprehending a spatial zone near the ground and bounded by the horizon. The horizonal area tended to be understood as heterogeneous, each location and object affording different "'values' and 'meanings' . . . in the environment" for the viewer, the text, and the audience.[1] According to devices and geometric writings in the mechanical arts, which formed a bridge between academic study and practical use, sighted objects, such as towers and pits, were the significant elements in an area. The viewer was also involved in a locale; the observer accounted for eye height, could lie on the ground, and sometimes lay an eye along an instrument to view another object. The body and the eyes could also be triangulated in various ways in relation to another object and the ground behind the body. The rhetorical trope of *topographia*, according to Matthew of Vendôme and Gerald of Wales, furthermore suggested that each of the objects or activities within an area, including humans and their practices, might arise as a kind of consequence or effect of an individual place. However, constant scale, the graticule, and other aspects of some cartographic practices indicate that the period could at moments abstract from individuated horizontal locales to apprehend space as homogeneous, in which locations could have equal weight as if evenly dispersed over a grid. This abstraction of space was sometimes perceived from, or was enabled or accompanied by, an aerial or overhead perspective, but these moments were exceptional, occurring in dream states, visions, and afterlife moments.

This chapter focuses on the varieties of physical areas presented in *The Book of Margery Kempe* and *The Book of Sir John Mandeville*. The two are distinct narratives written for very different purposes, but they both involve a great deal of travel, and an analysis of the spaces described in each reveals certain structural homologies. The events and objects in Margery Kempe's *Book* typically appear within discrete episodes, each capturing something different about her experiences in diverse locales in King's Lynn, in England, and beyond. Within these episodes, her *Book* provides a sense of the individual objects within locations, particularly buildings, but they appear without relative measurement between them. The reader gains an appreciation of the familiarity she had with the spaces in her town and how an interpersonal sociality could also attain across larger distances. The majority of *The Book of Sir John Mandeville* likewise collects the putative author's travels outside Britain into bounded scenes in discrete locations, each distinct from the other and object oriented with the narrator featuring in each. In both, then, space is most often presented as heterogeneous with individuated, intrinsically interesting items within it.

The Book of Margery Kempe and *The Book of Sir John Mandeville*, however, occasionally achieve abstract perspectives in which places are apprehended as, in some respects, homogeneous. Even though moments of this sort are rare and brief, and occur under unusual circumstances, at the risk of suggesting their greater significance, I linger on them in this chapter precisely because they are so exceptional and reveal a different sense of space from that of the more common horizontal perspective. In Kempe's visions, like others before and contemporary with her—including Dante, Boccaccio, and Chaucer, as I have noted—one glimpses a spatial overview. For example, her intimate experiences of heaven afford an extraordinary perspective that is qualitatively different from that of her typically horizonal perceptions, one that might be called in spatial terms *a visionary overview*. It is a trope that is common to many texts about intense religious experiences. There is, however, another set of elevated spatial perceptions that remain earthbound when Margery physically climbs to higher places, and I focus in particular on the moment when she climbs Mount Quarantania in the Holy Land. Such occasions are similar to the visionary overview; they complement them but also bring to light the general characteristics of unusual spatial perceptions. In Mandeville's *Book*, despite it being a compendium of distinct episodes grouped according to region, one also gains a sense of homogeneous space because of the author's anthologizing tendencies. This is not just to repeat the observation that the author Mandeville may not be real and the book's sources can be traced to written rather than experiental sources, but rather I wish to draw attention to the effect of the *Book*'s structuring organization of knowledge on the sense of space the reader gains. The sheer quantity of geographical information in the narrative, and the fact that the Mandeville author's models for his writing lie with medieval encyclopedias, mean that the author employs summaries, comparisons, lists, parallels, and other techniques that transform the individuated spaces toward an impression of geographical areas *as area per se*. The discrete horizontal perceptions are transformed into overviews of space in itself. However, something unusual happens in the moments when these rarer spatial hermeneutics come into play in *The Book of Margery Kempe* and *The Book of Sir John Mandeville*. They tend to give way quickly, and in the exceptional moments of abstraction when an overview is perceived, they turn inside out, or *evert*, so that the abstracted view is frustrated or put aside and perception returns to a horizontal position of perceiving individual objects in particular locales. The exceptional moments—whether of spiritual, physical, or textual elevation—undo themselves.

What is going on here? Why this occasional fluctuation between the horizonal and the homogenizing senses of space? Is there something historically significant that involves a change in the fundamental ways that people—from a relatively unlearned woman to an encyclopedic, composite man—think about space? One is tempted to think that the abstracting moments in the Margery Kempe and the Mandeville have similarities with, for example, developments in cartography: Ptolemaic measurements, constant scale, grid coordinates, and so on that tend to treat geographical space as a single and abstract entity. In the previous chapter I noted how Bertrand Westphal and David Woodward attempted to identify "a common interest in geometrically proportioned representations of nature among artists and engineers of the fifteenth century."[2] Part of the conclusion there was that spatial abstraction did not originate solely in a renaissance of Ptolemaic mathematical projection. The current chapter suggests other sources for the emerging hermeneutics of space in a more generalized sense. Margery and Mandeville show that an apprehension of homogeneous space does not, or does not only, arise in this period as a consequence of mapping, a relatively rare practice after all, but instead begins to emerge in different kinds of writings. The contemplative life and the compendious propensities of late-medieval encyclopedism afford understandings of space not only as an agglomeration of things but also as an entity in itself with areas and phenomena that have roughly equal weight and significance. This sense of space, however, appears only briefly. It emerges as a sort of potential, but Kempe and Mandeville promptly turn away from it, shifting their attention back to the diversity of objects within the world.

By way of an introductory example of these tendencies in the period, consider Geoffrey Chaucer's multilayered presentation of space in the opening lines of the Clerk's Prologue and Tale in the *Canterbury Tales*. In the Prologue and the beginning of his tale, an overview of an area that is taken as a whole *everts*, or in this case it may be more accurate to say that the overview *gives way*, to a distinctly heterogeneous topography in which singularity—of place, objects, characters, and their features—becomes the distinguishing characteristics. Because of this telescoping down from an overview to chorographical details, and because of Chaucer's divergence from his sources, the effect is for the marquis of the land—Walter—and his predicament to emerge as the sole focus of the story's narrative. As the tale's introduction continues, the horizonal topography of the local area foretells the theme of the Clerk's Tale as a whole. The topography shows a necessary match between Walter and the land, and it prefigures the narrative events of the Clerk's Tale to provide a necessary outcome to the problem the tale establishes—namely, that the marquisate of Saluce faces the imminent crisis that its rich countryside and many towers and towns, which have always been in Walter's family, will pass to a stranger unless he produces an heir.

Critical attention on the significance of spaces in the Clerk's Prologue and Tale should not come as a surprise. No less a work than the 1589 *Arte of English Poesie* recognizes Chaucer's description of Saluce in the Prologue as an example of *topographia*. The probable author, George Puttenham, defines *topographia* as the "counterfait" of "any true place, citie, castell, hill, valley, or sea, and such like" or else as a description "yf ye fayne places untrue. . . . [S]o did *Chaucer* very well describe the country of *Saluces* in *Italie*, which ye may see, in his report of the Lady *Grysyll*."[3] The Clerk of Oxenford's Prologue and Tale begin by offering the possibility of a larger view, a kind of scenic overview of the "country of *Saluces* in *Italie*," but it thereafter narrows down so the focus is on the immediate scene from one standpoint. The Prologue begins with an appeal to the Clerk's *auctor* Petrarch, drawing attention to Petrarch's rhetorical prowess and to the Clerk's own craft by association. It then continues with a combined but distinct pair of rhetorical figures, one half of which is *proslepsis*, a series of denials that he will discuss a topic.[4] The *proslepsis* soon works in conjunction with *topographia* for a longer description of the area of Saluce, both providing a sense of large areas. The Clerk notes that his *auctor* Petrarch, in his rhetorically "heigh stile," described the countryside of Piedmont, Lombardy, and Emilia-Romagna, and he says that he will not follow him in doing so. He goes on to state that he will not follow Petrarch in describing the source and course of the river Po but tells how it begins at a small well before swelling as it passes from the foot of the mountain eastward through Italy. He will not, he says, depict the lands from which the river springs,

> Mount Vesulus in special,
> Where as the Poo out of a welle smal
> Taketh his firste spryngyng and his sours,
> That estward ay encresseth in his cours
> To Emele-ward, to Ferrare, and Venyse.

He concludes by claiming that it is "impertinent" to describe anything but the core of his story, his "mateere."[5]

The Clerk ends his *proslepsis* when he moves on from the Prologue to the Tale, but the *topographia* continues until it transforms into a description of Walter himself, now focusing on a particular place and a particular person, with the place matching the person. Exactly how Chaucer accomplishes the transformation of place to person is curious. He diverges from his Italian and French sources to clarify and focus the tale's setting. At this point, Boccaccio had not mentioned anything about place. Petrarch, on the other hand, had noted that Walter reigned over his estate and the land, but he had also said that other noble marquises ruled many "villages and castles" (*vicis et castellis*) at the "root" (*ad radicem*) of Vesulus. Walter is the "first and greatest" among these others (*primusque omnium et maximus*). The anonymous French *Livre Griseldis*, Chaucer's most immediate source, also describes Walter among "many great lords and noblemen" as the "first and greatest" (*plusieurs grans seigneurs et gentilz hommes . . . le premier et le plus grant*) who "principally" (*principaument*) governed the land of Saluce.[6] In his version, Chaucer does away with any ambiguity and especially the relativity so that Walter is isolated and the current terminus of his ancestry. The Clerk states simply, "A markys whilom lord was of that lond, / As were his worthy eldres hym bifore." He then begins the tale proper with a smaller overview: the larger, interregional vista has now contracted to a "lusty [fruitful] playn" at the foot of Vesulus,

> habundant of vitaille,
> Where many a tour and toun thou mayst biholde,
> That founded were in tyme of fadres olde,
> And many another delitable sighte,
> And Saluces this noble contree highte.[7]

The effect of restricting attention to Walter as the sole lord narrows the focus on him and his lineage. Chaucer doubles down on topography and thereby significantly raises the stakes of the impending crisis of inheritance.

As critics have pointed out, the *proslepsis-topographia* of the Prologue and Tale gently mocks the Clerk's verbosity at the same time that it displays the Clerk's or Chaucer's rhetorical abilities, but the combined tropes also metaphorically prefigure the tale's narrative in two related ways. First, the geography can be read as standing for Walter's noble origins. "Mount Vesulus in special" is the "welle," "firste spryngyng," and "sours" of the river Po. At the "roote" of this scene are Walter's own beginnings, the "worthy eldres" that came before him. Second, the tale asks what will happen to this history in which land and lineage are intertwined. Will the pressure on Walter, the lord of Saluce, be resolved? The *topographia* appears to foretell the narrative outcome as a happy resolution. The river leaves its source and "ay encresseth in his cours" just as Walter's elders will see Saluce and their line "encresseth" through Walter. The "worthy eldres" will have a fecund future in Walter, who will pass on his lands to an heir. The introductory chorography suggests that origins and outcomes are naturally joined via Walter, who as it were resides in the middle; the transference of property from lord to lord in the country therefore follows the spatial logic of the geographical setting.

The main narrative of Chaucer's Clerk's Tale can be read as an amplification of this opening theme of the genealogical roots and the bequest of ancient property. The tale's narrative can be seen to connect the source of the land with its issue across the region of Saluce. However, unlike the prematurely foreshortened and economical path previewed in the opening and what is often described as the folktale-like structure of the Clerk's Tale, the narrative is anything but straightforward.[8] Though the river's path ultimately leads to "encresse" into the future when Walter's son "succedeth in his heritage / In reste and pees, after his fader [father's] day,"[9] Chaucer, sometimes following his sources and sometimes not, takes into account impediments to the natural bequest of this land. They might be thought of as a series of dams in the even "cours" of the river between its source and its generative fertility. Of key concern are Griselda's class (often associated with her father, Janicula) and her affect, especially at the moment at the end of the tale when her children return. Other factors certainly contribute to the effects created in the Clerk's Tale, but its opening provides an introduction to the story that will follow so that the larger landscape comes to bear on the characters and the plot, narrowing the tale's focus to the horizontal world of a marquisate potentially heading toward genealogical and sociopolitical crisis. The suitability of the characters to the land they inhabit is thereafter explored and challenged.

The Book of Margery Kempe and The Book of Sir John Mandeville are not constructed in the same way as Chaucer's version of the Griselda story. Their scenes do not telescope topography down to a locale, arranging a close fit among place, character, and narrative. Instead, the exceptional places in Margery's and Mandeville's books initially give rise to exceptional experiences and ideas (and vice versa), but then Margery and Mandeville draw back from those moments. The overview dissolves or *everts* to become a local and horizontal place with objects, including Margery and Sir John, emplaced relatively within it. The match between topography and character or incident is momentary. This topographical-narrative feature of Margery's and Mandeville's books may be read as modesty topoi on the part of them both, as though out of humility they will not continue in their exploration of the exceptional realms of the vision and a global geography. The eversion of an overview to local objects and earthly experiences is not, however, so strongly marked by humility but instead is attributable to the force of habitual spatial hermeneutics as local and horizontal.

Margery Kempe in Space

The Book of Margery Kempe is a text that needs some care in approaching, including in terms of space. There remains a tendency to forget it is not an unmediated narrative but a recollected one that, moreover, is in part shaped by more than one amanuensis.[10] These factors indicate some of the complexity with which the *Book* addresses the spaces Margery Kempe encountered in her life, with few easily falling into the category of solely horizontal and heterogeneous or homogeneous with contents of similar value. It seems to be much more an example of the former, a text in which space is not considered as such but instead individual locales and their objects, including Margery herself, figure in diverse scenes that often challenge Margery's intentional and unintentional presentation of herself as she encounters the world through her developing experiences. She and others in her text move around Lynn and its immediate area, she and her companions travel to other places in England, she journeys across the continent alone and with others, and she tours around locations in the Holy Land. Her *Book* often provides the qualia of the spaces, the way someone might typically apprehend the area around him or her in a horizontal fashion. The *Book* also describes another similar form of movement in which readers witness the circulation of her reputation. As it were, the sign of her name, "Margery Kempe of Lynn," is communicated among secular and religious people in fifteenth-century

England and elsewhere. Indeed, it sometimes moves across vast distances with surprising effects.

The fact that *The Book of Margery Kempe* is a retrospective text, one influenced by the memorial arts, themselves a subset of persuasive rhetoric, and particularly of *dispositio* (arrangement; itself a different kind of spatial understanding), complicates an unmediated presentation of the spatial areas she and others experienced in the past.[11] The circulation of her reputation and additional factors contribute to the sophisticated and diverse kinds of interactions with space in the *Book*, so it may not be possible in the end to classify the spaces in it as wholly or even primarily understood or presented as individual and unpatterned. Her *Book* indeed seems to challenge the distinction between individuated spaces and abstracted ones. It contains moments in which Margery and her readers attain an overview of a geographically particular landscape in which each location can have as much chorographical, experiential, and spiritual weight as another, and it includes other extraordinary moments in which Margery sees similar kinds of spatial areas in internally witnessed visions. But both kinds of homogenizing views are often complicated and even to an extent undone because an everyday perception reenters the narrative. Conversely, if we consider what appear to be direct apprehensions of everyday space in the horizontal, individuated sense that compose the majority of Margery's *Book*, even these episodes are not simple. This is because her encounters with individual locales, people, and objects are not only remembered phenomena, but they are also mentally and spiritually prepared for or anticipated. Further complications arise because, as the mechanical arts have shown in terms of a measurer being a meaningful part of a space, she is a feature in the topography of the *Book of Margery Kempe*. Her presence is objectified as a woman and as a potential visionary by the culture around her during a time of theological suspicion. In this sense, her name and reputation are also objectified, circulating in addition to her personhood. Her name and repute exist independently of her person, the sign "Margery Kempe of Lynn" moving freely around, and, in one or two instances, her name circulates to the extent that it re-meets her own person or persona in an uncanny fashion.

Consider one of the more multifaceted moments in *The Book of Margery Kempe*, a moment in which Margery has the potential to see over a vast area. In chapter 30, about a third of the way through the *Book*, while Margery is on pilgrimage within the Holy Land, she journeys to the mountain where it was believed Christ fasted for forty days and nights, and was tempted by the devil by being offered the realm of earth. As frequently happens in the *Book*, Margery's fellow pilgrims do not want to travel with her, but she follows along with them anyway, first to the River Jordan, then to nearby Mount Quarantania.¹² She asks her companions for help in climbing the steep mountain, which rises some 1,200 feet above the floor of the valley, but they cannot help, because they are unable to scale it themselves. In a moment of cross-cultural assistance, a "welfaryng" (good-looking) "Sarazyn" man happens by, and Margery offers him money to assist her, indicating by signs what she wishes.¹³ He takes her "undyr hys arme," and they climb the mountain, Margery complaining of being thirsty and having no "comfort" of the fellowship of her traveling companions.¹⁴ The occasion is an important instance of cross-cultural exchange and quasi translation as others have pointed out, but the moment may in fact be as outstanding spatially as it is culturally.¹⁵

The moment is a rare occasion in which Margery might achieve a physical overview of a large area and consequently an expansive understanding of where she is, but that potential is not fulfilled in the scene. Margery instead turns away from the view to a kind of pensiveness not dissimilar to Richard II in his reaction to seeing the fires in the West of London during the revolt. She makes no mention of what she sees from up there. The elevated position on Mount Quarantania, one on top of a tall cliff, does not offer a physical panorama of the area or even a moment of reflection on the areas in the Holy Land she has visited earlier in her journey or is about to see on descent. She does not discuss being able to look out to Jericho, or to the place where John the Baptist was reputedly born, or to Bethany, or to any other locale she has visited or will visit. In James Gibson's terms, the height does not afford "value" or "meaning" to the objects.¹⁶ The viewpoint's only meaning lies instead in its singularity, in it being the location where Christ was tempted. As critics have noted, the places Margery visits are primarily meaningful in the sense that she has embarked on pilgrimages in the spirit of *imitatio Christi*.¹⁷ On top of the mountain, she turns to contemplation and thinks about the scorn she has received from her traveling companions as well as the gifts of visions and tears that she continues to accept from God in the Holy Land.

Margery's more internal visions afford additional moments when she appears to apprehend a larger space. At two points in the *Book*, she recounts a spiritual experience in which she describes a large area in her soul over which she seems to have a sort of sight. One important moment is in Jerusalem at the Chapel of the Holy Sepulcher, where she experiences her first "cries" (as distinct from the roaring and other behaviors to which she has already been prone). The priests lead a cross around the temple, including to the fifteen-foot-high rock of Calvary, and they describe Jesus's suffering. Margery bursts out, "and cryed wyth a lowde voys as thow hir hert schulde a brostyn asundyr, for in the cite of hir sowle sche saw veryly and freschly how owyr Lord was crucifyed." The vision continues, and she sees John, Mary Magdalen, and "many other that lovyd owyr Lord" in this "city."[18] The scene may not in fact reveal an extensively vast area, but note similar uses of "citte" by Julian of Norwich (whom Margery has visited), which can stand synecdochically for the whole world in the long version of her *Shewings*.[19] As Virginia Raguin says, "Through [Margery's] references to buildings she transfers the images of the actual cities to her inner world, the 'city of the soul.' It is from this association of the personal self with the heavenly one that she builds her authority."[20]

The second significant moment occurs when Christ assures Margery that he is aware of her visions and how significant Mary Magdalene is to her. He explains that "sumtyme, dowtyr, thu thynkyst thi sowle so large and so wyde that thu clepist [beckon] al the cowrt of hevyn into thi sowle for to wolcomyn me." He continues, "I wot [know] ryth wel, dowtyr, what thu seist, 'Comyth alle twelve apostelys that wer so wel belovyd of God in erde [on earth] and receyvyth yowr Lord in my sowle.'"[21] This is perhaps not as large a space as the "city" of before, and after the "sowle so large and so wyde," and the whole court of heaven, the image constricts to including just the apostles. However, the scene continues on to describe others in the heavenly space. Also, there is a certain parallel to be made here with another moment slightly earlier in the *Book* in which an analogy is drawn between Christ's presence in her soul and the large realm of the weather. Here Christ points out how the planets are obedient to his will and that he sometimes sends "thundir-krakkys" that frighten people. He reminds her that he also sends lightning that sets fire to buildings (significant because of the burning of Trinity Guild Hall in Lynn in chapter 67) and "gret wyndys that blowyn down stepelys, howsys, and trees owt of the erde and doth mech harm in many placys." He also makes earthquakes and "gret reynys and scharp schowerys, and sumtyme but smale and softe dropis."[22]

These are large spaces and forces, moments in which Margery seems to be perceiving things not perhaps from Christ's perspective but at least in analogous ways to him, a kind of spatial *imitatio*. If the moments do not exactly offer a bird's-eye view, they afford a sense of indistinct locations in a large area. These are, however, not everyday experiences even if they are reasonably common for Margery. We should also note that the spatial images and metaphors in these moments of vision do not go further in abstracting space per se for, as in the Mount Quarantania episode, Margery is intent on the spiritual lesson and understanding they offer. In fact, what is notable about the vast majority of her visions is that they more often than not contain everyday senses of space. To take one example almost at random, at a certain point when she is having an idle moment and asks Christ what she should think about, he says to consider his mother, and she subsequently has a vision first of Anne giving birth to Mary, then Mary giving birth to Christ, and Margery participating in raising the child. The vision continues with Margery accompanying Mary and Joseph to visit Elizabeth, John the Baptist's mother, before going with Mary to Bethlehem, paying for her nightly accommodations and getting her food, and so it goes on.[23]

Yi-Fu Tuan's *Space and Place* reasons at one point that "'[d]istance' connotes degrees of accessibility and also of concern. Human beings are interested in other people and in objects of importance to their livelihood. They want to know whether the significant others are far or near with respect to themselves and to each other."[24] This is true of *The Book of Margery Kempe* which, despite occasional moments in which Margery perceives a larger area, more typically recounts horizontal experiences of space in which objects are noted from a ground-level perspective and Margery's own body is implicated. Margery's much more common horizontal experiences have different qualities. Consider in turn key moments in the four principal spaces the *Book* addresses: King's Lynn, England, the Holy Land, and the continent. Two of the most distinctive features of her impressions of King's Lynn, or Bishop's Lynn as it was then known, are that she records only the buildings and hardly mentions any other physical features of the town. She also rarely details the relationships among buildings, locations, or other items in terms of time, distance, steps, or any other measure, so in this respect she diverges from the relational veracity of local maps, which record places not always with descriptive detail but with some accuracy about where objects stand in relation to each other. The Church of Saint Margaret and other religious structures in Lynn loom large in her recollections, but secular structures such as the guildhall and other houses are also noted.[25] At one vivid moment in her narrative, for example, she visits an insane woman who has been removed to one end of town and locked up in a "chambyr," but no more details about its location are offered. This moment is illustrative in that besides making a rare note of relative location—the place is at the "forthest end of the town"—the *Book* is not particularly specific about any of the places where Margery lives or otherwise commonly visits.[26] For instance, early on in her narrative when she is tempted to sleep with a man, she first meets with him in an unidentified location in Lynn before they part for evensong at Saint Margaret; later on, she re-meets him to have sex, yet the location is not named.[27]

Perhaps this lack of specificity is not surprising given the clandestine and, to Margery, shameful nature of the meeting, but even her own home's external features and location are not noted. Later in the *Book*, when she and her husband have been living apart for some time and he falls down the stairs and severely injures his head, it is said that neighbors hear the fall, giving a sense of the severity of his accident and the proximity of houses. They send for Margery, but there is no sense of where she has been living. The narrative in fact notes that both Margery and John "dwellyd and sojowryd in divers placys" because people were suspicious of whether they were living up to their vow of chastity.[28] Locations are not mentioned, again likely because they are not significant for Margery's reconstructed narrative. The impression we are left with in terms of Lynn (and other towns such as Canterbury) is of significant structures but with very little description and almost entirely without relative distance or position. These characteristics paradoxically hint at a kind of pattern in her recollections, so we might say that her impressions strain at the sense of the horizontal. As Dee Dyas suggests, "Intention rather than distance seems to have been the defining characteristic" of Margery's and others' pilgrimages.[29] Local places are therefore all treated in similar ways instead of being individuated and spatially distinguished from one another.

Outside of Lynn or Canterbury, other English towns and locales might be noted in terms of measured relative distance but not in terms of cardinal direction, and this fact suggests a way of thinking about places, unconscious though it is, that is not exactly abstraction but again strains against the individuation of each location. Within Lynn—perhaps because it is more familiar to Margery, to her amanuensis, and possibly to an initial audience—distance is rarely mentioned. In contrast, outside a town, distance is often noted in terms of miles. Consider places around Lynn. At one important moment when two unnamed priests want to test whether Margery cries only in front of others and not alone, they take her, we are told, "too myle fro then [the place] sche dwellyd."[30] Another time, the priest who writes Margery's story is offered two books for sale, and one of the two sellers claims to be from Pentney Priory, "but fyve myle fro this place" (meaning from Lynn). The man is lying, so it is interesting to note that he is playing on Margery's priest's familiarity with a nearby place in an attempt to earn his trust, but the priest knows the inhabitants of the priory well enough to realize that the man is fraudulently presenting himself.[31] Other places in England, in addition to ones around Bishop's Lynn, are commonly noted in terms of distance but again not in terms of cardinal directions.

Locations outside England are described in similar ways. Towns and cities on the continent are rendered, like Lynn, in terms of their buildings and often in relation to Margery's own person. Larger distances commonly get some indication of measure. In Middleburg in the Low Countries, for example, she is warned by God that a thunderstorm is coming, and she and her companions quickly go back to their hostel, which is described as nearby.[32] During her pilgrimage to the Holy Land, the *Book* describes structures within Jerusalem in the same manner as places within Lynn and on the continent. In this case, not just buildings but also holy sites more generally receive some topographic emphasis and detail. The lack of description of the holy sites may be partially due to Margery's potential familiarity with replicas of them in England (although she does not mention them in her narrative). Or there may be some shared degree of similarity in outlook on the world and *contemptus mundi* between the Grey Friars who led pilgrims around Bethlehem and Margery herself.[33]

It is also worth noting how many times she re-meets people she has encountered earlier in her travels, especially British people, whether she is near or far from Britain. Such chance reconvenings quite frequently occur in romances, but in *The Book of Margery Kempe*, they signal the popularity of pilgrimages and the set routes that pilgrims traveled thanks to available roads and guidebooks. This known quality about pilgrimage routes might contribute to the occasional vagueness about location because Margery's narrative assumes readers will be familiar with the places it describes. That is, the individuality of pilgrimage routes is reduced because of people's knowledge of the locations and designated ways to them. Yet despite this potential sense of familiarity, Margery is forlorn when people leave her and sometimes (though not always) surprised when she sees them again. One of the more interesting moments is when she returns to England from the Low Countries and goes to Sheen (or Syon Abbey) on the Thames outside London. There she encounters the same hermit who had initially led her from Lynn to Ipswich when she had set out for Prussia fifteen months earlier.[34] Equally noteworthy is when she re-meets Richard, an Irish "broke-bakkyd man," in Bristol, whom she had initially met in Venice and had arranged to meet later in England.[35] This agreement shows people thinking not only across large distances but also across time to coordinate a specific place and location. This sort of casting of the mind across distances and time indicates another nuance in the horizontal.

These kinds of ecumenical movement, coincidence, and coordination also apply to Margery's reputation. Examining the sign "Margery Kempe of Lynn" further highlights the networks of in-person and oral communication that existed across spaces within England and elsewhere. Analysis of her name and reputation further brings to light a sense of relationships among locations and peoples, even places and people that are not physically contiguous. Kate Parker has detailed the remnants of economic power of the governing class within Lynn, of which Margery and her family had been an integral part. Despite the decline of her family's influence, her family name still empowers Margery in her travels and helps make her known among the clergy as well as secular people.[36] But this power, already being challenged on the economic and political fronts, and further questioned because of Margery's gender, was contingent on additional factors. We are told about her reputation in Lynn on several occasions, where some people sympathize with her experiences and understand them, while others follow the "jangelyng of other personys" and, we learn, "seyd ful evyl of hir [and cau]syd hir to have mech enmyte and mech dysese."[37] At one key point, a friar preaches against her in the town, or rather he does not name her but "so . . . expleytyd hys conseytys [explained his ideas] that men undirstod wel that he ment hir." The town knows from his descriptions that she is indicated, a technique that Chaucer's Pardoner also uses when he says that "though I telle noght" a person's "propre name, / Men shal wel knowe that it is the same, / By signes, and by othere circumstances."[38] The friar causes so much ill will among the people that they no longer communicate with Margery. Even the priest who had agreed to be her scribe is persuaded enough to eschew her.[39] Such a case strengthens the impression that Margery does not detail relationships among buildings in Lynn because of familiarity with them. The *Book* relies on the idea that its audience will understand Lynn, or a town like it, as an interactive network of spaces and people. The ability of these circuits of money, reputation, and affect to function successfully depends on a variety of factors, including power, knowledge, and in Margery's case in particular, trust in her experience.

"I leve ther was nevyr woman in Inglond so ferd wythal as sche is and hath ben."[40] As these words from Henry Bowet, archbishop of York, suggest, Margery's reputation also extended well beyond Bishop's Lynn. His words to those gathered at her examination in Beverley in Yorkshire point to how the sign "Margery Kempe" spread widely through rumor and reputation. Trouble with the Duke of Bedford had by this point led her to examination before Bowet, and here the archbishop expresses exasperation that he does not know what to do with her. Margery suggests that he release her with a letter stating that she is neither in error in her faith nor heretical, a powerful document that should allow her to travel freely thereafter. Also indicative of the way her reputation has traveled across distances is that the archbishop marvels that she is able to get money from people around her, but she explains that she can get funds if she promises she will pray for people. In addition, elsewhere in the *Book*, Philip Repingdon, bishop of Lincoln, has heard of her, and later the bishop of Worcester says he knows her.[41] Such is the power of her reputation and its ubiquity that when she is in Rome, she is given money from a man she has never met who has heard about her in England and has traveled to Rome to help her with funds.[42] In Bethlehem, the Grey Friars who guide the pilgrims seek her out. When she first meets the poor Irishman, Richard, in Venice, he knows of her reputation and the troubles she has had with her traveling companions.[43] All of these examples in England, on the continent, and in the Holy Land suggest the relative strength not only of pilgrimage but also of mercantile, ecclesiastical, and other networks. At times Margery intentionally taps into them, drawing on their prior establishment to her advantage, and at other times her name circulates along similar routes to her potential detriment.

Perhaps the most remarkable example of the circulation of Margery's reputation occurs late in the *Book* when she is back in London after returning from the Low Countries via Calais and Canterbury. She does not know her way to Canterbury, but she seems to know London quite well, and more pertinently, London knows her. The episode in London speaks to the proximity of people in the city, and it has an almost uncanny tone as "Margery Kempe" makes its way back to its owner. This is the only moment in the *Book* where we learn the name "Mar. Kempe of Lynne," so there is a kind of irony that we know her name today only because of the malicious intentions of those who want to use it against her. When she returns to London, "mech pepil knew hir wel anow" even though she has covered her face because she is seemingly embarrassed by the poverty of her clothing (a "cloth of canvas"). These and other "dissolute personys, supposyng it was" her, taunt her for her vegetarianism and her hypocrisy, attributing a saying to her that she had criticized people for eating meat when she ate the most expensive dishes herself. The *Book* goes on to recount that "sche many tymys and in many placys had gret repref therby," including "in many a place wher sche was nevyr kyd [identified] ne knowyn." What follows is even stranger. When she goes to a wealthy woman's house in London for a feast, she is among diverse sorts of people, including some of Cardinal Beaufort's retinue. There they repeat the saying about her as a hypocrite without realizing they are sitting at the table with the person they are laughing about. Margery reveals that they are saying it about her and takes the opportunity to reprove them. According to the narrative, they are abashed and ask her "coreccyon."[44]

Margery's more typical perceptions of horizonal local space, even from what we would consider vantage points such as up the mountain, are perhaps to be expected since we are not reading a modern form of travelogue in which views might be appreciated for post-Enlightenment senses of the sublime or an aesthetic appreciation of nature. The contemplative life and not the physical world from which she is often withdrawn is the experience that she values and that largely shapes her apprehension and comprehension of the world around her, so it is not a surprise that it tends to structure and color the recollections that are gathered in her *Book* and written quite some time after she experienced them. As Sebastian Sobecki says in relation to the names mentioned in *The Book of Margery Kempe*, the text attempts to draw away from specific people and "universalize" her experience.[45]

In addition to this spiritual focus and the memorial arts that have an impact on the text, it is also possible to consider each episode in the *Book* as shaped according to the conventions of *chorographia* and *topographia* in which individual localities are significant in terms of the relationship between places and the objects and events that occurred there. The *Book*'s opening proposes that it is a "schort tretys and a comfortabyl for synful wrecchys, wherin thei may have gret solas and comfort to hem and undyrstondyn the hy and unspecabyl mercy of ower sovereyn Savyowr Cryst Jhesu."[46] That is, it is written with a clear intent and set of topics or intended effects, especially to "comfort" its readers and offer understanding in the face of the "unspecabyl." It is, however, also tempting to read "comfortabyl" as "conforming" in the sense that the *Book*'s events match or suit the locations in which they occur. The *Middle English Dictionary* shows that *conform* has connotations of peace and harmony, and it notes an instance of *conform-comfort* word confusion in *conformen*.[47] When Margery's amanuensis later states that matters are out of chronological order, he says they are placed where they are for "convenyens" in that the events he describes are grouped with other similar occurrences.[48] As with "comfortabyl," there is a way that *convenyens* has a stronger ameliorated sense of *congruity* and *suitability* than today's idea of something requiring little effort.[49] He also notes in passing that the shaping of the *Book* could affect not only the order of the events but also identification of the actual places Margery visited when he excuses himself for possibly not correctly noting place names. In the appended second book about Margery's travels, particularly in the Low Countries, her priest-writer similarly says she may not be perfectly remembering the places she visited, and he has to rely on her memory of them because he has not been there himself.[50]

For Margery, the heterogeneous space is the more common one in which locations tend to be distinct and individually recognized depending on the size of the area she is perceiving and/or her familiarity with an area even if she has not physically visited it previously, as may be the case with some of the holy sites in and around Jerusalem. Only occasionally is she able to obtain a view of a larger area, although even the diverse spaces of Lynn, other towns in England, and the Holy Land share characteristics that make them somewhat abstractly characterized. Margery and her *Book* can measure distance at least outside Bishop's Lynn and other towns just as the astrolable, quadrant, and treatises on measurement did, with the measures being in relation to the viewer. This kind of horizontal relativity indicates the earthly, everyday nature of her experiences of physical space and locations. In fact, *The Book of Margery Kempe* employs a key spatial metaphor of measurement almost in a self-referential manner in its repeated insistence that she cannot internally "mesuryn hirself ne rewlyn hirselfe." Variations on "mesuryn" and the related "rewlyn" occur many times in the narrative, and the words are usually taken metaphorically to mean that Margery cannot appropriately moderate and govern herself, which goes to the heart of her challenges to social norms of decorum. The metaphors, however, also connote her inability to calculate, gauge, or determine her place in her geographical world, and she shares this sense of mismeasurement with others. At one point, a priest who challenges her is given proof of her sanctity when he is provided with so many tears that he, like her, cannot "mesuryn hys wepyng ne hys sobbyng." Immediately thereafter, her own priest-scribe also reconsiders her copious tears, which convinces him of Margery's own gift and encourages him so that he "drow ageyn and inclined mor sadly to the sayd creatur."[51] It is this inability to measure the spiritual against the physical, a mismatch between visionary perception and topographical place that, although brief, forces Margery and others out of the ordinary and into the extraordinary.

Mandeville's Approaches

The Book of Margery Kempe and *The Book of Sir John Mandeville* contain parallels not only in terms of each one's complicated relationships among their main characters (as we might think of Margery Kempe and Sir John), the writers of their texts, and the narratives, but also in terms of their treatments of space. The granular, detailed descriptions of distinct locations pair them in that they address horizonal spaces as composed of dissimilar, individuated places and phenomena located in sites that relate to the viewer. We have already seen that Margery is occasionally, albeit briefly, capable of seeing or thinking across larger areas, and she has the ability to understand areas of different kinds—King's Lynn, England, the continent, and the Holy Land—in some categorical fashion. Her foreknowledge of the spaces she encounters, and indeed people's general ability to think and act across large distances, also complicates a distinction between heterogeneous and homogeneous apprehensions of space because she anticipates what she sees, and therefore the level of detail pertaining to a particular location can be reduced. This effect is heightened because of the *Book*'s retrospective, remembered qualities, which tend to organize and pinpoint experiences on the basis of their rhetorical significance and effect.

CHAPTER 4

The Book of Sir John Mandeville is similar to Margery's *Book* in that it is largely composed of chorographical descriptions of individuated locations and the creatures and practices that are there, which often include Mandeville himself. Also, like *The Book of Margery Kempe* when it achieves an overview of space, Sir John's *Book* has the propensity to turn away from that overview and return to seeing individual locales in a heterogeneous fashion. Its sense of *topographia* is strong in that it frequently and simply remarks on what exists in a given location as though the objects and events are fitting to the place. It does not, however, often judge these places in an ethical fashion, a fact that is somewhat remarkable given that *topographia* often considered the ethical fit of, for instance, a practice in relation to a place. *The Book of Sir John Mandeville*, however, goes even further than Margery's text in everting its heterogeneous view of spaces so that they become somewhat generic.

It is possible to attribute these features of the *Book* to its putative and possibly composite author, to the content and the nature of the sources the *Book* uses for its information about locations, and to the scale of the areas its narrator describes. Analyses of the *Book of Sir John Mandeville* have explored these facets as well as its presentation of other cultures and peoples.[52] Anthony Bale describes how Mandeville was adopted by later readers as a "point of departure to build their own textual repositories of memories of Jerusalem and the East" because the fifteenth- and early sixteenth-century men he describes also traveled to the Holy Land. He points out that Mandeville's book presented to those readers a textual rather than contemporary account of the area because the author based his descriptions on older written sources rather than current accounts, so its value as a practical guidebook was limited.[53] Evidence such as this suggests that the characteristic ways in which the Mandeville author presents at least some locations in the *Book* may be attributable to his antecedents, but it more significantly indicates that a form of realist objectivity was not the only governing mode of presenting the spaces in the text.

The Book of Sir John Mandeville explicitly displays four kinds of spatially homogenous hermeneutics when its narrator discusses the diversity of locations, compares one place with others, uses lists, and employs spatial synecdoches. The *Book* also implicitly contains a fifth method of understanding space in that it tends to structure its episodes in parallel ways. First, Mandeville has a sense of "diverse" locales being in some way all the same. For example, in the opening pages he says he traveled through "manye dyverse londes and many provynces and kyngdomes and iles," followed by a list of lands from Turkey to Egypt, "Ethiope" (Africa), India, and more.[54] *Diverse* and its forms are common words in Mandeville's *Book*, allowing him to be particular and also to generalize across large areas and phenomena. It is a kind of compression or *abbreviatio* that is the opposite of what Mary Campbell noticed of Mandeville some time ago, namely, *amplificatio*.[55] He begins the *Book* by saying he will not describe all the lands and places in Europe but only select ones.[56] Examples of this technique of merely naming different areas are common; later on he says, for instance, that in the area of Ethiopia and "Libye" (North Africa), "many othere londes ther ben that it were to long to tellen or to nombre," so only "of sum parties I schalle speke."[57]

Second, the narrator employs comparisons between one place and another to draw parallels or to make contrasts, in both cases suggesting the ability to think of two or more places at once and to generalize about them. The *Book* opens with the Holy Land being compared with other lands; it is the "most worthi" because of Christ and the Virgin; other comparisons abound elsewhere in the narrative.[58] He compares not only places but also routes, as when he names two directions to Babylon, Mount Sinai, and Jerusalem, describing a passage back via Jerusalem as "the more ny weye" (the shorter way?) and "the more worthi."[59] Third, Mandeville turns to lists of various kinds to provide a shorthand overview of an area. He, for instance, merely registers all the islands around Greece.[60] Fourth, he often says that a place or a phenomenon is representative of a larger area of things in the same region and even elsewhere; this is synecdoche. For example, when he has described the divergent customs of the Eastern church, he states only that they are representative of the diversity of faiths throughout the world.[61]

In addition, a structure for presenting spaces soon emerges in which he offers an overview of a place before scaling down to smaller and smaller areas and features in a concentric, centripetal fashion such as we saw Chaucer doing in the opening of the Clerk's Tale. For example, Mandeville begins his description of Egypt early in the *Book* as "a long contree but it is streyt, that is to seye narow" and "sett along upon the ryvere of Nyle," and then after the overview he delves into its provinces and cities.[62] Another example from later on, among many others, includes his description of India as "fulle [of] manye dyverse contrees" with more than five thousand islands around it, not counting uninhabitable and smaller ones. He begins with that large area, then delves into each of its regions, peoples, and practices.[63] On a smaller scale, when he describes Jerusalem, he depicts its chorographical situation among hills, the history of the area's name, and the surrounding lands and locations before going on to describe the entries to the city, then the town's contents, and eventually returns to the surrounding lands (including Mount Quarantania).[64] This regular structure suggests an understanding of topographical features as contained within larger areas, a kind of nesting. It also indicates a spatial hermeneutics that involves scale as an important factor since each larger area recurrently progresses down in scale to smaller and smaller areas.

Two moments in particular in the *Book of Sir John Mandeville* take all these techniques to an extreme so that the spatial hermeneutics reach toward an abstracted overview, flattening individual differences and presenting an apprehension of space per se. One occurs right at the conclusion of the *Book* in the last chapter when he has gone, he says, as far east as he can, beyond the land of Prester John to the realms of the Great Khan, including its islands, where he remarks not only that there is so much diversity that he cannot describe it all but also that he will restrain himself from describing the places for the benefit of his readers. (There are also locales, he concedes, that he has not seen.) He asks of his readers' "noblesse" that what he has said should "suffise you at this tyme." He adds that he will not describe more of the region so that future travelers might "fynde ynowe to speke of that I have not touched of in no wyse."[65] After all, one of the key moments when Mandeville actually stops traveling has just been described, his attempt to reach Earthly Paradise. No mortal man can enter it, he says, "withouten specyalle grace of God, so that of that place I can sey you no more. And therfore I schalle holde me stille and retornen to that that I have seen."[66] Beyond Prester John's land, the sheer profusion of places is a kind of limit for Mandeville, one that suggests an almost intimidating though apparently acceptable sense to him that all places are in some sense similar, all worthy of description as if his descriptions, if not also his travel, could go on forever. Failing that, others can continue his voyages and accounts of them. They can continue this study of spatial plenitude.

The second telling moment when the Mandeville text describes space as a dispersed area in which everything is potentially the same occurs in chapter 20 of the *Book*, again when the putative Sir John gets far from England, out to part of Sumatra ("Lamary") where cannibals carry out their abhorrent practices. He takes a moment to meditate on what is there, what may lie beyond Lamary, and he thinks about other locations where he has been farther south (another example of comparison, here across very large expanses). He notices that the North Star is no longer visible, so the most important referent in navigation in the Northern Hemisphere is lost, but he claims the star is compensated for by an "Antartyk" one. This is his initial sense of the Southern Hemisphere as a logical mirror image of the North, gained from old authorities, and this recognition of symmetry prompts him to go on to reflect on the overall regularity of the earth to a degree that he imagines the whole planet as graticular with each place on the earth—whether Far East, South, North, or West—having equal value. People easily perceive, he says, "that the lond and the see ben of rownde schapp and forme" so that circumnavigation "alle aboute the world" is possible. He continues expounding on the roundness of the earth, the possibilities of circumnavigation, the relative positions of stars at certain degrees in the sky, the logical existence of the antipodes, and the fact that it is night in the land of Prester John (nearby) when it is day in the West.[67] To imagine the earth as all spatially the same is the largest sense of homogeneity. Margery might catch partial glimpses of the vastness of the spiritual kingdom, and elsewhere her *Book* may more simply note "diverse" places, but in Mandeville we have a full overview of the terrestrial planet.

We should not, however, disregard the fact that his vision at this point is unusual in its discourse, and it is momentary. Mandeville's discussion in the passage is geographical in a broad way, and it is that level of discourse, combined with his physical location far away, that seems to afford the understanding he has of the likeness of every place on earth. Where Margery's visions enable her sense of large areas, Mandeville's *scientia* and extraordinary location enable his perspective and understanding. What is nonetheless curious in his discussion in the chapter is that he concludes by critiquing the everyday forms of spatial hermeneutics that people hold. First he scorns "symple men unlearned" who think it is not possible to go "under the erthe, and also that men scholde falle toward the Hevene from under." The roundness of the earth means this will not happen.[68] But then he continues to a subtler spatio-sociological denunciation of people who have a limited view of the earth in which each place is relative only to where one happens to be currently situated. He says that "fro what partie of the erthe that man duelle, outher aboven or benethen, it semeth always to hem that duellen that thei gon more right than ony other folk."[69] This appears to be a critique of an aspect of the horizonal view, one in which the world would appear bounded by the horizon around the location where one happens to be at a given time, but of course the horizonal view does not necessarily mean that one is so "simple" as to believe the world ends at one's own horizon.[70] Mandeville instead implies that one should not think of one's own location as unusual or "more right" than anywhere else. We are again at a plenitudinal limit in which the fact of the multiplicity of spaces suffices, or else confronts, the traveler. Sir John turns, as if for comfort in the face of the multitude of locales, to Job 26.7, "Do not fear me who suspends the earth upon nothing" (Non timeas me qui suspendi terram ex nichilo). That is not, however, where the chapter ends, Mandeville continuing to imagine the slim chances of circumnavigating the globe without getting lost and engaging with Eratosthenes's and other scientific measurements of the earth.[71] Philosophical and encyclopedic ways of understanding the earth supersede, or at least encompass, theological ones. The chapter therefore ends with recounting "the opynyoun of olde wise philosophres and astronomeres" about the climes of the earth, and the next chapter starts with the island of Java next to Lamary. The *Book* continues thereafter with its episodic presentation of diverse locations.

It is possible to talk about *The Book of Margery Kempe* together with *The Book of Sir John Mandeville* because these very different works nevertheless reveal similar engagements with space. Both primarily concern themselves with objects in local spaces, although the former rarely acknowledges the distances between objects, including people's locations, in local and more familiar towns. Margery's *Book* examines in places the effects of her name circulating along defined pilgrimage routes, whereas the Mandeville author gives the impression that he has to discover his routes when he moves beyond Europe and the Holy Land. Margery's recollected narrative has moments when she removes herself from the horizontal view, or else she is removed by her immeasurable cries, and she achieves a sense of a larger spacious area. But the overview does not ultimately interest her, and she returns to a more homely situation. *The Book of Sir John Mandeville* is similar to Margery Kempe's *Book* except it exists more in a realm of secular or interreligious travel rather than solely Christian pilgrimage—well beyond Margery's sympathetic encounter with the Saracen—and Mandeville is more willing to extend his outlook to encompass the whole earth not, or not only, from a spiritual overview but from a worldly and scientific one. Its narrator remains on the earth and does not circumnavigate it or reach earthly paradise but instead turns back to return along diverse ways to his home. In both cases, rather than a Ptolemaic cartography enabling a geographical sense of a world that is large and composed of geometrically similar areas that "vibrate with curiosity and interest,"[72] other traditions embodied in Margery's spiritual experiences and Mandeville's encyclopedic aspirations enable a sense of the world as expansive and transiently homogeneous.

CHAPTER 5

The Science of Motion
New Ideas of Impetus and Measurement

Like space and place, motion remains a difficult phenomenon to describe scientifically and to understand more generally. What does it mean for something to move? Is motion relative to a fixed point, is it a quality obtaining to a moving object, or does it involve a force? Physics has to appeal to a number of theories: Newton's laws for the movement of large objects, special relativity for space together with time, and quantum mechanics for the motion of very small phenomena. In the Middle Ages, motion was not as it is today, a subfield of the larger discipline of physics. It was central to virtually all understandings of the natural world. This chapter focuses on this key characteristic of the world through an examination of the questions that occupied late medieval natural philosophy: What is motion, and what is the nature of continuous motion? Is motion something external added to a body, or can it inhere in a body? What kinds of objects can be affected by motion? How can motion be measured? Late medieval scholastics vigorously debated concepts of motion in part because they found Aristotle's ideas about change in his *Physics*, especially change in spatial position, wanting. Thomas Aquinas explained and generally followed Aristotle's ideas in the thirteenth century in his *Commentaria in octo libros Physicorum*, but in the fourteenth century, Jean Buridan, the Merton School of Oxford Calculators, Nicole Oresme, Thomas Bradwardine, and others challenged Aristotelian, Thomist, and each other's ideas about motion. They did so because the subject presented problems that fascinated them and that they wanted to solve, and they did so on new grounds. Where thirteenth-century natural philosophy addressed the origins and especially the destination of objects in motion, fourteenth-century philosophy largely put aside the question of destination and instead addressed motion in terms of internal motivation. This was a development that meant the causes of motion, moving objects themselves, and spatial change came to be seen differently.

An important consequence of what historian of science Anneliese Maier in 1949 called a "fundamental transformation" in the science of motion in the fourteenth century is that the significance of natural place is diminished.[1] *Locus naturalis* was a key idea in thirteenth-century and earlier mechanics, and an important way of thinking about the world more generally, intersecting as it did with broader ideas about physical properties in the world, the behavior of creatures generally, and theological ideas about the origins of the universe. Kellie Robertson describes *locus naturalis* as the "drama of inclination" in which all objects, "both people and things," advance toward a natural "home."[2] The concept of natural place continued to exert authority throughout the Middle Ages and beyond. However, that location—whether above or below an object, behind an object or in front of an object, in the past or in the future, or a place that something has left and to which it returns—loses importance in the late medieval mechanics I examine. The new concepts diverge considerably from a kind of onto-teleology in which all phenomena, including humans, move toward a place of rest or seek to remain there. The reasons that fourteenth-century natural science does this are varied but related. They include the idea that rather than natural place being a motivating or attractive location, the source of an object's motion came to be conceived of as embedded within the object itself. Jean Buridan adapts the word *impetus* to describe the quality of force within objects for the first time in history. Furthermore, because of the diminished significance of natural place and the embodiment of motive force, in terms of space, all moments in the existence of moving bodies, and all the positions of those objects, become as significant as each other.

The following discussion about these innovations in the nature of motion (a phrase medieval science would consider redundant since nature is largely identified with motion) has three parts. The first section explores the central role that motion plays in Aristotle and Aquinas's understandings of the cosmos in the thirteenth century. It also demonstrates how natural place functioned as a destination in traditional mechanics. The second section examines philosophies that sought to answer the question of where motion lies. Is it separate from a moving body, or is it part of it? It investigates the answers to this key question in the works of Duns Scotus, Francis de Marchia, and Jean Buridan, all of whom disagreed with Aristotle and each other. The most important theory about the location of motion is Buridan's impetus because it overturns a central Aristotelian point and proposes the idea of a force that inheres in an object. The final section addresses the measurement of speed and the acceleration of objects through an examination of Thomas Bradwardine, the Merton School at Oxford, and Nicole Oresme. Their analyses of how to measure speed may appear technical to the modern reader because they elaborate on all variations of changes in rate of motion, but their descriptions of measuring speed also demonstrate how differently these philosophers thought about space from a previous generation. Oresme is often credited with creating the first graphs of motion in history, and their significance is described.

The change from discussions of the origin and destination of motion in the thirteenth century to internal motivation in the fourteenth is a change in the kind of causation attributed to motion. Although Buridan, the Merton School, Bradwardine, Oresme, and others did not articulate it in these ways, even though the terminology is Aristotelian, the cause of motion shifts from a focus on final cause, as in the final natural place of movement, to efficient cause, or the source of motion regardless of position. As Maier argues in a discussion of natural place, a "final cause always presupposes a corresponding efficient cause. This, then, is the real problem [of late medieval mechanics], and it was considered one of the most difficult questions in physics: what is the active cause or, to phrase it differently, what is the motive force, the *movens*, that causes natural motion?"[3] Efficient cause in this case is the cause of motion in itself; it is not only the starting agent of motion that begins something moving but also the cause that keeps an object moving. Fourteenth-century science focused directly on the efficient cause of motion, and the final cause of a *locus naturalis* is de-emphasized. The import of this change in Aristotelian and medieval science is that, as in chapter 3 in which the prominence of a vertical hierarchy of cosmological spheres down to the earth is decreased in favor of a horizonal band or zone near the earth, here in this chapter, the new ideas about motion reduce another kind of hierarchy of place. Where before an object was thought of as out of place until it reached the end of its motion, an idea that therefore privileged its *locus naturalis*, now any place becomes as significant as any other. This is not a form of spatial relativity in the same spirit as Newton or Einstein, and not all late medieval science agreed on these new mechanics, but there was a general shift in the culture that nevertheless departed from the sense of natural place being the only way the culture thought about motion and location.

Traditional Ideas of Motion and Natural Place

"[A]nimals and their parts, plants, and simple bodies like earth, air, fire, and water . . . each has in itself a source of change and staying unchanged, whether in respect of place, or growth and decay, or alteration. . . . This suggests that nature is a sort of source and cause of change and remaining unchanged."[4] Aristotle's opening to book 2 of the *Physics* assigns a central role to change (*kinēsis*) and constancy in nature so much so that alteration is not merely a part of the universe but its central characteristic. He reiterates at the beginning of book 3 that if change "is not known, it must be that nature is not known either."[5] His ideas on change of a general kind had broad effects throughout the Middle Ages. For example, Averroës in the twelfth century follows Aristotle in a discussion of celestial bodies, stating that "if motion were destroyed, so would the heaven itself [be destroyed]. Indeed, the heaven exists because of its motion; and if celestial motion were destroyed, the motion of all inferior beings would be destroyed and so also would the world." This provocative hypothesis was condemned in Paris, yet later European philosophers continued to discuss it.[6] Trevisa's translation of Bartholomaeus Anglicus's *De Proprietatibus Rerum* provides one piece of evidence among many of Aristotle's effect in English when it states that it is interested in six kinds of "meovingis": "generacioun, corrupcioun, alteracioun, augmentacioun, diminucioun, and chaunginge of place."[7] Aristotle's idea of change was often taken to mean motion in the Middle Ages, and mobility consequently became central to the natural sciences. A thirteenth-century English textbook containing Aristotle states, "The natural sciences consider the mobile body as its subject" (In scientia naturali est consideratio de corpore mobili sicut de subjecto), while a contemporary Oxford master writes "the subject of natural philosophy is mobile bodies."[8] Anneliese Maier, John Murdoch and Edith Sylla, Marshall Clagett, and other historians of medieval science note the following: "The first and foremost problem of scholastic physics is the concept of motion," "In one sense, to speak of the 'science of motion' in the Middle Ages is to speak of all physics or even of all natural philosophy," and the "study of nature becomes the study of movement."[9]

For Aristotle and the tradition that followed Aquinas when his texts on natural philosophy were revived in the thirteenth century after they were banned in 1277, a tradition that included writers such as Dante, motion was of central importance because it was understood to affect all beings. Aquinas subtly changed the focus of the study of motion; unlike a figure such as Albertus Magnus, or the English textbooks that name the subject of natural philosophy as *corpore mobili*, Aquinas said motion was the study of mobile things (*ens mobili*) as in mobile being in itself:[10] "And because everything which has matter is mobile, it follows that mobile being is the subject of natural philosophy. For natural philosophy is about natural things, and natural things are those whose principle is nature. But nature is a principle of motion and rest in that in which it is. Therefore, natural science deals with those things which have in them a principle of motion" (Et quia omne quod habet materiam mobile est, consequens est quod ens mobile sit subiectum naturalis philosophiae. Naturalis enim philosophia de naturalibus est; naturalia autem sunt quorum principium est natura; natura autem est principium motus et quietis in eo in quo est; de his igitur quae habent in se principium motus, est scientia naturalis).[11] Following Aquinas, Jean Buridan, "the most distinguished and influential teacher of natural philosophy at the University of Paris in the fourteenth century,"[12] stated, "Every term pertaining to natural science, if it is perfectly defined as it pertains to natural science, ought to be defined by the term 'motion' or 'mutation,' 'to move' or 'to be moved,' or by another term including in its meaning the term 'motion' or 'mutation,' or equivalent terms" (Prima est quod omnis terminus pertinens ad scientiam naturalem, si perfecti definiatur prout pertinet ad scientiam naturalem, debet definiri per illum terminum 'motus' vel 'mutatio,' 'movere' vel 'moveri' vel per alium terminum implicantem in sua ratione illum terminum 'motus' aut 'mutatio' etc. aut per terminos aequivalentes).[13]

To study "meovingis" is to study the universe below, or within, the unmovable realm of "devyne substaunce," as Chaucer's translation of Boethius describes a fixed sphere.[14] As Boethius, who translated Aristotelian works, suggests, motion affects diverse things of different scales. The planets are subject to motion, as are the elements; heat, according to medieval lore, is the most moveable element of them all. Motion in this theory in large part defines terrestrial geographical features with the sea the most moveable, but the earth moves as well owing to the force of earthquakes and winds. When we get to the human world, it too is subject to motion, coinciding with, or otherwise subject to, the movements of fortune. Boethius also emphasizes a sense of natural versus unnatural motions across the scales, an idea with an important history and long-reaching effects. In the *Boece*, "Destyne is the disposicioun and ordenance clyvyng to moevable thinges."[15] Further down in scale, people move their bodies of course, in the sense that "the moevement of goynge" is "in men by kynde," as Boethius says elsewhere in Chaucer's translation. Indeed, Boethius privileges a close tie between, as it were, usual or habitual forms of human movement and a person who is "more myghty by right." A person who walks on feet, which are made for that "office" of walking, is "more myghti" than someone who is unable to use his or her feet and goes along with aid of the hands.[16] It may be said that in this theory, motion is the key to understanding one's place in the world, comprehending the world around a creature, and perceiving the more powerful—indeed, purportedly natural—ways of being and moving in space.

As is well known, the origin of all motion, according to Aquinas's reading of Aristotle, is the first unmoved mover. In his *Commentary on Aristotle's Physics*, Aquinas takes up Aristotle's observations about a "first mover" who "must be unmoved."[17] His argument is as follows: All motion requires an external force. There cannot be an infinite regress in time or in causal relationships as to where a motion or all motion begins. There must be an origin for all motion that is itself unmoving. Aquinas concludes that this first unmoved mover must have infinite power and be indivisible, and he identifies this force as God.[18] We see these ideas appear in literature in Dante, Chaucer, and many others.

Aquinas and later authors also posited that all objects in the cosmos have a destination toward which they move naturally. Aquinas might be taken as representative of this older understanding of motion that centered around the movement of an object not just to any place but to the ultimate location. The first part of his discussions about the properties of space in his commentaries on Aristotle's *Physics* and in the *De caelo* have quotations from Aristotle to the effect that "each kind of body should be carried to its own place," and bodies "which are in contact interact on each other."[19] Aquinas largely agrees with Aristotle on this topic, describing each being as having a *locus naturalis*; the earth, for example, has an *aptitudinem naturalem* to remain at the center of the universe.[20] He continues, using Aristotle's examples of a stone moving toward earth and fire appearing to be attracted upward toward the heavens, to posit that "it is clear that a natural body is moved toward its own natural place and rests there naturally on account of the kinship it has with its place and because it has no kinship to the place from which it departs" (Manifestum est enim quod corpus naturale movetur ad locum suum naturalem et quiescit in eo naturaliter, propter convenientiam quam habet cum ipso, et quia non convenit cum loco a quo recedit).[21] He also concurs with Aristotle in his description of the origins of animal movement as coming from within the body itself or external to the body, either of which can cause motion *a natura*, adding that "motion proceeds from the soul which is the nature and form of the animal" (quia est ab anima, quae est natura et forma animalis).[22]

The key element is, as Aquinas explains, "the fact that a body is naturally borne to its proper place" (Et primo quantum ad hoc, quod corpus naturaliter fertur ad proprium locum). A *proprium locum* is an unfamiliar concept to grasp for people brought up with modern physics, and I believe we can see

Aquinas struggling to understand the idea, but his explanation is as follows. The first thing to note is that it relies on Aristotle's definition of place as "the boundary of the container": "The reason why each body is naturally borne to its own place can be given: it is because the containing body (which is next to the contained and located body, and which is touched by it so that the boundaries of both are together not by compulsion) is akin to it in nature" (quia illud corpus continens, ad quod consequenter se habet corpus contentum et locatum, et quod ab eo tangitur terminis simul existentibus, et hoc non per violentiam, est proximum ei secundum naturam). Why are bodies "akin" to their places? Aquinas does not explain the idea directly but instead appeals to nature for an explanation: "The order of *situs* in the parts of the universe follows upon the order of nature" (Ordo enim situs in partibus universi attenditur secundum ordinem naturae). As if sensing that this explanation of an "order of *situs*" (position) is not entirely adequate, he offers a second explanation based on the "qualities" of each object: "[W]hen distinct things are in contact, [they] mutually interact on account of the contrariety of their active and passive qualities. Therefore it is the kinship of nature existing between the container and the thing contained that explains why a body is naturally moved to its own place" (dum tanguntur distincta existentia, propter contrarietatem qualitatum activarum et passivarum, sunt activa et passiva ad invicem. Sic igitur proximitas naturae, quae est inter corpus continens et contentum, est causa quare corpus naturaliter movetur ad suum locum). Aquinas then continues his explanation by reinforcing his propositions via an analogy with the trope of synecdoche. Given that "place is the boundary of the containing body," then "the contained body is related to the containing body after the manner of a part to a whole—a separated part, however" (corpus locatum se habet ad corpus continens sicut quaedam pars ad totum, divisa tamen).[23]

Aristotle and Aquinas after him furthermore note a distinction between "natural motion" and "violent motion," also familiar concepts to most readers. In Aristotle, motion "by force and contrary to nature" is when an object moves "contrary to the normal positions and manners of movement. The latter are more clearly perceivable as being moved by another." So stones falling downward and fire rising are moving toward their "natural places."[24] Otherwise, objects might be violently, and even unnaturally, separated from their natural place and can move in a direction contrary to that place. *Motus violentus* can occur in animals because of their overall composition or a part thereof, or because of another body. While Aristotle is not evaluating natural as better or more effective than violent motion, the emphasis on "natural" location and "normal" movement implies an evaluative perception. For example, Aristotle posits that objects are "moved to the opposite places by force, but to their proper places by nature," and "change contrary to nature is secondary to change according to nature."[25] As the editor and translator Edward Hussey notes, "natural movement" has "ontological and other priority . . . over unnatural movement."[26]

These theories of natural place were used to account for acceleration as well. Aristotle in *De caelo* proposes that "earth moves more quickly the nearer it is to the centre, and fire the nearer it is to the upper place."[27] While the reason this is so may be "unclear," as Marshall Clagett points out, in terms of heavy objects, Aristotle "appears to have believed that acceleration depends on the increasing proximity of the body to its natural place, i.e., [in the case of heavy objects] the center of the world. This increasing proximity produces additional weight, which in accordance to its dynamical rules . . . would bring about the observed quickening."[28] Aquinas explains the concept, first thinking of a falling body, that the "earth, the closer it approaches the middle, the more swiftly it is moved" and "something is more strongly impelled by the heavy in its fall as it nears the terminus of its motion. . . . The same holds for fire whose motion is swifter, the closer it approaches an upward place" (terra, quanto magis appropinquat ad medium, velocius fertur . . . a gravi cadente fortius impellitur aliquid iuxta terminum sui motus). He goes on to clarify that "to the extent that a heavy body descends more, to that extent is its heaviness the more strengthened on account of its proximity to its proper place" (quanto corpus grave magis descendit, tanto magis confortatur gravitas eius, propter propinquitatem ad proprium locum).[29]

In sum, Maier states that in the thirteenth century moving objects were thought to have two basic propensities in terms of a final place. She writes, "The first of these internal tendencies that manifest themselves as resistance is the striving of every material substance to reach its natural place and to remain there once it has attained this goal. . . . The second internal tendency is the striving of every body already situated in its natural place to remain at rest or, if set in motion, to return to a state of rest."[30] Aristotle and Aquinas after him are clear that an object cannot act on itself, because then the motive force, which is also a thing in the world, would be in a sense the same as the moving object, collapsing a distinction between two beings and making them one. Natural philosophy after Aquinas would come more and more to focus not on the question of natural place and its interactions with an object via resistance, but instead on the relation between the motive force that propels an object and the thing itself.

Motion: Part of a Moving Object and Apart from a Moving Object

The ideas that developed out of Aristotle—the first unmoved mover, natural place, and the distinction between natural and violent motion—are familiar, but to simply reiterate them would be to overlook the significant deviations from Thomist and other earlier theories of motion as well as to disregard fourteenth-century innovations in science. Natural philosophers in later years continued to question the very idea of what causes motion. Rather than focusing on the role of natural place as destination, fourteenth-century natural philosophy looked more horizontally at local motion. Instead of considering the first unmoved mover and a *locus naturalis* as destination, natural philosophers fundamentally altered the relationship between motive force and moving object. They did so because they noticed contradictions in Aristotle's theories, and they objected to the ways that motion had been theorized by authorities such as Aquinas. They asked: Was the origin of motion indeed separate from an object, and if it were, did that motive force always have to be in contact with the object? With the examples of stone and fire, what was the motive force since none seemed to be present while the object fell or rose? And what about a projectile that does not have a source of motion in contact with it after it has left, for example, a person's hand? These fundamental questions gave rise to one of the most innovative ideas in the late Middle Ages: impetus.

Aquinas's conclusions identifying the first mover with God were not particularly debated, but his first and most significant premise, which followed Aristotle—namely, that all motion requires a separate force—was. In book 8 of the *Physics*, Aristotle notes that while some things are obviously moved by an external thing or force (for example, an arm throwing a discus or the wind toppling a tree), a separate motive power is less clear in the case of things that appear to move themselves, such as when a creature is walking, talking, falling, growing, and so on.[31] Aquinas followed him and reasoned that this should not lead to the idea that things can move themselves. The separate force, for example, in the case of animal motion or change is either a distinct part of the animal or a force that is not easily discerned but is still separate. Some medieval philosophers, however, perceived a contradiction in Aristotle on the issue of whether motion was caused by a separate mover, despite Aquinas's pervasive influence. The reason was that Aristotle actually "vacillated, suggesting first that natural motion may result from an internal cause, the nature of the body, but arguing later that the nature of the body cannot be the whole story and that the participation of an external mover is required."[32] In book 7 of the *Physics*, for example, Aristotle writes, "Everything that is moved locally is moved either by itself or by something else. In things that are moved by themselves, the mover and the moved are obviously together; i.e., without intermediate."[33] The consequence of the Aristotelian ambiguity was that "[t]he debate over this issue continued through the later Middle Ages, with no clear victor."[34]

A clear departure from Aristotle on this topic occurred around 1300 when Duns Scotus questioned the idea that the agent that produced motion had to be external to a body "and argued instead that light and heavy inorganic bodies are self-moved by an internal principle."[35] Scotus reinterpreted Aristotle to claim "that self-change is a widespread feature of the physical world, where the agent and the patient involved in self-change are really the same."[36] It has been proposed that "though variations on this theme were formulated," Scotus's idea "became one of the basic positions in the later Middle Ages."[37] One of the most significant late medieval developments to arise out of this subject and out of Scotus's proposition was the idea of impetus. Aristotle had stated that the mover has to be in contact with the moved, but philosophers were interested in the problem of projectiles because they continue to move after they have been thrown or otherwise propelled, meaning after they appear to have broken contact with the mover. Aristotle's solution was to posit that a mover imparts motion to the surrounding air or water, which passes it on to "successive portions of air" with decreasing influence.[38] There were thus potentially three elements relevant to motion: the moving force, the moving object, and the medium through which an object moves.

Philosophers such as Robert Grosseteste, Richard Rufus of Cornwall, Francis de Marchia, and Jean Buridan found Aristotle's explanation inadequate, and the fundamental way of understanding motion began to change. As Maier outlines, "[T]he concept of motion caused by an external mover (*ab alio*) was gradually replaced by a concept of intrinsic motion (*ab intrinseco*). This shift in accent is not a mere nuance of interpretation, as may appear to be the case at first glance, but rather a fundamental transformation: the external mover, different from the object and in contact with it, is replaced as the principle of motion by a force that *inheres in* the moving object."[39] Grosseteste, who wrote an extensive commentary on Aristotle's *Physics*, proposed that a mover created a "disposition" in an object. Avicenna had influentially proposed that bodies could act on each other without contact. Albertus Magnus, inspired by Averroes, posited a force entering an object, describing more specific examples such as "the crossbow, the catapult, and the siege engine," but his ideas were rejected by his contemporary Aquinas. Richard Rufus of Cornwall, who lectured at Paris and Oxford in the early thirteenth century, excluded the idea of influence from the medium surrounding an object and instead went on to reason that the thrower made an "impression" on the object.[40]

It was left to William of Ockham's fellow Franciscan, Francis de Marchia, and after him, Buridan in the fourteenth century, to fully develop the idea of impetus as a force fully imprinted on a projectile. Francis de Marchia, following others, is credited with rejecting the idea that a force is transmitted to the air, water, or some other medium; instead, he posited that it was "a force left or impressed" (*virtus derelicta*) in the object itself.[41] Fabio Zanin emphasizes that Francis de Marchia's idea means that "in effect, *virtus derelicta* makes it possible to analyze a broad spectrum of non-violent motions that nevertheless are not natural motions, having no natural tendencies or aims," and "*virtus derelicta* is a form lacking any natural inclination or tendency; there is no teleological imperative connected to it."[42] Buridan may have been influenced by Francis de Marchia when he applied the word *impetus* to describe the force in an object. He adapted the word from general usage (from *petere*, meaning to seek or go toward something), which formerly described any impulse, vigor, and force. Buridan radically redirected its meaning away from any end point, indeed away from space itself and to embodiment within the object. For Buridan, "motion is not a mode of being of the moving object but a qualitylike factor that inheres in it."[43] In his *Quaestiones super octo libros Physicorum Aristotelis* of the 1350s, Buridan describes *impetus* as a "motive force . . . of the moving body" (*virtutem motivam illius mobilis*) that the motor gives to an object when it sets it in motion. Motion is thus a series of ratios: "[B]y the amount the motor moves the moving body more swiftly, by the same amount it will impress in it a stronger impetus. It is by that impetus that the stone is moved after the projector ceases to move. But that impetus is continually decreased (*remittitur*) by the resisting air and by the gravity of the stone. Thus the movement of the stone continually becomes slower" (quanto motor illud mobile velocius movet, tanto imprimit ei fortiorem impetum. Et ab illo impetu movetur lapis postquam proiicens cessat movere. Sed per aerem resistentem et per gravitatem lapidis inclinantem ab contrarium eius ad quod impetus est innatus movere, ille impetus continue remittitur; ideo continue fit ille motus tardior).[44]

In his *Quaestiones . . . de caelo et mundo* and his other lectures on Aristotle, which were also influential, Buridan argued further that since gravity is a constant, natural place cannot account for the acceleration of a falling object. It is not an external entity that can fully explain an increase in the velocity of a falling object. Having named the moving force an object acquires *impetus*, he states that impetus increases as a heavy object falls toward earth. First, he discusses the location of that force, putting aside the older explanations. The other force that moves an object "is not the place which attracts the heavy body as the magnet does the iron; nor is it some force (*virtus*) existing in the place and arising either from the heavens or from something else, because it would immediately follow that the same heavy body would begin to be moved more swiftly from a low place than from a high one, and we experience the contrary of this conclusion" (non est locus, qui attrahat grave sicut magnes ferrum; nec est aliqua virtus existens in loco sive a caelo sive ab alio, quia statim sequeretur quod idem grave inciperet velocius moveri a basso loco quam ab alto, cuius oppositum experimur). Impetus, he argues, increases as an object falls. Impetus is "acquired" (*acquiritur*) in the object as it descends.[45] The same rules apply in an inverse form in the case of objects such as projectiles that are propelled horizontally. The impetus initially communicated to the object gradually diminishes the farther away it moves from the motive force as well as due to the resistance of air or another medium, which "inclines it [the object] in a direction contrary to that in which the impetus was naturally predisposed to move it" (*inclinantem ad contrarium eius ad quod impetus est innatus movere*).[46]

Impetus continued to be an unsettled subject after Francis de Marchia and Buridan from the mid-fourteenth century on. Nicole Oresme, for example, came to a different conclusion about the idea of it as a force given to an object even though he appears to have been a student of Buridan. Oresme's writings on the motions of the heavens seem to have had "a sizeable and prestigious audience."[47] His work of interest here, variously titled in manuscripts *Tractatus de configurationibus qualitatum et motuum*, *Tractatus de uniformitate et difformitate intensionum* (confused with another work), or *Tractatus de figuratione potentiarum et mensura difformitatum*,[48] was likely composed at the College of Navarre in the 1350s or early 1360s, where he became grand master of the college, and versions of it were read at the University of Paris and in Italy, Vienna, Heidelberg, and Cologne by Albert of Saxony, Marsilius of Inghen, Pope Alexander V (the third rival pope), and later Galileo, among others. Charles V, the Wise, who became king in 1364, would reward Oresme for services to the crown as ambassador, councilor, chaplain, and translator. Jean Gerson would take up some of his ideas, and Henry of Hesse would adapt and apply his ideas to natural phenomena, like Oresme, not restricting himself to physics.[49] Oresme rejected the idea that impetus caused the motion of heavenly bodies, but he appears to rely on the idea in his description of acceleration and elsewhere.[50] The Merton School also rejected Buridan's theory. Indeed, it was not until 1600 that "the common scholastic position was that impressed impetus explains projectile motion."[51] The possible influence of these and other theories on Galileo has been detailed.[52] Whether rejected or not, however, the focus had changed to varied reexaminations of how a *virtus* interacts with a phenomenon rather than a natural place having a kind of pull on an object.

Speed's Measurement

Another set of problems to engage scholars concerned the measurement of motion, a seemingly technical subject that actually goes to the heart of understanding movement. There were three major innovations in measuring motion, each of which built on the new attention paid to embodied force in an object and all emphasizing the idea that any place in an object's movement is as significant as any other. In fact, they draw particular attention to each precise moment in the motion of an object. First, Thomas Bradwardine—Oxford philosopher, chaplain to King Edward III, and archbishop of Canterbury—identified an error in Aristotle's law of motion, this time to do with measuring it. Bradwardine's insight is considered important enough for Ernest Moody to describe it as a "radical shift from Aristotelian dynamics to modern dynamics, initiated in the early fourteenth century."[53] Second, another new and far-reaching development in motion's measurement, also from Bradwardine, is what he called the "quality" of motion at any point in time—soon known as "instantaneous velocity"—instead of motion over a longer period of time, during which the motion might have different rates of constant speed, acceleration, and deceleration.[54] Third, philosophers at the Merton School at Oxford built on Bradwardine's findings to calculate acceleration, formulating the mean speed theorem, and Nicole Oresme would take up the ideas of the Oxford Calculators with contributions of his own, including giving them graphic representation. Oresme's graphs are believed to be the first graphs of motion in history.

In modern terms, Aristotle's theory of motion in the *Physics* involves force, resistance, time, and distance. Aristotle had stated that a force applied to a moving object will make an object have velocity if the force is greater than the resistance to the object. Since an object moves a given distance in a certain time, Aristotle reasoned that the proportion of force to resistance is equal to the proportion of distance to time. So, for example, if a force doubles and the resistance remains the same, the distance will double in the same amount of time, or conversely the distance will remain the same but the time for an object to travel that distance will be half; the object will have twice the velocity.

Aristotle's theories appear logical, but they were not fully worked out, so Bradwardine explored them further in his *Tractatus de proportionibus velocitatum in motibus* of 1328. His innovations are still counted as of prime importance in the history of science; J. D. North describes them as nothing less than a "mathematization of Aristotle."[55] Joel Kaye states that "Bradwardine sought to create a mathematics of proportion proper to his perception of the world as he saw it, a world composed of qualities and quantities undergoing continual and continuous expansion and contraction."[56] The first point Bradwardine noticed was that Aristotle's measurement of motion as distance over time takes into account the whole time and distance of an object's movement and therefore is a calculation of average speed. The problem with Aristotle's formula was that it did not address "the whole question of how moment-to-moment velocities are related *within* the whole time of the movement."[57] If one object moves at a constant velocity over a period of time, and if a second object varies its velocity over the same time, both could have the same average speed. Bradwardine's explanation offers a solution to measuring what is called *velocitas instantanea*, the speed of an object at any moment, which he described as the "quality" or "intensity" of a motion (*qualitas motus, intensio motus, intensio velocitatis*).[58]

Bradwardine also found a contradiction in Aristotle's account of motion between his proposition that a force has to be greater than its resistance in order to move, and the "proportion" (Bradwardine's word; we would say *ratio*) of force to resistance equaling the proportion of distance to time. Bradwardine's solution in *Tractatus proportionum seu de proprortionibus velocitatum in motibus* was to reject Aristotle and propose instead (in modern terms) that the rate of velocity is the ratio of an exponential increase in force to resistance.[59] In Bradwardine's formulation, if a force applied to an object is doubled, and the resistance stays the same, then the object's velocity will double. If the force is raised to the power of three with the resistance remaining the same, the velocity will be three times faster. If the force is raised to the power of four with the same resistance, the velocity will be four times faster, and so on.[60] While Bradwardine's explanation is not correct in terms of modern understanding of rates of motion, he achieves what he sets out to do, which is to reconcile Aristotle's claims, and he is the first in history to introduce an exponential function into equations of physical motion.

His contemporaries and successors, now known as the Oxford Calculators or Merton School, which included Bradwardine, further developed his ideas about measuring motion. As Edith Sylla describes them, the *Anglici* or *Britannici*, as they were known on the continent, comprised a group of scholars at

Oxford in the fourteenth century, many of whom were fellows at Merton College, who developed insights into medieval physics, particularly kinematics. Their perceptions were widely influential, taught throughout Oxford, and immediately taken up at the University of Paris, and they continued to exert influence into the sixteenth century. This group included William Heytesbury ("the Calculator"), Richard Kilvington, Walter Burley, John Dumbleton, and Roger Swineshead.[61] Following Bradwardine, they measured motion not in terms of the total distance traveled as Aristotle had, which will not work if the speed within a given time varies, but in terms of extrapolating from a speed at one point in time.

Constant motion involves a uniform speed, and that problem appears to find resolution in Bradwardine's reformulations. What was more difficult to calculate was acceleration or deceleration in which the "quality" of the speed as Bradwardine had identified it, the velocity, intensified or weakened. The Calculators proposed the following solution: to measure the velocity of a uniformly accelerating or decelerating object over a particular time or distance, one takes its average or its midpoint over that time or distance.[62] This is the mean speed theorem, also called instantaneous velocity, because one is not calculating the whole length of movement but instead motion at one point in time. As Heytesbury explains, "such a nonuniform or instantaneous velocity is not measured by the distance traversed, but by the distance which *would* be traversed by such a point *if* it were moved uniformly over such or such a period of time at that degree of velocity with which it is moved at that assigned instant" (quod huiusmodi velocitas difformis seu instantanea, non attenditur penes lineam pertransitam, sed penes lineam quam describeret punctus talis, si per tantum tempus vel per tantum uniformiter moveretur illo gradu velocitatis quo movetur in illo instanti dato).[63] If one chooses the "instantaneous velocity" at the middle point of a uniformly accelerating or decelerating object, then one knows the average speed.

We are today accustomed to graphs of motion, but they are rare in medieval natural philosophy. The exception here is Oresme, whose manuscripts of *Tractatus de configurationibus qualitatum et motuum* habitually contain illustrations of his theories. The graphs are revealing not only because he was likely the earliest to use graphs to represent mechanical qualities, especially "intensities," but also because of his ability to see a relationship between being and spatial motion. Gilles Châtelet writes how Oresme was able to work out and display a special connection between being and motion, such that "ontological degrees and extensive sizes cooperate without becoming merged."[64] Take a physical quality first since it will make his presentation clear. Consider a moving object. Oresme uses a horizontal line at the base of a figure to represent the distance an object moves and/or the time it takes to travel a certain distance. This is the object's "extension." Perpendicular lines along this baseline can then be erected. These lines represent the object's velocities at points along the distance or time of the object's movement. If all the vertical lines are put together, then an object that moves uniformly at a constant speed over a certain distance or time will be represented by a figure that looks like a square or (more likely) a rectangle. Other rates of motion, including changing rates, require other graphs. A "uniformly difform" motion, or constant acceleration, can be represented by a right-angled triangle. More sophisticated graphs are necessary for more complex difform kinds of motions and for combinations of different rates of movement.

Figure 11 is a manuscript page of Oresme's treatise on the configurations of qualities of motions. The top of folio 72v explains "simple difform difformity" of any quality, with the diagrams in the first four lines illustrating varieties of them. Below that, the new chapter (1.16) indicates that there can be sixty-two (should be sixty-three) species of composite difform difformity in which different kinds of qualities are combined. Oresme provides some examples. The subsequent figures within the lines of writing below the center of the page show a variety of combinations of kinds of qualities. The larger left-hand upper marginal figure (keyed to the text with a red checkmark) also combines uniform qualities in a figure called a "graduated quality or difformity" (*qualitas seu difformitas graduata*). The second, lower left-hand marginal figure (keyed with three red dots) shows a combination of different kinds of uniform and uniformly difform qualities.[65]

The genius of Oresme's graphs is that they allow for calculations so that the area of the figure equals the total velocity in a given distance or time, what Oresme calls its *configuratio*. This can be as simple as multiplying the base of a rectangle by its side for constant motion or calculating the area of a right-angled triangle in the case of uniform acceleration. More complicated motions that vary over time require more complicatedly shaped graphs, such as we see lower down on the manuscript page and in the margins. Oresme's graphs make clear that one can still measure the total motion of an object even if its rates (or "intensities") of motion vary by dividing the graphs into squares, rectangles, and triangles, measuring each part, then combining them to get the total velocity. Oresme was interested in much more than the calculation of motion, delving into "intensities" of many kinds (as we will see in a later chapter), but it is fairly simple to understand that he can apply the same methodology to nonspatial qualities. For example, an object that is hot over a certain period of time can have heights on a graph representing degrees of hotness. If an object stays at the same temperature over a period of time, the area will

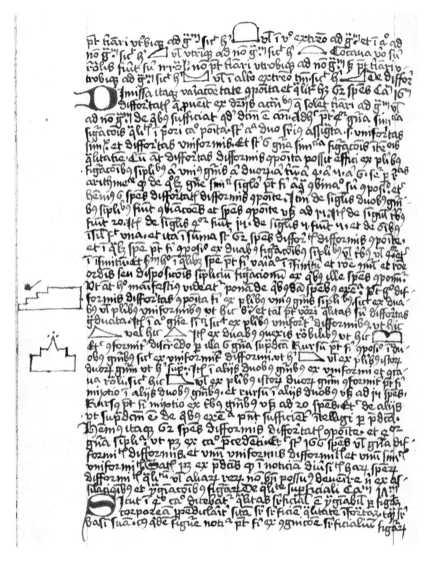

FIGURE 11. Graphs of motion in Nicole Oresme's *Tractatus de configurationibus qualitatum et motuum*. Bibliotheek der Rijksuniversiteit, Groningen, MS 103, fol. 72v. Reproduced courtesy of the Bibliotheek, Rijksuniversiteit, Groningen.

have a rectangular or square configuration; and if it cools in a uniform manner, its configuration can be represented as a triangle showing the decrease.

The graphs and the theories together suggest not just a different attitude toward motion but also a different approach to moving objects and to space. Motion is not directed toward a destination, nor is a propulsive force communicated to a medium, such as the air or water surrounding an object. The object instead embodies a motive force in itself. In fact, the medium now supplies only a fraction of what it did in the Thomist theory, namely resistance. Spatial distance and time are also altered in their fundamental natures, becoming continua rather than passages toward a natural location. There is an evenness and equality to places. The "naturalness" of some motions as opposed to others also disappears in this new way of seeing the world. Instead, all motions become equal, their complexities a key part of the normal wonder of the universe.

Chapter 6

Motion in Literature
Place and Movement in the *House of Fame*

The fourteenth-century philosophers who rethought motion—Jean Buridan, the Merton School, Nicole Oresme, Thomas Bradwardine, and others—proposed in essence new ways also of thinking about space and objects. Their science of motion redirected attention away from a *locus naturalis* with an attractive force for a moving object and toward a motive force embodied in an object with all locations in the transit of an item having equal weight. Also, a normal or habitual match between an object and a particular place of rest was no longer emphasized. Motion was now attached to an item, which could be impelled in any direction, each place in its movement as significant as any other. In working out these ideas, the philosophical writings exhibit important formal or stylistic traits. They may have been, in Edward Grant's phrase, "empiricism without observation" in the sense that they were thinking about the behavior of real objects in the world—fire, stones, and so on—except those items are traditional and found in their authorities, so they seemed to be reconsidering authoritative sources instead of actually experimenting with objects.[1] But at the same time they often disagreed with Aristotle, Aquinas, and each other because it appears as though they were at least thinking about observable phenomena acting in a way that was contrary to how they had previously been theorized. In either case, their writings deliberately and incrementally build on and diverge from their authorities because they are invested in internal consistency and extensive thoroughness in their texts. Propositions are made, and all varieties of results are explored. Every permutation of different species of motion is detailed—uniform, difform, uniformly difform, difformly difform, and so on—through straight, accelerating, circular, and other kinds of movement. It has also been noted that in terms of subject matter, the natural philosophies on motion show little to no interest in human behavior. They instead explore change on such a fundamental level that, while examples of human behavior might appear, they are not marked as special cases of movement. The overwhelming bulk of proofs and examples they offer are meant to apply to any thing, and so they discuss objects on a universal level.

These scholastic structures, styles, and generalized objects of attention in fourteenth-century scientific writings should not, however, imply that the new philosophies had little effect on culture outside the pages of the tracts. There is evidence of the rapid and widespread impact of their ideas. I have already provided some evidence for their dissemination, and I will lay out further evidence. It is not so much in the area of direct or practical application that we can see the effect of the theories. Weapons involving projectile propulsion were not changed, for instance. Taken as a whole, Middle English literature also does not echo that many specific, concrete examples from the scholastic treatises on motion or the practical geometries. There nevertheless is indirect evidence of the central innovations the scholastic philosophers made on the topics of natural place, the new nature of motion, and the measurement of all places and kinds of movement. Literature's engagement with the subject of motion also diverges from the philosophical discussions of the topic in that literature tends to single out the implications of motion for human beings. All moving objects in the world are considered, but literature more often addresses motion in relation to human behavior, or if not directly human behavior, then on other actions in the world that affect characters and how they apprehend the world around them.

This chapter focuses on structural and thematic elements in Geoffrey Chaucer's *House of Fame* in order to contend that it is a poem about a narrator who does not know where he is. The narrator, Geffrey, travels without stop in the dream vision, and his motion seems to promise answers to the questions the poem raises about inspiration, the acquisition of knowledge, poetic production, and more, except that the poem fails to answer them and, indeed, the origin of the questions is unclear. The cause of these obscurities is that the *House of Fame* appears to register the debates in fourteenth-century mechanics in a manner that is central to the whole poem. First, like the new science of motion, Chaucer's dream vision recognizes natural place as retaining some significance as a final cause, but the narrator also has impetus, which Buridan had redirected away from a destination to be embodied instead within an object. The presence of these two ideas is therefore similar to the state of science at the time, except that the poem obscures the origin and nature of that impressed force. Part of the narrator's problem stems from the fact that he is unmotivated in the sense that he does not have a clear reason for moving in the first place. Chaucer takes the idea of impetus but complicates or even removes its function as an efficient cause, the main reason for the motion. The narrator moves but without clear motivation. Second, with a present but reduced sense of *locus naturalis*, and with an embodied but occluded motivating force, the narrator's movement through space is potentially errant in a spatial sense. The etymology of *errant* traces back from Middle English to Anglo-French and Latin in two senses of *errer*: from a French and Latin deformation of *iterare* and *itinerare*, meaning to journey or travel, and Latin *errant* from *errare*, meaning to err, stray, and wander.[2] The narrator does both, and the *House of Fame* blurs a distinction between purposeful and undirected travel. It is not merely, as the new philosophies had suggested, that each location has equal weight. At least in their formulations, as we saw in Nicole Oresme's graphs, even if motion can run at greatly varying speeds and degrees of acceleration and deceleration, motion is still along a line. Something more radical occurs in the *House of Fame* because of its horizontal spatiality. Any location in any direction comes near to having equal significance. Motion becomes directionless.

The *House of Fame* takes up these aspects of unmotivated motion and directionless movement in an exploration principally of the narrator's mentality as well as his often fraught attempts to comprehend himself and the often strange spaces in which he finds himself. The poem's meditations on motion are reflections on the possibility of knowing about the self and the world. Jeff Malpas proposes that "one might say that to have beliefs (or any other mental state) is to have a capacity for embodied, oriented agency that is most directly expressed as a capacity for a certain sort of complex, organised movement; and to have knowledge or understanding of the world in general, then, is fundamentally a matter of being able to act and to move, in a certain complex and organised fashion, within that world." He suggests that "epistemology is itself properly an inquiry into the possibility of movement and agency."[3] The narrator's movements in the *House of Fame* are certainly, in Malpas's terms, "complex" in the sense that they are quite often unwilled, but they are not "organised," or at least any systematicity is suppressed to the point of a certain aimlessness. Geffrey appears to embody the new philosophy's ideas that tend to flatten a hierarchy of places, especially the sense of a special class of place, namely a destination. The problem then becomes where to go. What determines or at least guides a figure in motion in a world in which all objects may embody a force and move in any direction?

Let us begin with a preliminary example of another poem that addresses motion and space in ways that in some respects parallel Chaucer's: Robert Henryson's late fifteenth-century poem of 633 lines, *Orpheus and Eurydice*. Henryson's adaptation of the Orpheus and Eurydice story, based on Boethius's *Consolation of Philosophy* and Nicholas Trivet's moralizing commentary on it, seems to afford an uncommon overhead perspective on geographical area when its hero, a demigod, flies over the earth, ascends into the heavens, and travels down into the infernal regions in his search for his love. Henryson and his sources give Orpheus, the son of the two gods Calliope and Phoebus, supernatural powers that allow him to move freely across the earth, throughout the heavens, and among the infernal regions. It appears that he is a character who has the potential to perceive space from a bird's-eye view and that Henryson's audience could thereby also vicariously experience it in this way and gain a vantage point of understanding as well. The viewpoint seems to enable an abstraction of space that allows for mastery. Henryson is, however, characteristically more ambiguous than this might suggest, offering a complex version of the Orpheus myth. In terms of motion, Orpheus has a locus in the sense that his goal is seeking Eurydice, but that destination is in a sense obscured because he does not know where she is for the majority of the poem. His motion seems directed, but he is lost for most of the narrative. He does not know where he is or where he is going.

In some ways, *Orpheus and Eurydice* is similar to *The Book of Margery Kempe* and *The Book of Sir John Mandeville*, discussed in chapter 4. Indeed, analysis of the hero's interactions with the spaces in the poem reveals a chorographical understanding that seems initially to take up something similar to Margery's vision of larger spaces and Mandeville's apprehension of the globe in which all places are geographically equivalent and potentially all navigable. Yet both Margery and Sir John did not sustain their visions of the whole world. Margery turns back to homely and familiar scenes. Sir John's story of a man who circumnavigated the earth but was unable to recognize his own home goes on to explain an image of the earth suspended over or in nothing by noting that even though it is possible to travel all around the earth, not one person in a thousand would be able to find his or her way back. The world is full, too full, of "weyes."⁴ It is as though Henryson's Orpheus is Margery or Mandeville in pagan and demigod form, a character who has the potential to perceive and/or travel over the whole earth, but who in effect loses himself in its dissimilar, heterogeneous spaces.

Henryson does not exhibit the specific language of the fourteenth-century mechanical sciences, but he certainly renders Orpheus's movement in ways that appear to reflect recent developments. One authoritative print from Scotland's first published book of about 1508 titles the poem as the [T]*he traitie of Orpheus kyng and how he yeid to hewyn and to hel to seik his quene*.⁵ The topic it announces is twofold: Orpheus and his movements. *Yeid* means "traveled," existing in Middle English and Middle Scots only as a participle. Although a common word, it has some technical connotations, appearing in relation to celestial bodies, road travel, plows, ships, and waterways.⁶ My reading of the poem suggests that the hero's motion has a goal, and he certainly has cause for his movements, but neither this motivation nor his demigod status affords him effective motion through space. He travels across and within horizontal planes of space, with each place seeming to appear alike, as in being homogeneous, but this perspective does not allow him to attain a different understanding. The tone is the "grim" Henryson that Jill Mann and others have noted.⁷

In the beginning of *Orpheus and Eurydice*, when Eurydice is taken away to Proserpina's realm, Orpheus seeks her everywhere with a freedom of movement that his divine nature appears to enable, and we, the audience, witness that equality of each place within the spaces he searches. His search is organized into the three realms of the earth, heavens, and underground, and each is distinct, but each of those vast regions at least begins as equally available. The first is the realm of the earth, and the space Orpheus goes to there is the wilderness, a familiar place to readers of the poem in the sense that Eurydice's potentially dangerous encounter with a herdsman and poisoning by the snake have already occurred there by this point in the poem. The woods are of course also familiar to readers of Middle Scots romances and romance more generally: potentially magical, they are areas that reflect but also distort and challenge the hierarchies and conventions of courtly spaces and society. In Henryson's poem, however, the woods also hypothetically offer a route to where Proserpina rules, meaning that they are not only located away from the earthly court but also suggest a middle space between the living and the dead.

The woods also provide the setting for Orpheus's own transformation into someone insane with grief, a king who can abandon his court and live in—or more accurately, with—the woods, his songs there causing the birds to sing and the trees to dance. He complains:

> "Fairweill my place, fairweill plesance and play
> And wylcum woddis wyld and wilsum way.
> My wicket werd [fate] in wildirnes to ware [endure],
> My rob ryell and all my riche array
> Changit salbe in [shall be into] rude russet and gray."[8]

Orpheus welcomes the "wyld," and it is also worth pausing to note the significance of the "wilsum way" in this scene. The *Middle English Dictionary* tells us that *wilsum* is a word whose stem derives from the Old Norse adjective *wil* and whose meaning ranges from "wandering, straying, lost," to "perplexed, uncertain," and also "errant, wayward, and false." *Wilsum* is typically applied to diverse spaces, including internally undifferentiated areas as in "pathless" and "desolate" spaces but also locations that have moral connotations of being misleading or "devious." The shepherds are "wilsom" at the beginning of the York Second Shepherd's Play, and Gawain rides "mony wylsum way" when he leaves Camelot in *Sir Gawain and the Green Knight*.[9] In the first book of Gavin Douglas's *Eneados*, rocks, dangerous shores, forests, deserts, realms, and Aeneas's own path is "wilsum." Dido asks Aeneas to describe his travels: "Thi wandring be the way . . . / Wilsum and errant, throu euery land and see."[10] Ian Johnson attributes Henryson's usage of the word to Nicholas Trivet, where "wilsum" is a quality of wild and savage men.[11] Orpheus willingly, and willfully, chooses a formless, directionless way that exists there in the wilderness. "Wilsum" is repeated like a refrain as the hero wanders, and Henryson will moralize about that choice at the conclusion of the poem.[12] At this point in the narrative, Orpheus's loss of his former identity within civilized space, and his embrace of whatever fate may hold for him, are implicated with a potentially deceitful and aimlessly wandering movement.

The freedom of motion that the immortal Orpheus possesses in *Orpheus and Eurydice* is therefore not positive; the "wilsum way" implies that he does not know where to go. Not finding his love on earth, Orpheus flies up to the heavens, but this ascension fails to find her there, nor does it enable a masterful overview, such as a recognition of Platonic truth.[13] In Henryson's interpellation in the Ovidian story, the hero travels to the Milky Way, Saturn's realm, Jupiter's sphere (his grandfather who orders it to "be socht fro end to end" for Eurydice[14]), Mars's arena, his own father's—Phoebus's—place, but all for nothing. He has to leave and go to Venus's realm for the answer, but the response he receives is merely that he has to seek elsewhere farther down, so he passes through the realm of Mercury, past the moon, and back to earth. Along the way he hears the music of the spheres, which itself is not locatable in any one place. It is "armony throu all this mappamound," not bound to one sphere but occurring in the relations among different spaces.[15] The spheres are in a sense uniform, and together they compose one celestial harmony, a distinctly nonearthly mode of being. But Henryson does not allow Orpheus to linger there too long and, after the descriptions of the music his hero learns while there, he returns to earth to continue his search. His mortal wife could not be found in the heavens, so he must return to the ground and the "wilsum wone [area]."[16]

It is not so much the desolation of the earth that frustrates Orpheus, but rather that the earth is without clear markings to provide him direction or else with so many asymmetrical marks or routes that it is bewildering in the sense that he has become both wild and lost. Upon his return to the ground, he seems able to skim across the earth, seeking throughout "grey groves" and desolate places, yet the potential for mastery that a privileged view might offer is absent because on earth "ay he fand streitis and reddy wayis."[17] Whether the "streitis" are actual paths or just narrow passes of some sort, the significance is that there are too many ways that are all equal to him. The poem emphasizes that he always—"ay"—comes across these diverse places and passages; they are all too "reddy." Orpheus continues on his labyrinthine way on earth in the poem, "Fer and ful fer and ferrer than I can tell,"[18] until he eventually reaches the gates of hell, and, despite the aura of finality when he gets to the underworld, his route in hell is like those in the heavens and his two travels on earth. In the underworld, he goes from place to place, first crossing a "flude," then "Wepand allone a wilsum way," and down another "streit."[19] Orpheus in this version of the story eventually helps some of the tortured souls as he descends through hell to find Eurydice, and the rest of the story, with Proserpina's condition that Orpheus not look back at his beloved, is well known.

In Henryson's moralization, which immediately follows the tale and which draws on but also diverges from Trivet, we are explicitly told that Orpheus is a symbol for the intellectual part of a person's soul and for understanding "fre / And seperat fra sensualitie."[20] The problem, the moralization goes on to explain, is that the soul is attracted to Eurydice, who here stands for "effectioun," which has the potential to climb to the heavens of reason but also can desire "the flesche." The metaphor and characteristics of the earthly and other spaces that Orpheus explores therefore become clear. Eurydice can symbolize "nocht bot gud vertew," but when a person is stung by the snake of bodily fondness, "we fle outthrow [throughout] the medow grene / Fra vertew till [to] this warldis vane plesans, / Myngit [mingled] with cair and full of variance."[21] Orpheus's desire for Eurydice has misled him into the unclear, mixed ways. At the conclusion of the poem, attention turns directly to the "wilsum ways" once more when Henryson discusses the self, which "stammeris" (stumbles) after affection[22]:

> This ugly way, this myrk and dully streit
> Is nocht ellis bot blinding of the spreit
> With myrk cluddis and myst of ignorance,
> Affetterrit [Fettered] in this warldis vane plesance
> And bissines of temporalite.[23]

The moralization ends with the idea that eventually we may be granted freedom from earthly distractions in order to look up to, and ascend to, reason so long as we do not look back to fleshly lusts.

Henryson's *Orpheus and Eurydice* is a poem that in no small way depicts its main character via the spaces its hero encounters and the movements he enacts. Its *moralitas* associates heavenly spaces with reason, far away from the earthly, bodily temptations that can lead a soul astray. As critics have noticed about other properties of Henryson's moralization, the allegoresis only partially matches the poem's narrative.[24] This incomplete morality is also expressed spatially. The moral is that even when Orpheus ascends to the heavens in his search, which is the realm of divine harmony and of his ancestors (the latter evoking genealogy at the beginning of the poem), he does not find an answer to his quest because he is still seeking Eurydice, the flesh. In spatial terms, his spiritual or godlike qualities give him the potential to transcend the bewilderment of grief and earth's countless paths, but even the realm of heaven does not allow a masterful understanding of area. Nor does his status as a demigod take precedence over the aimlessness of his motion. Even in the heavens, the hero is unable to view all, his body leading him on what is potentially a path with an oriented direction, but one that in fact pulls him back into the bafflement of limited outlook and endless wandering.

The Culture of Motion in the Fourteenth Century

Geoffrey Chaucer's experimental poem, the *House of Fame*, is similar in some ways to the later *Orpheus and Eurydice*. In Chaucer's poem, the central character, Geffrey, who is also said to be on a quest concerning love, passes through some spaces that are similar to those in *Orpheus* in that both main characters go from earth up into the heavens. Like Orpheus, Geffrey also experiences spatial disorientation despite the potentially comprehensive perspective over a whole realm that transportation into the heavens might make available. However, unlike Orpheus, who finds nothing in the "wilsum ways," Chaucer's character is potentially led to greater understanding most when he is taken to smaller byways, but those constricted spaces nevertheless frustrate any understanding that the character seeks. Geffrey is also less certain in his exploration than Henryson's hero and less motivated too since he is often led places and astonished by them rather than actively seeking a definable goal. Even though Orpheus might despair, at least in Henryson he is driven in his search for his true love to look (twice) all over earth, in the heavens, and then, tragically, in hell. Geffrey in the *House of Fame* has a more complicated, reduplicated set of motivations and goals. He needs to comprehend the spaces he encounters and the journey he is on in order to understand where he is or at least be able to talk about where he might be. The stakes are high for the narrator because, as Jeff Malpas points out, "place is . . . that within and with respect to which subjectivity is itself established—place is not founded *on* subjectivity, but that *on which* the notion of subjectivity is founded. Thus one does not first have a subject that apprehends certain features of the world in terms of the idea of place; instead the structure of subjectivity is given in and through the structure of place."[25] Without knowing where he is, the narrator of the *House of Fame* risks not knowing who he is. The poem shows him as having impetus, but the motivation for his travel is not clearly defined, and it is not evident where he is heading.

Let us pause for a moment to assess the evidence of the influence of the fourteenth-century scientific innovations. In chapter 3 I discussed the evidence for the influence of mechanical ideas outside of universities. The impact of scientific concepts on literary writings, especially Chaucer's, has been well researched in other studies. In terms of the science of motion, while specific scholastic terms for new ideas about acceleration and other mechanical processes may not explicitly appear in vernacular literature, Chaucer and his contemporaries employ other words that signal interest in similar topics. The scholastic term *impetus*, for example, seems not to appear in poetic works with the same meaning, and synonyms like *impress* do not appear to suggest a sense of Buridan's new physics. Measurements or descriptions of instantaneous velocity or of uniformly accelerating motion, or a register of graphic presentations of motion, likewise do not show up explicitly in literary writings. But when the reach is extended beyond the specialized language of the scholastics, it is possible to perceive interest in the kinds of problems they were exploring. For example, a sense of the new idea of impetus shows up in Middle English texts, although these are admittedly difficult to identify clearly. When a ship is described in the *Gesta Romanorum* as "drivinge withe a grete ympet," "ympet" seems simply to mean "speed" with little, if any, connotation of a scientific sense of embodied force, yet the ship seems to travel under a force that it incorporates.[26] Borrowings from Latin and French of words meaning *impetuous*, which also do not show a scientific influence of this sort, further confuse matters in this case. Natural phenomena, especially winds and waters, are described in ways that likewise seem to imply a notion of force, but here too the specifics of impetus are not easily seen.

Despite the challenge of discerning the direct influence of new mechanical terms, other evidence indicates that perceptions of the physical world were essentially changed by new ideas such as impetus. Some evidence is clear. For example, Bartholomaeus Anglicus's popular treatise *De Proprietatibus Rerum* demonstrates an interest in motive forces, acceleration, and resistance, which are the common topics in science, and it seems to make distinctions that were new in mechanical writings. It is careful to distinguish between an object that moves "kyndeliche" and one that moves "by strengþe."[27] In the former case, Bartholomaeus posits that a falling stone gathers speed because the "mevinge wexiþ strengere," while the other object that moves "by violence" (his example is an arrow) clearly outlines the effect of resistance while also implying that an embodied force of movement reduces over a distance. He says that "whanne a þing meoviþ by violence, þe strenkþe is in þe biginnynge of þe meovinge, and þe furþer it is fro þe biginnynge, þe feb[l]ere is þe mevinge, as it fareþ in arowe þat comeþ out of bowe, þat may flee so ferre þat at þe laste he þurleþ [pierces] noþing nouþir greveþ."[28]

These examples might simply be traditional, reflecting an older Aristotelian influence since the example of the stone can be found therein, but acceleration, resistance, natural versus violent movement, impetus, and so on are precisely the same subjects that science took up. The fifteenth-century English *Vegetius*, a treatise on warfare, for example, depicts the effectiveness of projectiles as in part attributable to the force with which they are propelled and partly because of added weight in the object itself: "There is also anoþer maner of fiȝt þat Romayns used . . . þat is, wiþ dartes and speres ledede above þe heued . . . and þat schoot, what for strengþe of schete or þe castere, what for peys of þe leed, þere was none armure þat miȝte wiþstonde þe strook."[29] There is no medium mentioned here, no sense of Aristotle's description of movement as relying on the subtraction and addition of the air in front of and behind a moving projectile. Instead, the weight of the lead head of the dart affects its motion in addition to the force of the shooter of the arrow. Martial examples of this kind, incidentally, are common; the later Wycliffite Bible translates *impetus* simply as "asawt" (assault).[30] These military objects and actions can then appear in metaphorical situations.

Other historical evidence provides proof of the swift and relatively widespread promulgation of new physical ideas in late medieval culture, often with implications for literary writings. Bradwardine's innovations, some of the more significant examples of developing ideas about motion in the late Middle Ages, present a good case. Bradwardine was confessor to Edward III and was made archbishop of Canterbury in 1349 but died in the same year. His *Tractatus de proportionibus* from 1328 is from earlier in his career while he was a fellow at Merton College in Oxford. It survives in at least thirty manuscripts and "quickly dominated the whole mathematical discussion of motion in the 14th century."[31] The work of Bradwardine and the Oxford Calculators was typically summarized "in elementary outlines and manuals" while "[p]resumably, most arts students at Oxford in the last half of the [fourteenth] century were exposed to the Merton studies in one form or another." Among those who show the influence of Bradwardine are John Wyclif, Ralph Strode, and Chaucer.[32] The new mechanics became part of the required texts for an undergraduate degree at universities in Paris, Italy, Vienna, and Freiberg by 1400.[33] Bradwardine is mentioned for his theological work on simple versus conditional necessity in the Nun's Priest's Tale, and the chronicler Henry Knighton describes him as "celebrated beyond all the scholars of Christendom, in theology pre-eminently, but also in all other liberal learning" (*Hic erat famosus prae caeteris clericis totius Christianitatis, in theologia praecipue, similiter et in caeteris scientiis liberalibus*).[34]

Historical work on the sciences, while tracing many of these influences on individual figures, institutions, culture, and literature, also concentrates on the larger context for the new scientific developments, and it is the parallels and broader influences that one might seek in studying literature in relation to the new physical science on motion. Pierre Souffrin and Joel Kaye after him provide examples that show how historical changes affected the scholastic philosophers and, in turn, how the new physical scientific study influenced a wider culture. Kaye's larger project in *Economy and Nature* traces parallels between "the rapid monetization of European society" and developments in late thirteenth- and fourteenth-century natural sciences, while his *History of Balance* also explores common threads among social, political, economic, and technological developments in the same period.[35] Souffrin, Kaye, and Lynn Staley have perceived extensive exchanges between philosophical ideas and economic developments and vice versa, including the speculation that, for example, Oresme's graphic forms describing motion derive from developments outside the academic sphere in "everyday practices and instruments of measurement."[36]

Unmotivated Motion in the *House of Fame*

Middle English literature explores philosophical ideas, but it also supplements the science with its own insights and developments. Sometimes the topic of motion is presented in literature in ambiguous terms, where it is difficult to judge the effects of the physics. At other moments, however, a sense of irony enters into literary presentations of motion, indicating a kind of self-consciousness about the topic. The *House of Fame*, in my reading a poem about a dreamer-narrator questioning where he is, explores the reasons for Geffrey's own motion and other movements in the poem in terms of the origins of motion that are sometimes external to the narrator and sometimes embodied within him. In the science of Francis de Marchia and John Buridan, a force "impresses" (*imprimit*) on an object; impetus is the *motor* that is left behind in the object once the mover has transmitted that force to the item. The reasons why the narrator of the *House of Fame* does not know where he is are due in part to the conventions of the dream vision and Chaucer's innovations on that genre, but they are also attributable to the new science. Without a clear origin to the motive force, whose own nature is also ambiguous, the narrator has an embodied motor but is unable to recognize it. The poem fully takes on this opacity so that the topic of the poem is also lost, yet the subject keeps moving.

Previous criticism of the *House of Fame* has addressed the ambiguity of the narrator's persona, the aimlessness of his journeys, and the poem as a whole in relation to medieval science. Sheila Delany in a chapter titled "The Limits of Science" posits that the eagle's entrance at the end of book 1 "provides a motive for the dramatic action of the second and third books of the poem."[37] Delany traces the origins of the ideas about natural place, including the extended discussion of sound in book 2 of the poem, to Aristotle. She denies that Chaucer was familiar enough with the new natural philosophy of Buridan and others to be parodying it directly, but instead posits that Chaucer is offering a "burlesque of scientific method" through a "cumulative critique" of "an empirical view of the world."[38] In her reading, the science in the *House of Fame* is ultimately "limited," and the narrator of the poem instead makes a "leap of faith" into the heavens, transcending the mundane world of physics.[39] Neil Cartlidge emphasizes Robert Holcot's ideas from his popular Wisdom commentaries of about 1335 about sound being like circles on a body of water after a stone is thrown in. Holcot was cast as the antagonist in Thomas Bradwardine's *De causa Dei*, but Cartlidge's exploration turns to Holcot's "eclectic" techniques as offering a model for Chaucer's eagle in the poem.[40] Kathryn Lynch's reading of the poem is darker than Delany's and Cartlidge's. It takes up John Fyler's interpretation of the narrator's frustrated journey toward clarifying vision in the tradition of Saint Paul, Virgil, Scipio, Martianus Capella, Boethius, Dante, and others. She, however, broadens Fyler's reading of the poem so that it is not about the narrator's journey alone but about larger thematic issues, and she also reads it in relation to the scholastic discipline of logic rather than the natural sciences. She nevertheless does not give up the idea of travel and offers a sense of the poem as about "the underlying difficulties that open up for all people, and especially writers, when the path from this world to the next is no longer laid out with complete clarity."[41] Kellie Robertson adds another sophisticated reading of the poem that incorporates insights into its structure with thematic issues that return it to the context of late medieval physics. As with the other works she examines in her *Nature Speaks*, she reads the *House of Fame* as a "drama of inclination" in the sense that the eagle in the poem describes the "kyndely enclynyng" of objects toward their natural places. In her reading, the second book of the poem is a kind of "fulcrum" that moves the discussion from one about paganism and determinism in book 1 to Fame's arbitrariness in book 3. The poem brings together the conundrum of the presence of free will within a world seemingly governed by natural forces that "incline" an object in one direction.[42]

These readings have in common an interest in the motion of the narrator and the related development, or lack thereof, of the *House of Fame*. Robertson aptly describes the poem as a whole as a "series of narrative non sequiturs" and as having an "appositive form."[43] The structure of the poem is a key element, and my spatial reading of the *House of Fame* also notes the significance of the journey for the structure of the poem. Like Robertson's reading, it takes up the discussions of the natural sciences, but unlike prior interpretations, it sees key changes in the science of mechanics that alter the characteristics of the narrator's and the poem's movement so that the issue is not so much about a clash between the way and the will but about more fundamental problems with motion in the first place. Structurally, the narrator and the subjects of discussion in the poem move through different *topoi*, but that movement and the discovery of the new space do not advance toward a destination.

Any spatial reading of the *House of Fame* might strike readers as unusual because the poem begins with a conspicuous lack of physical location. In its opening, it also differs from *Orpheus and Eurydice*, which starts with a structured genealogy of Orpheus's ancestors in Greece before proceeding to the love and marriage of the hero and heroine of the story in Thrace. It is also unlike Chaucer's earlier work *The Book of the Duchess*, in which the narrator describes fairly early in the poem where he is: in bed. The *House of Fame* does not begin with a defined, concrete locale but instead discourses on the topic of interpreting dreams, reviewing theories of what they mean, and the narrator further claims in a voice familiar to readers of Chaucer that he possesses no special or authoritative understanding of dream lore. His invocation at line

69 of the god of sleep as his guide is also highly problematic, perhaps as troublesome as invoking the Furies for inspiration to guide *Troilus and Criseyde*. Not only is Morpheus traditionally an ambiguous and malevolent figure, but he is also central to the topic, Chaucer conjuring the god to help him interpret his sleep while awake.[44]

With Morpheus's introduction, we get the first attempt to ground the narrative in time if not also in space. We are told that the narrator had his dream on the tenth of December,[45] and we are also given spatial details but not about the narrator; instead, the reader learns about Morpheus's location in a stone cave on the river Lethe. It is not until over a hundred lines into the poem that the first location associated with the narrator's physical world is introduced, and while the location is specific, at the same time it is only an analogy. The narrator begins (again) his account of his dream by saying he fell asleep "wonder sone, / As he that wery was forgo [weary was who went] / On pilgrymage myles two / To the corseynt [Saint] Leonard, / To make lythe [easy] of that was hard."[46] The simile between the narrator and the weary traveler has been taken as comic or ironic in that a person may be worn out by such a short "pilgrimage" to the Benedictine house of Saint Leonard at Stratford-at-Bowe, which was only about two or three miles from London, where it is thought Chaucer was likely residing at the time he was writing the poem.[47] Whether ironic or not, the simile continues a certain vagueness about the narrator's own location, and, while it is not unusual for a dream vision to be imprecise about a narrator's precise place, the as-yet-unnamed narrator in the *House of Fame* is strongly associated with dislocation before he describes his dream.[48] The first place directly involving him that he is precise about occurs about line 120: the temple of glass.

But where is that temple and what exactly is it? The poem focuses pressure on the temple because it is the first locational content of his dream. Questions about the temple's disposition and composition therefore are likely to engage the narrator in a strong way in Malpas's sense—namely, as "the sort of creature that can engage *with* a world . . . that can think *about* that world, and that can find itself *in* the world."[49] The introduction to the temple, however, first offers a path that is potentially as endless as in Henryson's *Orpheus* because the sleeping narrator is overwhelmed in his dream by the images and statues in the temple, more than he has ever seen before. This overload of "olde work" of a High Gothic kind ("ymages . . . in sondry stages," "ryche tabernacles," "pynacles," "curiouse portreytures," "queynte [curious] maner of figures"), instead of providing information that can help place the dreamer, exacerbates his inability to know where he is, and he concludes, "For certeynly, I nyste never / Wher that I was" until he comes to know "wel" "Hyt was of Venus redely [truly]."⁵⁰ At this later point, even though he realizes it is Venus's temple (or rather, in a temple "of Venus," but seemingly not just one of her temples nor a temple merely to her), he does not know where the temple is or its significance.

He has moved, but why? We are told thereafter that in the temple he "romed up and doun," and soon he will encounter a small fence, but the reason for his motion is unclear.⁵¹ The verb *romen* is a distinctly neutral word, employed in Chaucer, *Piers Plowman*, and other Middle English poems to mean simply to walk about, sometimes with purpose and sometimes without any strong motivation.⁵² To a reader familiar with late medieval mechanics, the narrator is definitely moving, but the *fors* or *virtus* that impels him is not apparent. He has not been impelled to move somewhere by an external source, so he is moving as if with an embodied power. He discovers the place he is now in is Venus's temple, but that is not a destination to which he has naturally been led.

The poem presents an additional and particularly literary challenge in terms of the structure of the motions within it, because the narrator and the locations he encounters are presented metonymically. At this first place, the dreamer appears in a temple of glass, then it is by a kind of analogy again that he can discern that it is a temple of Venus because he sees a portrait of her. The dreamer does not linger on her image, but continues on in metonymic fashion to look at another image, a "table" of brass in which he reads Virgil's *Aeneid* and sees the story of the hero's exploits in that poem.⁵³ The narrator surveys history, which is not a spatial object, but at several points he seems to look over a large area or at least a large section of history in the brass engravings on the walls of Venus's temple. It is not always easy to distinguish between view and *historia*, and it seems that they in fact inform each other. For example, he first sees from a nonlocalized perspective the destruction of Troy as an image or series of images that are arranged more in terms of narrative than space: the wooden horse, the destruction of Ilium, and so on. A spatial element nevertheless soon appears more clearly, for when he views the story of Aeneas, for example, a kind of combined historico-geographical perception comes into being. He sees Aeneas and his company sailing to Italy "as streight as that they myghte goo" but diverted to Carthage, where Aeneas is to reunite with his "folk," and the story of Aeneas continues for some 225 lines.⁵⁴

But where is the narrator? Has Virgil's *Aeneid* somehow helped him discern where he is or understand the significance of the temple of glass and of Venus? He concludes his story of Aeneas by explicitly drawing a contrast between the marvelous and unusual nature of those sights and where he himself is despite— or because of—this unusual vision. When he finishes viewing the engraved images, he says he is unaware who might have made them, and he reiterates that he does not know where he is: "But not wot I whoo did hem wirche, / Ne where I am, ne in what contree."⁵⁵ It is as though his lack of knowing who made the brass table (not presumably who made the poem) is somehow connected to not knowing where he is, so the images have not enabled him to know any more about his situation. Either the images have been pointless or the poem is suggesting that without the kind of knowledge that would tell not only who made the images but also why they were made, his viewing of history and ekphrasis (in reverse) as well as being able to recount Virgil (in compressed form) lead to no greater knowledge. The question remains about where the narrator is and therefore also the meaning of where he is, of his dream, and of his role in the motion he is undertaking. He has impetus, but the origins of the propelling force that have somehow been transmitted to him are obscure, and each new location and event neither furthers his knowledge nor has much significance in itself.

Before continuing to examine the results of this lack of knowing, it is worth pausing to meditate briefly on the fact that the narrator is spatially dislocated in such a way that even though his understanding is not furthered, the lack of specific place affords him (albeit in some second- or thirdhand way via the images from Virgil) an overview or series of overviews, whether of history or spatial area or both. Such moments are significant because they speak further to the structure of the *House of Fame*. At this point in the poem, Geffrey has been able to look over the Trojan, Carthaginian, and Italian history and geography. Similar viewpoints appear occasionally at various moments in the poem in which the narrator is able, and sometimes forced, to get an overview that allows him to see all the objects before him as (literally in the case of the brass plates) flat, rendering a space that seems in some respects homogeneous. In book 2, for example, the eagle explains to him why Jupiter has sent the eagle, namely, that "no tydynge" of love "cometh to thee," whether those stories come from "fer contree" or "thy verray neyghebores / That duellen almost at thy dores."⁵⁶ Near and far distance is irrelevant; scale is flexible in the aural-literary-mythical source evoked here, but the strongest sense of an overview across a vast area occurs later in the same book. From his talons, the eagle asks the dreamer to look "yond adoun" to see if he recognizes any place at all:

> And y adoun gan loken thoo,
> And beheld feldes and playnes,
> And now hilles, and now mountaynes,
> Now valeyes, now forestes,
> And now unnethes [barely] grete bestes,

> Now ryveres, now citees,
> Now tounes, and now grete trees,
> Now shippes seyllynge in the see.[57]

The list of places is presented as though purposely random, the lexical equivalences among objects and the repetition of "now" emphasizing the availability of all things to Geffrey: a mix of valleys, forests, beasts, rivers, cities, towns, large trees, and sailing ships. Following the passage, the narrator continues on and soon achieves a Troilus-like view, indeed, an intensification of that into a world that "[n]o more semed than a prikke" so that the dreamer-narrator cannot recognize any place because he is so high.[58] This presentation is similar to a moment later in the poem, when Aeolus trumpets people's reputations "thrughout every regioun" to "the worldes ende";[59] it is as though everywhere and everything is equal.

Chaucer's *House of Fame* therefore displays an abstract sense of space, but such an overview does not provide any more understanding of where one is or indeed why one is where one happens to be. The poem in fact also presents the other end of the spatial scale—namely, familiar, horizonally perceived locations, each of which is presented in a plain voice such as we find in *The Book of Margery Kempe*, *The Book of Sir John Mandeville*, and other narratives. As Ruth Evans explains, and as Kellie Robertson observed of the poem's "apositive" structure, Chaucer's poem is made up of parts,[60] a reconstruction of memorized parts just as much as Margery Kempe's *Book*, though those pieces tend to have origins in different places: historical lore in Chaucer's case and vision in Margery's. Consider even the repeated, simple way the dreamer said he witnesses the scenes in the *Aeneid*: "sawgh I," "And next that sawgh I," and "I saugh next," then "Ther sawgh I" repeated five times in a row and more.[61] It is difficult to characterize the structural relationship between the abstracted and horizontal-heterogeneous modes of apprehension in the poem. The *House of Fame* seems to oscillate between two extremes, one more unusual in which a large area is flattened and presented without divulging anything more about where oneself might be or the significance of one's viewpoint, and the other more commonplace apprehension of space in which places are familiar and presented in a way that is unremarkable. Despite this going back and forth, and the narrator's somewhat comical and largely passive series of unmotivated movements, the poem still holds out the promise of understanding where he is in the sense of someone who will be able to engage with the world and even find himself in it.

Another spatio-hermeneutic problem, however, arises for the narrator and the audience because the horizontal view of everyday spaces is also complicated.

To return to the moment at the end of book 2 where the narrator has been in the temple of glass and Venus, and he has looked over the history of Aeneas, he admits for a second time that he does not know where he is, nor does he know who made the images, "not wot I whoo did hem wirche, / Ne where I am, ne in what contree."[62] Yet he arbitrarily decides to continue moving. He will go out, he says, and before him is a small gate, a "wiket." Where did that come from? A partial answer may lie in a source for the poem, the *Roman de la rose*, where the narrator also notices a gate. In that poem, the dreamer, standing outside the garden and hearing the birds singing, wishes to enter its delights. Alone, he circles the garden, not finding any way to go over or through the wall (which, like the brass plates in the *House of Fame*, presents images) until he finds a small "wiket" by which to enter.[63] The *House of Fame*, however, inverts this trope, which had by Chaucer's time become a tradition of garden entranceways leading to further knowledge, so that instead of entering a garden, he leaves where he is for something quite different.

The poem at this point refuses to meet readers' expectations, because when the narrator leaves the temple of glass by means of the little gate, he encounters a very unusual space—a desert—one of the strangest settings in all of Chaucer's poetry. It is as though he has not entered a garden but left one, or else the garden has become decimated, dessicated. He finds himself in front of a large "feld" and can see to the horizon:

> When I out at the dores cam,
> I faste aboute me beheld.
> Then sawgh I but a large feld,
> As fer as that I myghte see,
> Withouten toun, or hous, or tree,
> Or bush, or gras, or eryd [cultivated] lond;
> For al the feld nas but of sond
> As smal as man may se yet lye
> In the desert of Lybye.[64]

The scene the poem presents has sources in Lucan, Virgil, Dante, and French poetry. Its meaning and significance have been read as a kind of existential barrenness or, as Steven Kruger astutely says, at least "stripping away of appearances."[65] It is certainly a contrast to the crowd of images and information that has preceded it; Delany notes, "a new spaciousness, striking after the enclosure of the Temple and its limited visual field; now the Narrator is able to look 'As fer as that I myghte see.'"[66] The gate is so simple and the desert so plain and unhistoriated, yet the stark contrast and the openness and expansiveness of the new view of the desert still do not help the narrator understand where he is. Also, his motivation is again lacking; he has gone out through the wicket for no reason except that he was struck with the idea of seeing if there is some "stiryng man / That may me telle where I am."[67] Why specifically a "stiryng" man? Is it because the former scenes have been without anyone alive, because the previous descriptions have not included any movement, or is there an affective quality to his efforts, the narrator seeking someone who can move (stir) him? Whatever the possibility, the answer is that no one is there. The parallel in the passage between the clauses "As fer as that I myghte see" and the sand "As smal as man may se" draws the dreamer and the reader together in their ability to perceive and to understand the view in such a way that the analogy between the grains before the dreamer and the sands of Libya becomes important. However, deserts in Middle English literature are often "wilsom" like the constricted and fruitless paths in Henryson's poem. The alliterative poem the *Wars of Alexander* tells of its hero who drives his heathen enemies "in-to desert landis," "A wilsom wast & a wild," and in the N-Town Play of the Baptism, Jesus looks ahead to his time in the wilderness: "In whylsum place of desertnes. . . . For man thus do I swynke [labor]."[68]

The scene presents a kind of microcosm of the poem's structure so far and provides a fitting conclusion to book 1. Notice its contrast and contradictions. The narrator, we are told, upon emerging out of the gate, looks "faste aboute" himself as though he looks nearby, but "faste" also suggests a certain eagerness and alacrity, and connotes a sense of fearful guardedness. Only then does he perceive a large area, yet even this he describes as "but a large feld," certainly implying the plainness of it as "nothing but a large field" but also suggesting a sense of disappointment with it being nothing more, especially given the implicit contrast with the French tradition of enclosed gardens. It appears he sees far even though he looks nearby, yet either way he does not see any recognizable topographical feature, and his view comes to focus on the fineness, I am tempted to say *finitude*, of the grains of sand, "as smal as man may se." And no one is there to advise him, so the real area he perceives, which contrast with the images of the brittle temple of glass, is also potentially illusory or fruitless. He concludes by asking God to protect him from "fantome and illusion."[69]

The desert may remind us of Margery Kempe up Mount Quarantania, where she does not contemplate the view or find anything particularly revelatory about having an outlook over a larger area, in that the narrator in Chaucer's poem also does not seem to gain anything from coming upon his large and simpler view. Or consider where Geffrey gains another overview later in the *House of Fame* in book 3, in which he describes Fame's house as the highest of all. He says the house "stood upon so hygh a roche" that "Hier stant ther non in Spayne." He begins to climb this vertiginous peak, but his "intent" as it is described is not to gather a view or even to find out where Fame is located, but instead to "powren wonder lowe, / Yf I koude any weyes knowe / What maner stoon this roche was."[70] Like Margery, he turns away from the view, but unlike her, he does not turn to a spiritual contemplation but instead to a stonily material one. His emergence through the wicket to the desert is an earlier moment of this sort in the *House of Fame* in that Geffrey is led, without a motive force, to concentrate on the smallest grains of sand, which do not explain anything. Perhaps there is a sense of the sand not as singular and tiny but as enormous and vast, the desert the most homogeneous space, but even that does not help the narrator.

A pattern of a kind therefore establishes itself in the *House of Fame*, one in which, despite moments that might afford an entry to a new area or an overview of a space that has the potential to help the narrator locate where he is, what the place is, and why he is moved there, the poem turns away from and frustrates those expectations. Such a scheme or at least a concentration on spatiality and motion may help explain the compelling discussions of sound that cluster within book 2. In book 2, sound appears to have the potential to be pervasive and even natural in contrast to the manner in which the poem has so far failed to concentrate on the natural or Nature or, if it has, has rendered nature in strange and defamiliarizing terms. It would be reassuring to Chaucer's readers to encounter the older idea that everything, sound included, has a natural place. The objects raised in the *House of Fame*'s lines on sound—fire and a stone—come up again and again in traditional philosophical analyses of motion, and so do the forces in the poem. The eagle starts with a natural metaphor when he explains that each sound "Moveth . . . into his kyndely place," but he goes on to render that "place" in much less philosophical and less explanatory terms. The eagle tells how Fame's place stands "in myddes of the weye / Betwixen hevene and erthe and see" with a "way thereto . . . so overt" and also that her house "stant eke in so juste a place / That every soun mot to hyt pace."[71] Chaucer's repetition of "weye" and "way" in four lines is a metaphor that is not in his Ovidian source, and while a "way" can be just an abstract course in Middle English, it more commonly has a literal sense of a road or physical route in Chaucer. Here it provides a concrete sense of the House of Fame's location that the also very Chaucerian metaphor of walking suggests with the word "pace." Sound does not seem to move in an aerial medium to its natural place but instead travels quite materially along a road as if embodied in the human world of walking.[72] Objects will move to their places unless they encounter resistance: the eagle explains that a light thing moves only while it is "at his large," meaning unrestrained.[73]

Alexander Gabrovsky's *Chaucer the Alchemist* accurately notes the significance of the medieval physics of motion in the *House of Fame* more than scientific theories of sound or optics. He argues that "Chaucer's dream vision complicates the physics of movement by incorporating fourteenth-century ideas of relative motion, especially as it relates to uncertain knowledge. In the *House of Fame*, poetic imagination accomplishes in one artful dream what the terms of logic and science never communicate in full." Yet despite an extensive and valuable discussion of the poem in relation to motion, Gabrovsky ultimately turns away from physics to argue that Chaucer's poem depends on "human psychology and imagination" more than the physics in itself.[74] In my reading, the poem does not come to focus on the fallibility of perception but instead maintains attention on physical actions and motions that are at least in part independent of the actors involved. The poem is about more than the narrator and extends to the nature of places and motion as well.

In the eagle's ongoing explanation, objects move through a medium in the poem, echoing the Aristotelian and some medieval examples of natural media: air and water. Were someone in Chaucer's day who was familiar with contemporary science to read the poem, however, he or she would be made starkly aware that the Aristotelian idea that a motive force was communicated to the medium surrounding an object was old-fashioned and incorrect. Francis de Marchia, Buridan, Richard Rufus of Cornwall, and almost every other fourteenth-century scientist had turned to the idea of impetus instead and had given up on the idea of a force being communicated to a medium. After explaining that sound moves naturally to Fame's palace in the middle of the heavens, earth, and sea, the eagle turns to the discussion of sound as "nought but eyr ybroken," in part a pun on farting.[75] To make a claim about the physics of motion, however, is to confuse the moving object (the sound) with the medium (the air), and the eagle reiterates this idea by saying that the blowing of a horn "twists" the air "with violence" and so "rends" the air, and a harp struck also "breaks" the air.[76] Natural scientists would say that the separate mover would be the blower or harp striker, the moving object would be the horn's or harp's sound, and the air the medium through which the sound moves and which provides resistance. The musician communicates the force to the instrument, which produces a sound that then embodies the force, but in the poem, the last two are conflated: sound is a broken form of air.

The eagle continues his explanation with the analogy of a stone thrown in water, a parallel that aligns more with contemporary ideas about natural science and physics, up to a point. The analogy is that the stone thrower is the mover; the sound of a horn, harp, and speech is the stone or moving object that embodies the force of the throw; and the waves across and into the pond are the medium. Every word "Moveth first an ayr aboute, / And of thys movynge, out of doute, / Another ayr anoon ys meved."[77] Cartlidge attributes the origin of this idea to Holcot and a discussion of sound rather than motion in itself, but Aristotle and others preceding Francis de Marchia and Buridan describe motion in just this way, as an object causing a wave-like effect in the air or another medium.[78] Of course, the analogy only goes so far; the eagle is conflating a discussion about a stone, which belongs, on the one hand, within a discussion of heavy objects and air (and gravity), with, on the other hand, a discussion of light objects moving more horizontally (and with analysis of projectiles). The stone is not carried by the medium of the water to the edge of the body of water. But the analogy is understandable. Where the explanation however again diverges from discussions in contemporary physics is when the eagle describes each ring or wave of air "stirring" each other "[m]ore and more," whereas contemporary physics suggests that the medium provides resistance, and the object or its waves would slow down.[79] Sound, in this theory, would decrease as the distance from its producer increases. The difference here that may, in part or in whole, explain the eagle's description is that movement to a natural place is a motive force in the moving object that may increase as an object moves to its place, but this applies to objects affected by gravity. In horizontal movement, an increase in velocity or acceleration would require the addition of more force.

The explanation of sound is intriguing; it is as though a confluence of older ideas about motion from late medieval mechanics were pushed together in an anthologized or encyclopedic fashion while lacking some of the precise thinking we find in the scholastic philosophy. The eagle would have been seen as conflating the vertical and "kyndeliche" motion of stones with the violent impetus of a horizontally projected object such as an arrow and, what is more, transferring the whole to a discussion of sound. One part of physics on motion is transplanted in another area. It requires a stretch of the imagination, or recollection, on the part of Chaucer's audience to bring these two distinct discussions together, however confusedly, a conflation that only becomes more challenging when one tries to figure out how the widening pattern of waves of sound will end up in the middle of the heavens, earth, and sea as we are soon told. The eagle goes on, repeatedly emphasizing the natural movement of the sounds upward and to their natural place, and, at the conclusion of his explanation, the eagle says they travel to that middle location because it is "most conservatyf" of sound.[80] Centrifugal and centripetal forces are compounded.

The point of this analysis is not to discern whether Chaucer's eagle explicitly uses or does not use certain mechanical theories, nor is it meant to imply that Chaucer's audience would have uniformly perceived the eagle as philosophically old-fashioned, confused, or ignorant. The bird, after all, admits, with a certain irony perhaps, that he has attempted to present the ideas "symply, / Withoute any subtilite / Of speche, or gret prolixitie / Of termes of philosophie," and he offers that what he has said will be proved "by experience."[81] It would take a sophisticated audience to have felt, or to have perceived, the confusion in the discussion and therefore to have seen the comic absurdity in the situation of Chaucer being led, and misled, by the eagle. Sound, the passages seem to suggest, is not going to reach anywhere or anyone naturally in this manner despite the eagle's assertions. They may in fact be Chaucer indicating that scientific discourse might be appealed to in order to make a point but that it can be as mistaken as any other form of authority. Either way, book 2 goes on to end with a discussion of sound that is not about medieval physics or sonics but is reminiscent of the Franklin's Tale. The eagle asks Geffrey to listen to sound as it approaches the House of Fame, and we are told that it is inauspiciously like the sound of waves breaking on rocks that threaten ships or the tail end of thunder, and the dreamer is afraid. The book concludes with the idea that, instead of the abstraction of sound and its passage to Fame's house, each sound eventually has to be made even more specific and embodied again, reappearing as the person who spoke it.[82] The physics have come full circle back to the simplicity and individuality of those who produce each sound. The narrator is again left at the end of the book still looking to learn something of his physical world and where he is.

    The familiar "wicket" and the flattened landscape, the ability to see over vast flattened areas and the mundane local street, the natural and expansive proclivities of sound and their reindividuation: these eversions feel similar to the plastic structures of Fame's house, of Fame herself, and of Ovid, all of which/whom dilate in size. In book 3 Geffrey views the house through the distortions of glass and ice, and he travels from the outside with its musicians upon the walls "That shoone ful lyghter than a glas / And made wel more than hit was / To semen every thing" to the inside of the temple through a gate, and there Fame, like Philosophy in *Boece*, seems only a cubit tall but then stretches herself like some godly rubber band so that her feet reach earth while her head touches the heavens. When the narrator looks upon Ovid on his pillar, Ovid also seems to extend extremely high because the interior hall also appears to dilate "Wel more be a thousand del [times] / Than hyt was erst [before]."[83]

    These torsions of physical space, of figures, and of the poem's subjects find a parallel in the remainder of book 3 in which the narrator encounters lies and truth compounded. The Socratic feature of the poem, from Dante, also continues in book 3 with the narrator next meeting an anonymous man who serves as a guide and interlocutor, and who seems to intuit what he desires. It would appear at this point that the narrator is learning where he is and why he is traveling. He says he has found a "cause." The man asks him "what doost thou here than?" And the narrator replies that he will tell the man "The cause why y stonde here."[84] Given that the *House of Fame* opens with questions of cause (the words "causeth," "causes," and "cause" occur seven times in the proem), this scene promises a sense that the narrator has progressed, which he has done, except when he answers the man's question, his response points to a lack of motivation except for a kind of stock of dream poetry, one that we have already learned in the poem, namely that he wants to learn of "love or suche thynges glade." Otherwise, he replies that he does not seem to care what he hears, "Somme newe thinges, y not what," he says, "Tydynges, other this or that" except he adds that these "tidings" in Fame's house are not what he came to hear. The guide responds he can perceive what the narrator "desirest for to here" and leads him to the house of Daedalus and rumor. The man's promise to the dreamer that he will take him to a place where he will hear "many" more tidings is not auspicious.[85] The problem again is not one of paucity but of plenitude, and the guide's offer will take the narrator to a place that is all surfeit.

The man bids Geffrey to go with him to another place, in a "valeye," nearby.[86] We return to the physical metonymics of earlier in the poem, the anonymous guide offering an answer to where the narrator is and what he seeks, and offering a solution in the form of moving to another locale. Place and the promise of destination are intertwined, but the motion is merely to another site. These complex imbrications of space and cause or result have certain parallels in Dante's *Commedia*. In a discussion of the measurements in the *Inferno*, John Kleiner analyzes the fact that the mathematical descriptions of Dante's hell become more and more prevalent and precise as the poem progresses. The dark woods at its beginning are in stark contrast to the end, in which "Hell's terrain is measured out numerically and the travelers' position plotted precisely." That precision is, however, ironically rendered in the *Inferno* because the measurements that excited early modern and later scholars about Dante, including those who created graphic maps of hell according to the measurements, become contradictory and confused. Kleiner suggests that "when Dante's measurements of the final circles are analyzed objectively, they reveal a terrain *dis*ordered by number and measure."[87] He concludes that these phenomena are Dante's ultimately self-conscious parody of his own enterprise in creating the *Inferno* in that "it is not God's Hell that is deformed, but Dante's counterfeit."[88]

Chaucer is not so self-critical. The narrator's search is fruitless but not ironic. He needs to know where he is. The narrator and the action of the *House of Fame* move, but neither he nor the narrative has a natural place to which to travel, and, moreover, reasons for moving are obscure. It is as though the spaces and the narrator's movements are necessary for finding something out, but the locations he encounters repeatedly forestall any development in his understanding because they offer, in the end, just other locations. The end of the poem contains a kind of speeding up of these kinds of oscillations, what Rebecca Davis calls "a dialectical course": "the movement between enclosure and open space, fixed form and flowing matter, tradition and improvisation.... Every turn away from authority generates a corresponding impulse to return to it."[89] Except, given the metonymics, the poem does not so much oscillate as move laterally. When the narrator hears the clamor from the House of Rumor, physical distance is again contorted and appears to be stretched over a vast area. He stands next to the house and regards the spinning structure, from which comes "so gret a noyse / That, had hyt stonden upon Oyse [in Northern France], / Men myghte hyt han herd esely / to Rome."[90] It is shaped like a number of commonplace baskets, or a "cage," yet it is full of holes large enough for sounds to enter and exit. It is also full of doors, which stand open during the day and are closed at night, but either way, we are told that no porter stops any sounds. It seems round as it spins, but it has seemingly many "angles." It is large, sixty miles "of lengthe," but it appears flimsy.[91]

The dreamer reencounters the eagle and pleads with him that he might remain where he is to see what "wondres" are there in the spinning wicker house and also to learn something "good" or at least something that he wishes to hear.[92] The desire to learn some "good" reminds readers of the beginning of the poem ("God turne us every drem to good!"), but the accompanying desire simply to see "wondres" or to find something he wishes to hear undercuts an ethical impulse and redirects the dreamer's request to stray toward a less clear goal. Yet the appetite to remain in place and to learn is significant because it indicates a yearning to comprehend where he is, but the eagle does not allow him to remain there. Indeed, the eagle suggests that he is going to take him into the House of Rumor for no better reason than because it is the place he will hear the most "tidynges." The proliferation of sounds and information from earlier in the poem finds its match here at the end of book 3 as Geffrey continues to seek out knowledge, including "a tydynge for to here, / That I had herd of som contre / That shal not now be told for me."[93] A "contre" is evoked again, and this awkward observation that he wishes to hear something either from or of the other country but then immediately says he will "not" hear it is only partly explained when he goes on to say that it is not necessary for him to hear the tiding now because it will all come out at some time, and anyway, others can "synge" it better than he can.[94] Whatever it is, the cause is weak, confused, and multiplied to the extent of meaninglessness. It is not really possible to tell whether the tiding "of som contre" is significant, but it offers a conclusion to this part of the dreamer's search into something "good" and refocuses the poem back on a question associated with a place. Not finding an answer about what he seeks, or even just a report from where he has been led, it would seem that news from another, farther away country would be welcome. This exotic promise is interrupted with a "gret noyse" and people running to a corner where stories of love are told.[95] This concluding echo of one of the narrator's initial reasons for his dream search into love promises just as much as the rest of the poem. As the others in the House of Rumor climb over each other to hear the latest tiding, the sense at the end of the poem is that more will happen, sound, or be heard, but nothing will be learned.

The *House of Fame* urges answers about sleep, love, history, fame, sound, and rumor. The poem's journey to find out about these topics is bound up with the fundamental questions of where the narrator is and why he is moving from one place to another. If he can understand where he is in the sense of what a place is and what it means to him, and if he could discern his motivation, then the poem promises an answer as to why he is dreaming what he dreams. The problem is that motivation is absent in the poem. The narrator dreams what he does, but why? The narrative moves from local, intimate places to spatial overviews, but in each case, the overview seems to be unrevealing and local places expand and proliferate without offering comfort or leading to a greater revelation. A question is frequently posed in the poem, but it is never quite clear what that question is, and, instead of an answer, the dreamer is simply led to another place, which also does not answer anything about the previous location. The poem ends with a kind of crescendo in which the extremes of scale, as it were, speed up into a series of unresolved spatial contradictions. Understanding has shifted in location but has not progressed. The narrator has traveled and sought, and he has found more and more without sufficient reason for it.

Rebecca Davis's reading of the poem astutely suggests that motion is key to understanding Chaucer's "fugitive poetics, a way of making poetry in a world in which 'every kyndely thyng that is' reveals itself in transit."[96] However, for Davis as for Delany and Robertson, motion is subject to a "law of natural propensity" in the spirit of Augustine, Dante, and Boethius, a world full of movement in which things incline to their natural places. Davis reasons that Chaucer lights on permeable forms—the House of Fame, the House of

Rumor, a basket eel trap—that can "gather" moving "contents in passive fashion" so that each house and other structures function not so much as an end point of motion as "deferred *telos*."[97] In fact, in their discussions of motion, the new physics had already spoken of such a "deferral" or, more strongly, had begun to turn away from destinations altogether. The *locus naturalis* was still there as a kind of remaining promise, but motive force was now embedded in each object, an origin or cause that impelled an item on in exponential fashion until resistance countered it. The ending of the *House of Fame*, with people clambering over each other to reach the "man of gret auctorite" in one angle of Rumor's house, may evoke the older teleological scheme of the universe, but in the new physics, that location is merely one of many along the dreamer's and the poem's uneven, sideways passage of motion.[98] The narrator has already dreamed and desired things that have sent him on his way. Each location he encounters should be significant and provide a measure of the ratio of force to resistance, and in a way Chaucer hints at those measures when the dreamer is offered new tidings that propel him. The poem, however, tells a narrative of motion without clear motive. Its impetus arrives as a series of obscure causes that bring about lateral movements one after another.

CHAPTER 7

Intense Proximate Affect
Nicole Oresme's *Tractatus de configurationibus qualitatum et motuum*

In Herodotus's *Histories*, *geōmetría* is an art the Greeks learned from the Pharaoh Sesostris, who, after building temples and digging canals in Egypt, divided land into plots for taxation. Disputes over land ownership needed resolution because of flooding, and geometry was the solution.[1] Aristophanes's *The Clouds*, which contains one of the earliest references to a map, refers to *geōmetría* as the art of measuring large areas.[2] While geometry in these works may appear to imply divided space on the ground that is perceived from above, the more typical horizontal perspective in the late Middle Ages means that "geometrical" space was registered as Hugh of St. Victor defines geometry, as *mensura terrae* (earth measurement).[3] A fifteenth-century English treatise on heights, breadths, and depths notes that the word "geometri" means "erthly mesure."[4] In all cases, the difference between *gē* or *geo* and *geōmetría* is the addition of measurement, the sectioning of spatially continuous area into separate but contiguous plots. In this study of late medieval space, I have examined how the science and literature of the time approached space to gain insight into how people apprehended the areas around them. I have analyzed their perceptions as to what composes a local area, the habitual ways they saw the spaces as horizonal and its alternative in overhead views, and the qualities of objects as they move through local areas. The next spatial degree is one out from a single one. This chapter examines discussions in late medieval science on proximity, the distance between objects.

Medieval natural philosophy of the thirteenth and fourteenth centuries was interested in what it called *propinquitas*, the quality of nearness that expresses the relationship between one place and another, one entity and another. Nicole Oresme's name stands out once more in the natural philosophy, with his *Tractatus de configurationibus qualitatum et motuum* the object of central attention.

Propinquity was a significant topic in diverse kinds of writings in the late Middle Ages, including ones we would today group as sociological. Genealogies comprehended proximity in terms of nearness to a person, whether an immediate ancestor or a direct descendant, laterally as a peer, or even more distantly yet incrementally as in the genealogical rolls that trace histories of family members or historical figures back to creation. Writings on behavior and etiquette often worked in parallel with genealogy, translating genealogical-historical distance into a more physicalized set of spatial guidelines for relations between people—for example, describing in more concrete detail where and how to stand near a person of higher rank. The young addressee of the poem "The Babees Book," for example, is instructed to remain in front of his lord, not leaning on a post or other part of a building. The shorter poem "How the Good Wiff Tauȝte Hir Douȝtir" counsels a young woman how to behave with a man who "biddiþ þe worschip," advising that she "Sitte not bi him, neiþer stoonde, þere synne myȝte be wrouȝt." It also instructs her not to be near a man on a street.[5] Julian of Norwich meditates extensively on the vision of a servant standing quite precisely "full nere the lorde, not even foranenst him but in perty aside [not exactly in front of him but partly to one side], and that on the lefte side" in chapter 51 of her visions.[6] In a larger sociological context, when it came to the "dense sociability" of the more concentrated late medieval urban areas, it has been argued that while physical "proximity did not always wear" a "seductive face," it nevertheless "assured some fundamental functions" in establishing and maintaining social relations.[7] Reginald Pecock's theological tract for the laity, *The Folewer to the Donet*, examines one of the limits of the human social world when it considers, but rules out, moral obligations to nonhuman animals even though he notices that we work closely with them.[8]

This range of social factors is significant for the topic of proximity, but social ideas about being close to another were not the only rubric through which people considered the subject. Astronomy and medicine included but also extended beyond the human world to propose that when celestial and other objects came close, they could affect each other in a number of possible ways. Bartholomaeus Anglicus's encyclopedia says of cosmological activity that "whan a planete comeþ wiþ a planete in þe same signe ascendent oþir

[or] in þe nexte signe þeretofore oþir bihinde, þan hit is iclepid coniunctioun. And þis coniunccioun may be good ʒif þe planetis be goode, and aʒenwarde [on the contrary] iuel ʒif þe planetis beþ euel."[9] An early fifteenth-century translation from Latin of Guy de Chauliac's *Grande Chirurgie* describes how in astronomy "Þe more coniunccionʒ [among celestial objects] . . . signifieþ wonderful þingeʒ stronge and ferdful and mutacionʒ of kyng-domeʒ, comyngʒ of propheteʒ, and grete mortaliteeʒ." For example, "þat grete coniunccioun [of Saturn, Jupiter, and Mars] signified a meruelous mortalitee and a ferdful, for not al-only it was of þe more, ʒe, bot as it was of þe moste."[10] Bartholomaeus's encyclopedia also addresses physiognomy. His work attributes "spewinge" to "pressi[n]ge and reringe of þe stomak by noyous [harmful] companye of oþir membres" while also observing that "by ferenesse and nyʒnesse of þe sonne mennes face and bestis bodies ben disposid in strengþe and hete."[11]

Critical work on the uncanny, neighbors, and the touch in relation to literature and other kinds of writings has addressed additional medieval understandings of proximity. Among psychoanalytic approaches, some have studied, for example, uncanny characters in works such as *Sir Orfeo* or the uncanny proximity of geographical areas in Chaucer's poetry.[12] Others have looked to legal and other records to examine the history of neighbor relationships, and still more have applied ideas about "neighborliness" and vicinity to manuscript relations.[13] Work on the touch, especially in queer studies, has considered moments when bodies come so close to each other as to come in contact, with the potential to challenge conventions of sexuality and gender.[14] Note at this point that a linguistic analysis of vernacular senses of proximity is not particularly revealing because of the vast array of words in English, including prepositions, that denote spatial relationships between objects. Whether inanimate or animate, things could be *ner*, *neigh*, *next*, *bi*, *biside*; they could *ajoin*, *approchen*, *attenden*, *resten*, be *at on*, and be *withouten*; they could move *to*, *toward*, *unto*, *aboute*, *bifore*, and *apart*. The words do not tell much about the relationships between objects, nor do they help distinguish whether an association is social, astronomical, medical, literary, or of an even more abstract kind.

My intention is to focus on spatial proximity in its constitutive physical form, and late medieval mechanics offers a lesser-known lens through which to consider the primary sense of the moments in which objects are near each other. There is some overlap with sociological, astronomical, physiological, and critical senses of propinquity as the language in one area of knowledge was sometimes applied to another, but a principal part of the aim is to avoid shading off into metaphorical uses of nearness or associations of other kinds. Mechanics examined the idea of nearness in relation to all natural and supernatural objects and events. It complemented and informed the astronomical and medicinal sciences, and it appears to provide some of the underlying philosophical thinking on social bodies when they come close to each other.

The topic of proximity is related to the previous chapters on place and motion because all of mechanics is concerned with motion. The chapter on late medieval innovations in the science of motion noted that because of a reduction in the significance of natural place, all locations in the movement of an object were scientifically meaningful. The hermeneutics of motion and space largely changed from a teleological physics to a content- or extent-based understanding. Like ideas about motion, late medieval notions of *propinquitas* also drew on ancient traditions and made a number of innovations. One main idea about proximity concerns what came to be known in the fourteenth century as "the intension and remission of forms" (*intensio et remissio formarum*). By *forms* here are generally meant qualities, and change was thought to come about because of an increase or decrease in what was known as the "intensity" (*intensio*) of one or more qualities. An intensity could increase, or gain "intension," or it could decrease, or undergo "remission." Behind these ideas lay the law that even though two objects, two qualities, or an object and a quality might be contiguous to the extent that, at first glance, they appear indistinguishable, they remain separate. This is an Aristotelian principle, but the new physics found itself needing to reiterate and rethink the point because the new notion of impetus seemed to imply that a motive force and a moving object were integrated into one entity. The intension and remission of forms also gave rise to the idea that because a qualitative change—an intension or remission in a quality—came to be thought of as divisible into degrees of intensity, the change could be measured quantitatively, although this idea was subject to some disagreement. Also, it was theorized that all phenomena have intensities, and indeed most have more than one at the same time, such as color, temperature, velocity, and so on. The ways that these intensities vary in degree, alter over time, move across a distance, or otherwise change according to another measure were thought of in spatial terms as a "configuration" (*configuratio*).[15] Close objects affected each other by means of an item's configuration of its intensities coming near to another's. Interactions between proximate intensities could take place among configurations in an individual body, and they could occur between bodies if the entities were near each other. One may note in addition that the mechanics of *propinquitas* can contribute to studies of affect in the history of feelings. In the late Middle Ages, mental states, dispositions, and emotions are included in the consideration of intensities and configurations.

Nicole Oresme's three-part treatise from the mid-fourteenth century, *De configurationibus qualitatum et motuum*, is examined here as the central case study in order to describe his ideas about intensities and to investigate how the qualities of objects can interact with each other within an individual item and between different items due to the intensities and configurations of their forms. Previous historians of science—Pierre Duhem, Anneliese Maier, Edward Grant, and especially Marshall Clagett (who produced and analyzed an edition of Oresme's writings)—have tended to focus on the fundamental physics of *De configurationibus*. I examine the more neglected chapters of the treatise because what appears remarkable about it is that Oresme is rigorous in applying his mechanical view to all phenomena in the world, all qualities. He considers the intensities of the qualities of geometrical entities: points, lines, planes, and solids. He addresses beauty, amity and hostility, imagination and affection. He is profuse: the passions; the lion, eagle, and falcon; a tin vessel; a cure for a disease; noble, beautiful, and perfect corporeal figures; magnets; the soul; motion; pain and joy. All objects are composed of qualities in Oresme's flat, intense world. In this mode of thinking, he is representative of other natural philosophers at Paris and Oxford in the fourteenth century, although he developed the ideas to the greatest degree and completeness in ninety-three chapters in the *De configurationibus*. This way of considering the whole world as composed of complexly interacting qualities in configurations, all of which could be measured, became dominant in this period. A second relatively neglected aspect of his mechanical theory is the idea that parts of each configuration of an item are potentially interactive when one quality nears another within one item or across physically separate components. *Propinquitas* is necessary, and the tensile distance between qualities, a spatial distance, becomes significant. The discussions of the natural philosophers on this topic concern what today is called *affect* except their analyses go far beyond a focus only on human subjects to include a world in which everything is intense.

Intensities and Configurations

Even though a work such as John Wyclif's *De officio pastorali* suggests that "preyeris & many oþere gode dedis ben as wel don afer as neer," practically all Latin and Middle English texts suggest that proximity is necessary for an object or action to have an effect on another entity.[16] Pecock's *Folewer*, for example, states quite plainly that, except for sight, "eche of þe out-ward wittis knowiþ and iugiþ oonli of þe þing which is immediat to him."[17] Wyclif's and Pecock's topics are different of course; Wyclif is describing the efficacy of prayer and Pecock the more modest human senses, but one might be wary of relying too much on theological hierarchies in this case. The philosophy behind observations such as Pecock's stems from no less an authority than Aristotle, who notes in several places in his discussion of motion that "movers and the moveds must be in contact with each other." At one point Aristotle says, "It is necessary . . . for the self-mover to have a part that causes motion while it is unmoved and a part that is moved but does not necessarily cause motion, with either both parts touching each other, or one part touching the other."[18] The efficient cause of motion depends on contact of some sort. In the fourteenth century, Francis de Marchia, John Buridan, and other philosophers challenged key aspects of this idea, positing impetus as an embodied force or *virtus derelicta* that an impelling body transmits to a moving object. The difference in terms of Aristotle's last statement on one part not being moved and the other moving is that for the fourteenth-century philosophers, the motive force can move along with the moving item. But they did not disagree that there remains a distinction between the force and the object. Otherwise there would be a collapse between a thing and its attendant qualities, which would be to make a phenomenological or ontological mistake.

Other philosophers would repeat and alter Aristotle, insisting on an intimate proximity of objects, whether moving or not, to affect each other, but the objects and the forces governing their interactions still remained separate entities. Thomas Aquinas comments on Aristotle about moving objects to the effect that when the philosopher says in book 5 of the *Physics* "that things in the same place are together," Aquinas says "he adds that 'together' is not taken here in the sense of being in the same place, but in the sense that nothing is intermediate between the mover and the moved. It is in this sense that things in contact, or things that are continuous are together, because their extremities are together or are one and the same" (dixerat ea esse simul quae sunt in eodem loco. . . . ideo ad hoc excludendum subiungit, quod simul dicit hic, non quidem esse in eodem loco, sed quia nihil est medium inter movens et motum; secundum quod contacta vel continua sunt simul, quia termini eorum sunt simul, vel quia sunt unum).[19] The idea is to clarify that no two objects can occupy the same place but that contact is required for an efficient cause of motion. The thinking is similar to Aristotle's definition of platial location in terms of a relationship between a container and a contained object: "the limit of the surrounding body, at which it is in contact with that which it is surrounded."[20] *Propinquitas* is necessary for place, but there too items are distinct even though they touch.

The theory that came to be known in the fourteenth century as the intension and remission of forms had several parts that accounted for the behavior of all things in the world. It was associated with the discussion of motion, but it was applied beyond moving objects to encompass change more generally. New thinking about intensities is said to originate with Godfrey of Fontaines and John Duns Scotus in the late thirteenth and early fourteenth centuries. The central problem was seen to concern changes in the qualities of objects, such as something becoming redder, hotter, faster, and so on. This was a change in intensity, but what is the actual nature of the change? Godfrey of Fontaines sharpened the focus of discussions about changes in intensities "toward concentration on concrete individual qualitative forms."[21] Other natural philosophers, including Walter Burley, would attribute to Godfrey the further idea of successive qualities, meaning that when the quality of an object undergoes change, a new quality or "form" replaces the old one. So an increase in heat would mean that a current intensity of heat is supplanted by a new, more intense heat. Scotus and others after him, especially the Merton School at Oxford (Burley excepted), posited instead of a successive theory of forms an additive idea of change, meaning that the old intensity is not erased and replaced by a new one, but instead the quality is divisible into individual parts, as Godfrey had posited, so that an increase in intensity is therefore a quantitative change. The intensity of heat is added to with more intension of heat, the intensity of a shade of red becomes less red with the remission of redness, and so on. The additive and subtractive theory of forms meant that change could be measured mathematically with numbers. The most common method was to use ratios in the ways I have discussed in Mertonian and other ideas about motion.[22]

Nicole Oresme starts out *De configurationibus qualitatum et motuum* by saying he has written his tract "so that the treatise might be useful not only as an exercise but also as a discipline" (*ut iste tractatus non solum exercitationi prodesset sed etiam disciplinae*).[23] He uses graphs to represent his measurements of phenomena, for example, the duration of motion on the horizontal axis and the intensity of motion on the vertical axis. As explained, a rectangle would represent uniform motion, that is, constant speed. A right-angled triangle would show constant acceleration or, as Oresme calls it, "uniformly difform" motion. (His treatise was also called *Tractatus de uniformitate et difformitate intensionum*). The area of each figure is the quantity of the quality over a period of time or another measure, and this Oresme calls a "configuration." This method of measuring change is applicable to all qualities, not just motion. Everything in the universe has qualities that have distinct intensities. He writes:

> It is manifest from natural philosophy and experience alike that all natural bodies . . . determine in themselves their shapes, as, for example, animals, plants, some stones, and the parts of these. They also determine

in themselves certain qualities which are natural to them. In addition to the shape that these qualities possess from their subject, it is necessary that they be figured with a figuration which they possess from their intensity—to employ the previously described imagery. It is necessary, therefore, that the aforesaid natural bodies, or their forms, determine in themselves a certain figuration of their fundamental, constitutional, or innate qualities.

Ex philosophia naturali et per experientiam manifestum est omnia corpora naturalia . . . sibi determinare figuras, sicut sunt animalia et plante et aliqui lapides et partes istorum. Determinant etiam sibi certas qualitates eis naturales quas quidem qualitates preter figurationem quam habent a subiecto necesse est figurari figuratione quam habent ab earum intensione secundum ymaginationem premissam. Oportet igitur quod predicta corpora naturalia vel forme ipsorum determinent sibi certam figurationem suarum qualitatum radicalium seu complexionalium et sibi naturalium.[24]

Oresme appears in places to reject the additive and subtractive theory of intension and remission of forms or qualities, or he at least rejects the reality of the additive theory for a species of qualities that are continuous. His different writings seem to disagree on the subject. The problem lies in the fact that "successive" qualities (the word is the same in Oresme's Latin), like light or heat, do not really divide into small units of measure. When someone measures a change in temperature by numbers of degree, for instance, he or she is using a convenient but inaccurate word since a change in temperature is successive. Indeed, Oresme begins his whole theory of the configuration of qualities and motions by describing it as a way of "thinking" about change rather than a truly accurate way of capturing change that matches its real nature. He uses the word *fingere*, meaning "to make up" or "imagine," when he describes his methodology. He goes on to depict his way of measuring change as a process of analogy: measure "is recognized [or known] by similarity" (*cognoscitur in similitudine*). A qualitative temperature change, for example, has to "be imagined" (*ymaginatur*) as a quantitative one in order to represent it even though it is successive, but there is not "real multitude or superposition of degrees, as some people think" (*Nec est ibi realis multitudo sive superpositio graduum prout aliqui opinantur*).[25]

Given that in reality, some objects and qualities are continuous and indivisible—his examples include light, hotness, and whiteness, among others—change is not a matter of infinitely divisible parts being added or subtracted.

Oresme in this treatise posits instead that for an indivisible object to change, it must already be nonuniform or difform. Hotness, for example, is difform in the sense that it is the same quality over time, but its intension and remission, its degrees of hotness we say, differ. He offers the example of an object that constantly increases in heat for one hour, remains at constant heat for the second hour, and then constantly increases again for the third hour. Such an object is uniformly difform, uniform, and then uniformly difform again; a graph of these changes would have a right-angled triangle showing the increase, a rectangle, and another right-angled triangle, the total area of which could easily be calculated in Oresme's graphing method. Motion is similarly subject to "the division and extension or continuity of the mobile," time, and intensity (*secundum divisionem et extensionem seu continuitatem mobilis*). The mobile object can be measured according to *uniformitas* and *difformitas*, and time according to *regularitas* and *irregularitas*. So movement can be uniform and regular (*uniformitas et regularitas*), uniform and irregular (*uniformis et irregularitas*), difform and regular (*difformitas et regularitas*), and difform and irregular (*difformitas et irregularitas*).[26] A graph of the hotness over the hours or a graph of a mobile object over time or distance would present a total area, and it is the total relation of changing intensity to the time or another measure that he calls a "configuration" (*configuratione*). This configuration of intensity in relation to time, distance, or something else is the conclusive aspect of his theory. He gives, for example, the case of changes in the seasons, which he says are known to produce diseases. Great seasonal change can produce great diseases. But "sometimes a minor change causes more diseases than a major change produces at another time" (aliquando minor mutatio plures egritudines causat quam alia vice faciat mutatio minor). This is because diseases are attributable not only to the magnitude of the change but also to the "configuration" of the change. Multiple diverse changes in a season are "difform" and produce more severe diseases. Presumably, a large and difform change might produce even greater disease.[27]

Oresme's ideas about propinquity build on the theory of the intension and remission of forms, Oresme drawing on earlier philosophers in describing the ways that qualities can affect each other due to proximity. Pierre Duhem points to the origins of Oresme's ideas in William of Auvergne, Roger Bacon, Albert of Saxony, and Marsilius of Inghen in the thirteenth and fourteenth centuries. Thinking about propinquity in relation to the intension and remission of forms, however, became more refined because of Oresme, who in effect added "the concept of internal configurations of intensity to the earlier ideas of the varying ratios and dispositions of qualities."[28] His ideas also become necessarily more complex when he considers proximity, because an intensity of a quality and its configuration are now put in relation to another quality, or more than one quality, which also has its own configuration. As Clagett writes, "Differences in internal configuration may explain many physical and even psychological phenomena, otherwise not simply explicable on the basis of the primary elements that make up a body. Thus two bodies might have the same amounts of primary elements in them and even in the same intensity but the configuration of their intensities may well differ, thus producing different effects in natural actions."[29] Any sense of nearness can cause effects, but Oresme habitually describes physical—in the sense of spatial—proximity. He largely discusses natural bodies when they are "mutually united" or "brought together" (*corpora naturalia advincem comparata*).[30] At one point he offers an example of medicine in which a cure might be particularly effective when it is applied, in part due to "contact" (*contactus*) with a body and because of close parallels between "the figuration in intensity of the quality of the thing . . . applied" and the body or smaller object affected (*figurationem in intensione qualitatis illius rei que applicatur*). Symmetry may not be the only answer when it comes to medicine or other near effects, for dissimilarity can conversely give rise to the "flight" or the "corruption" (*fuga vel corruptio*) in the affected body.[31]

Oresme, like other natural philosophers of his day, is systematic and exhaustive in describing the phenomena that contain configurations of intensities of qualities. Because of this rigor, he does not particularly distinguish between a situation in which the intensity of a quality within one body (say the weight of a tin vessel) interacts with another quality within the same item (for example, the color of the tin) and a situation of one body affecting another distinct entity. He would argue that the proximity of one quality to another is what matters in terms of affect, and the physical incorporation of the quality in one body or the existence of qualities in two (or more) separate bodies is not particularly important. Propinquity applies in either case. The idea of how one quality affects another is the key thing.

Proximate Affect

Affect theory is a useful lens through which to understand the interactions of proximate configurations of intense qualities, and Oresme's *Tractatus de configurationibus qualitatum et motuum*, conversely, is a text that can inform ideas about affect. Indeed, it appears that Oresme has already had some influence on affect theory. Mary Beth Mader has traced the impact of Aristotelian science and its medieval engagements, especially Oresme's ideas about intensities, on Gilles Deleuze and Félix Guattari, who in turn are a source of inspiration for work in affect studies.[32] Brian Massumi, who translated *A Thousand Plateaus* (one of two Deleuze and Guattari works that refer to Oresme) and who is also a major figure in affect studies, discusses the complex affective relationships between bodies and within each body, all in the context of a discussion about motion. He follows Leibniz in asking, "Can we think a body without this: an intrinsic connection between movement and sensation, whereby each immediately summons the other?"[33] Massumi does not explicitly mention Oresme or other medieval natural philosophers, nor does the influential anthology *The Affect Theory Reader* (which includes an essay by Massumi) address anything from the Middle Ages. The introduction of the *Reader* nevertheless describes affect as arising "in the midst of *in-between-ness*: in the capacities to act and be acted upon. Affect is an impingement or extrusion of a momentary or sometimes more sustained state of relation *as well as* the passage (and the duration of passage) of forces or intensities. That is, affect is found in those intensities that pass body to body (human, nonhuman, part-body, and otherwise), in those resonances that circulate about, between, and sometimes stick to bodies and worlds, *and* in the very passages or variations between these intensities and resonances themselves."[34] One of course wants to be careful and not make affect an ahistorical phenomenon, as Holly Crocker, among others, reminds us,[35] but the passage sounds very much like Nicole Oresme, for whom everything is intense and can interact via proximate intensities. Oresme is, after all, much more systematic in working out the permutations of intensities and configurations.

Book 1 of Oresme's discussion in the *Tractatus de configurationibus* starts out with the geometrical presentation of configurations and the total number of configurations of uniform and difform qualities, which is sixty-three. It then progresses to particular examples of physical, aesthetic, and psychological phenomena that do not occur in time or across a distance, or for which the time or distance is too small or difficult to measure, but he retains the idea of configurations when he goes on to apply his theory to natural bodies, which include people and all their constituent parts: their bodies, souls, intellects, emotions, imaginations, and so on, all of which can affect each other. The first animal quality he discusses are passions, and we can see the transition where Oresme begins to apply what he has been saying about geometrical figures and inanimate objects to animate ones, although he speaks in such a way that all objects, inanimate and animate, fall under his purview:

> Just as differences in actions arise from diversity in the configuration of qualities, so accordingly one can assign the reason for certain differences in passions. Thus it might be said that just as bodies are rare and porous according to quantitative figuration, so they are more easily and quickly affected, other things being equal, than bodies disposed in a different way.

Thus, other things being equal, those bodies, whose qualities in the figuration previously posited are as if porous to the insertion of the contrary quality throughout the imperceptible particles of the subject, will, in place of others, be more capable of being affected quickly and will be more susceptible to penetration by alteration.

Sicut ex diversitate configurationis qualitatum provenit differentia actionum, ita ex hoc potest assignari ratio de quibusdam differentiis passionum ut dicatur quod quemadmodum corpora que sunt rara et porosa secundum figurationem quantitativam, ipsa quidem facilius et citius patiuntur, ceteris paribus, quam corpora aliter disposita; ita illa corpora, ceteris paribus, erunt pro allis velocius passibilia et alteratione magis penetrabilia quorum qualitates erunt secundum figurationem prius positam quasi porose per interpositionem qualitatis contrarie secundum particulas subiecti imperceptibiles.[36]

Book 2 of the treatise is similarly structured. It begins as a discussion of the topic of motion in an abstract sense—its "double difformity," on the quantity of its intensity, on succession—but then moves on to more concrete examples of motive objects: celestial bodies, sound, and so on. Of most interest is his discussion of qualitative "accidents of the soul" (*accidentibus anime*) as he defines them: "apprehension, cognition, or imagination" (*apprehensio aut cogito seu ymaginatio*) and their "accompanying desire or passion" (*appetitus concomitantis vel etiam passionis*), such as revenge, excitement, fear, or joy.[37] In this turn to the "sensitive soul" here and elsewhere in his writing, the same physics applies: "Accidents of the sensitive soul are, in accordance with the extension of the subject, figured in the organs with respect to uniformity and difformity in completely the same way as are sensibles or the other qualities of which we have spoken before" (Accidentia anime sensitive secundum extensionem subiecti eadem modo penitus figurantur in organis quantum ad uniformitatem et difformitatem sicut sensibiles vel alie qualitates de quibus dictum est ante). He concedes that the intellect is both "indivisible and inorganic," and "hence is not figured properly either according to body or quality," but nevertheless "still there can be improperly imagined in it, by some means, a certain spiritual configuration corresponding to the configuration of the sense, because its intellection depends on the sense" (indivisibilis et inorganica . . . et ideo nec est corporaliter neque qualitative proprie figurata, tamen aliquo modo improprie ymaginari potest in ea quedam spiritualis configuratio correspondens configurationi sensus, quoniam eius intellectio dependet ex sensu). So "just as the interior sense sometimes approaches uniformity and evenness and sometimes is figured with great difformity and, as it were, unevenness," so too can we consider the "power of the intellect" (Sicut ergo sensus interior quandoque ad uniformitatem et planiciem accedit quandoque etiam valde difformiter et quasi aspere figuratur, ita conformiter suo modo diceretur de potentia intellectus).[38] In his admittedly adapted reasoning involving a certain degree of imagination, quantitative figuration can be applied to qualitative configuration and so to sensitive and intellective power. Therefore, for example, "the soul occupied by many thoughts and affected by many passions has been made as it were rough and difform" (anima enim multiplicibus cogitationibus occupata et passionibus affectata facta est quasi aspera et difformis). He goes on to explain how someone may remain obsessed about a topic so they cannot rid themselves of the "difformity," or a person might be able to cast off the "difformity" by moving on to another idea by forgetting the original cause of torment or because of the simple passage of time.[39]

Oresme's discussions of the relations among intensities are especially fascinating when he talks as he does about animate objects and human qualities. His understanding of the universe is in some sense parallel with astrological writings and humoral theory in that those areas of study picture a world in which each element can be strong or weak, dominant or deficient, and affect another part of a system, except Oresme's theories are rigorously and methodically worked out in terms of proportions and ratios. Oresme critiques astrology in other tracts.[40] It has been emphasized (arguably at the expense of his other observations) that one of Oresme's particular criticisms in *De configurationibus qualitatum et motuum* is reserved for magic. Oresme does not deny the possibility of magical effects, such as the conjuring of spirits, but he is impatient with accounts that attribute effects of the supernatural on people and things when his proportions and configurations of intensities provide a natural and more realistic explanation of them.[41]

The second part of book 2 of the *Tractatus de configurationibus*, for example, explains away the claims of magicians who exploit or deceive susceptible people by making demons seem to appear or an object look as though it contracts and dilates at a person's will or command. Oresme is not saying that other more honest and informed people would be unable to point out that—even where—demons actually exist; he is not suggesting that demons and other seemingly supernatural phenomena are merely a consequence of a mind playing a trick. They exist. What he does is to emphasize that propinquity and a natural series of causes can usually account for the ability of one body, soul, or mind to influence another rather than magic. For instance, he describes how easy it is for a nonphysical entity to affect another person: "The body or member of the person so altered [by the imagination, the affection, and so on] sometimes can alter the medium and a body near to it at a certain distance, and most of all if that body is disposed to receive easily such an alteration or impression" (Corpus autem vel membum hominis sic alteratum aliquando potest alterare medium et corpus sibi propinquum secundum distantiam certam, et maxime si illud corpus fuerit dispositum ad faciliter recipiendum talem alterationem vel impressionem).[42] Indeed, even though he is skeptical about magic, he is even more doubtful about the ability of something to affect another object at a distance. He concludes his discussion of the potential for magic to have effects by saying that "it seems much more incredible and remote from reason for the imagination of the soul to be able to act in a place where it is not than for demons or spirits to exist and, being present, to produce" miraculous sights or actions (si velimus iudicare recte, multo incredibilius et a ratione remotius videtur ymaginationem anime posse talia agere ubi ipsa non est quam demones vel spiritus esse qui presentes faciant ista).[43]

Of special interest are his discussions of *amicitie et inimicitie naturalis,* "natural friendship and hostility," which lead to analysis of friendship and love.[44] In his philosophy, the combination of intensities and their configurations lead to natural amity and enmity between members of different species, an idea that one can find in other philosophers such as Aquinas but that is put in Oresme's terms here: "Thus it is that one cause of the natural friendship between man and dog can be the fitting accord between the ratio of primary qualities in the human constitution and the ratio of the same qualities in that of the dog. Another cause can be the fitting accord between the configurations of the primary or other natural qualities in each of these species" (Inde est quod una causa amicitie naturalis inter hominem et canem potest esse convenientia proportionis primarum qualitatum in humana complexione ad proportionem earundem in canina. Et alia causa potest esse convenientia figurationum earum primarum vel aliarum naturalium qualitatum in utraque istarum specierum). The same is true of natural hostility between certain animals, his example being a sheep and a wolf.[45] He then applies the idea of configuration to members of the same species and "natural friendship" (*naturalis amicitie*). Here he divides the kinds of friendship into three species. First, "one individual naturally likes another similar to it more than it does one dissimilar" (unum individuum naturaliter diligit sibi simile in specie magis quam dissimile). Second is a "special mode" (*modus . . . specialis*) between males and females not because they are similar but because their ratios and figurations conform. He concedes that procreation and "mutual assistance" (*adiutorium motuum*) are also causative in friendships between men and women, but even these can arise from the conformity of ratio and figuration. The third *modus* of friendship, which Oresme calls "more special" (*specialior*), is when two individuals (Oresme's example is two men) "are impelled as if by a certain natural attraction towards mutual goodwill and love" (*quasi quodam tractu naturali impelluntur ad benivolentiam mutuam et amorem*), while others are never friends without knowing why, but the cause is attributable to their lack of conformity.[46]

Oresme evokes aesthetics in these discussions because a species that one finds more agreeable will also be considered more noble and beautiful.[47] He also addresses aesthetics in his work on the circular motion of celestial bodies, *Tractatus de commensurabilitate vel incommensurabilitate motuum celi*, a book that takes the form of a dialogue between arithmetic and geometry. His ideas about configurations of intensities are similar for aesthetics. A match between the ratio of qualities with their configurations and the "species" to which the object belongs will create a more beautiful animal. This appears to be a slight departure for Oresme, whose system does not necessarily require the suitability of an object's ratios and configurations in terms of a "kind" or an Aristotelian category, but the authority for such a statement lies, he says, in "natural philosophy and experience alike" (*Ex philosophia naturali et per experientiam*), namely in Aristotle's *De anima* as well as observation.[48] He says that individuals within each group will be more noble if their "fundamental quality possesses more fittingly the mode of configuration due its species" (*radicalis qualitas propius habet modum configurationis debitum seu speciei*).[49] Oresme's examples come from the animal world and apply to the human world within it. He connects beauty to perfection and nobility when he writes that "those qualitative configurations which are similar and proportional to nobler and more beautiful or more perfect corporeal figures are simply better or nobler" (videlicet configurationes earum que sunt similes et proportionales figuris corporeis nobilioribus et pulchrioribus seu perfectioribus sint simpliciter meliores seu nobiliores).[50] Some species are therefore by their configuration more noble than others. But he also suggests that some objects are simply more beautiful because their configurations of qualities may be "mutually conformable or fit together better while others do not fit together well" (*alique sibi invicem sunt conformes vel convenientes et alie disconvenientes*).[51] These ideas differ from statements about beauty in the *Tractatus de commensurabilitate vel incommensurabilitate motuum celi*. There geometry proposes that more beauty may occur when there is a mix of "commensurable" and "incommensurable" motions.[52]

Oresme's image of a world of different degrees of intensities that are configured and relate to each other in proportional ratios at different times due to propinquity is a necessarily complex one. Even on affects, however, he is systematic in that his discussion works through the many kinds of combinations of feelings that might occur. Some things near to others can influence each other's emotions because there is a symmetry in the figuration of intensity between the two objects, while other nearby objects can give rise to discord because of asymmetries between them. His discussion of pleasure and displeasure in relation to "sensible" things may stand as representative of his all-inclusive thinking about intensities and affect:

> Should we hasten through the exterior senses first to touch, we find it certain that some things are pleasant to the human touch while others are unpleasant, not only because of the [intrinsic] excellence of the touchable things but also because of something arising from another source. Therefore, certain things are unpleasant to the human touch which would be pleasant to the touch of another animal. And similarly, within the same species, some touchable object displeases one individual but pleases another. The same thing is true of taste. For example, thorns [unpleasant for man] please the ass, while pottage [tasty to man] displeases the ass. Also, something pleases a well man and is displeasing to a sick one, and there are many such examples. It is the same for something that can be heard. For certain animals like music and others do not, and men differ much in this regard, as has been noted. And so universally, in regard to all kinds of pleasurable and unpleasant sensibles, it is certain that different animals differ in many ways. And although many causes may concur in producing this diversity, still it perhaps seems probable that . . . two . . . [causes] are the most important, namely (1) the accord or discord between the ratios of qualities, and (2) the accord or discord between their configurations.

> Si per sensus exteriores velimus discurrere primo in tactu, certum fiet quod aliqua sunt tactui humano delectabilia, aliqua tristabilia, non solum propter excellentias tangibilium sed etiam aliunde, et ideo quedam sunt tristabilia humano tactui que forent delectabilia tactui alterius animalis. Et similiter in eadem specie aliquod tangibile displicet uni individuo quod placet alteri. Similiter in gustu sicut asino placent tribuli et pulmentum displicet. Aliquod etiam placet homini sano et displicet egro, et sic multipliciter. Consimiliter de audibili, nam quedam bruta delectantur in musica et alia non, et homines differenter se habent quantum ad hoc et multipliciter, ut notum est. Et ita universaliter de omnibus sensibilibus delectabilibus vel tristabilibus, certum est ea diversimode ad diversa animalia se habere quantum ad hoc. Et quamvis ad istam diversitatem multe cause concurrant, forsan tamen et probabile videtur quod due predicte sint principaliores, videlicet convenientia vel disconvenientia

proportionis qualitatum et convenientia vel disconvenientia configurationis earum.[53]

He goes on to single out pain and joy as two "passions of the soul" and to apply his ideas about intensities and configurations to them. Pain can be measured according to its intensity and its extension because it takes place in time. A short, intense pain can therefore be represented geometrically as a square if it is of a certain duration and intensity, or as a rectangle if it is longer. The usefulness of his graphic representations of qualities becomes clear when one realizes that the areas of these two very different experiences of pain could be the same, so one could say that the pain was of the same quantity even if each had a different configuration. Also, however, even though two pains can have the same extension or length in time, they can be different in that one can be uniform over the period of time and the other difform. Oresme proposes that the uniform and more "beautifully" or "harmoniously" (*pulchre, armonice*) configured one will be more tolerable than the difform one. In hell, it is possible that "any infernal pain is difform with a difform difformity and [varies] according to more and less" (pena infernalis difformis est difformitate difformi et secundum magis et minus).[54]

His last chapter of the second part of *De configurationibus* is on joy, which is like pain in its configurations and intensities. Some joys are "preferable and better" because their difformity is "more nobly and better figured" (*eligibilior atque melior . . . nobilius et melius figurata*), meaning that a difform joy can be preferable to a uniform one "because it is difform with a certain fitting difformity" (*eo quod ipsa est difformis quadam difformitate decora*).[55] For example, he explains why a person might act by distinguishing between a kind of predisposition one might have and the emotions or imagination at the moment when a person is urged to act:

> [I]t is possible that movement or passion in the body is varied not only because of greater and lesser intensity of imagination or affection but also because of a diversity as to difformity in the figuration of the aforesaid accidents of the soul. For example, if someone imagines [something] with affection or thinks about revenge, and the difformity of this cogitation or imagination is duly figured, then the act will duly carry out the commands and he will be as one particularly fortunate in carrying out or executing his intention. But if the imagination or affection is not duly figured, then the act will not be duly performed even though the imagination or affection is sufficiently intense.
>
> Possibile est . . . ut non solum propter intensionem maiorem et minorem ymaginationis vel affectionis varietur motus seu passio in corpore

> sed etiam propter diversitatem figurationis predictorum accidentium anime in difformitate. Verbi gratia, si quis cum affectione ymaginetur aut cogitet de vindicta et istius cogitationis vel ymaginationis difformitas fuerit debite figurata, tunc ipse actus imparatos exercebit debite et erit in prosecutione seu executione intentionis sue quasi bene fortunatus. Si vero ymaginatio sive affectio indebite figuretur, ipse operabitur indebite, quamvis ymaginatio vel affectio fuerit sufficienter intensa.[56]

Of particular interest for the topic of affectivity and propinquity is his discussion of "the causes of certain effects arising in an alien body" (*de causis quorundam effectuum in corpore alieno*), in which he analyzes the effects that the soul can have on another being.[57] One can, he says, move another body physically, with the hand, for instance, but one can also be the cause "of the transmutation of an alien body . . . without any exterior application, but [rather] by a certain latent action" (*transmutationis corporis alieni . . . sine tali exteriori applicatione sed quadam actione latenti*).[58] *Latent* here means simply that the action is hidden or obscure, something one does not usually notice. Oresme again evokes the occult in this section when he entertains the idea that someone's thoughts can directly affect another, but he rejects the idea as too much of a simplification. He entertains the possibility that "the intention of the imagination combined with a certain figuration of it" (*intentionem ymaginationis cum quadam figuratione eiusdem*) affects an alien body, but what he is particularly keen on eliminating as "excessively absurd and irrational" (*nimis absurdum et irrationabile*) is an effect that "arises from the soul by thought alone" (*fieri ab anima per solam cogitationem*), adding that such an idea was also condemned in the articles of Paris of 1277.[59] The "more rational way" (*rationabilior*), he counters and concludes, is that the

> imagination or affection can be so increased intensively and its difformity figured in such a way that it acts to change significantly the body or some member of the person who is so imagining: i.e. towards health or sickness, or towards some other special disposition. . . . Then the body or member of the person so altered sometimes can alter the medium and a body near to it at a certain distance, and most of all if that body is disposed to receive easily such an alteration or impression, [e.g.,] in the way that the eye of a menstruating woman infects the air and by means of the air infects a clean mirror. And this kind of alteration principally arises where visual rays are directed [at someone].
>
> potest tantum intendi et eius difformitas taliter figurari quod ipsa agit et immutat notabiliter corpus vel aliquod membrum hominis sic ymaginantis, videlicet ad sanitatem vel egritudinem aut ad aliquam

aliam dispositionem specialem. . . . Corpus autem vel membrum hominis sic alteratum aliquando potest alterare medium et corpus sibi propinquum secundum distantiam certam, et maxime si illud corpus fuerit dispositum ad faciliter recipiendum talem alterationem vel impressionem, eo modo quo oculus mulieris menstruose inficit aerem et mediante aere inficit speculum mundum. Et hec alteratio principaliter fit ubi diriguntur radii visuales.[60]

Although the ideas about infection and menstruation's effects on mirrors are voiced in other texts, such as Aristotle's, some new and interesting facets in terms of Oresme's ideas are introduced here. The first is that the imagination cannot affect someone or something else directly but instead first alters the body, or part thereof, of the person who imagines, thinks, feels, and so on. Second, the person's altered body or body part can then change the medium of the air, water, or whatever lies between the bodies. Third, and a slight alternative, is that the altered body or body part can then affect another body or part so long as it is nearby. Oresme singles out the eyes as significant in this process because they are openings in the body that lead more directly to the internal senses and the soul. He concludes the chapter with an important caveat that the power to affect someone via these processes "rarely operates externally in the aforesaid way, [and if it does, it does so] weakly and with difficulty, and in the presence [of many people], [operating] for a short distance and only on a body which is very disposed and apt for receiving an action of this sort" (Ipsa etiam raro operatur ad extra modo predicto et remisse et cum difficultate et in presentia et ad certam distantiam et non nisi in corpore multum disposito et apto ad suscipiendum huiusmodi actionem).[61]

In the previous passage, Oresme is not dismissing the idea that one body can affect another, for that would be to undo his physical theory of the intense world. He is merely expressing skepticism about magic, which can instead be explained through natural processes, and he is de-emphasizing the significance of interpersonal intensive affectivity because human-to-human influence is merely similar to all other proportional affect. The theory is nevertheless clear that one can consider affect as subject to physical laws. Given sufficient proximity, one can inquire into the accord or discord between qualities and configurations of qualities within and between people, or between people and (other) objects. In such a case, the topic must be addressed with some sophistication because the issue is complex. It is not enough just to identify the proportion or intensity of a certain quality in a person, such as courage or wisdom or jealousy. One also has to consider the relations of those qualities to other qualities and their particular configurations within that entity, then compare people; only then may the theory explain amity, hostility, and other affective relations among humans.

The medieval physics of Nicole Oresme about the necessity and dynamics of proximity provides useful ways of thinking about many kinds of relations, including ones between human beings. It also intersects with contemporary mechanical observations about motion. Beyond the simple observation that objects must be near each other to affect each other, late medieval physical science following Godfrey of Fontaines, Scotus, and others teaches that things are not naturally attracted to an end point, a resting location. Instead, all kinds of spatial relations are significant along a continual, though only rarely uniform, grade of acceleration because of accumulating impetus. The complexities of relations between objects, including between people, are many but can be considered physically in the sense Oresme and others mean when they discuss ratios, configurations of qualities, and their interactions.

Marshall Clagett's introduction to the edition of the *Tractatus de configurationibus qualitatum et motuum* addresses certain "efforts . . . made to foist on Oresme a proto-mechanistic view of the universe." Clagett argues that such a view is inaccurate in terms of Oresme's views of the heavens and large forces on the earth, including inertial ones, but he does not really address the topic of a potentially mechanistic view of other phenomena.[62] While it is not true that Oresme posits a mechanistic view of the world in the *Tractatus* in terms of proximate intensities necessarily having to affect each other (a certain configuration is not bound to influence another), he does insist that all the various phenomena in the world are accountable in terms of the configurations of intensities. They are certainly more attributable to his theories than to magic or unknown forces. Indeed, it seems from a reading of the treatise that a close-up understanding of individual entities interacting in the world as in a true investigation into their makeup that was able to perceive all their intensities in their proportions, strengths, and arrangements would be adequate in explaining interactions. Such observations would include all phenomena, so that even seemingly deeply subjective qualities—likes and dislikes, love and hate, for example—would be accountable in his scheme. Among other things, the significance of the will is reduced, or, rather, the will becomes a part of the world that is also subject to natural forces. Oresme's world may not be necessarily mechanistic, but it is one in which configurations of intensities are the significant factors when two objects near each other, and everything is an object, everything has intensities, and, by a species of imagination, everything can be measured and graphed.

CHAPTER 8

Proximal Literature
Nearness and Distinction in the *Legend of Good Women*

Like late medieval natural philosophy and Nicole Oresme's *Tractatus de configurationibus qualitatum et motuum* in particular, Middle English literature also engages with the subject of proximity. In poetry and prose of many kinds, characters, objects, ideas, emotions, and other phenomena near each other, move away from each other, and come together in telling ways so much so that the writing becomes about the tensile distance between beings. Sometimes, as with lovers, desire is a governing factor in efforts to reduce, if not eliminate, the space between persons, while other forces, including their own doubts, concerns, or repulsions, pull or push in the opposite direction to maintain separation. The narratives work in the proximal gap where these tensions reside. At other times, more purely mechanical laws increase and decrease the proximity between objects. For example, in Chaucer's most patently scientific poem, "The Complaint of Mars," planetary acceleration and relative speed govern the permutations of Mars and Venus. "Conjunction is the point," as Jessica Brantley has argued.[1] Elsewhere, Chaucer's Canon's Yeoman's Prologue and Tale also center on two relative rates of progress that bring the Yeoman and his master together with the Canterbury pilgrimage company. One rate of progress is physical and earthbound in the sense that the Yeoman and his master desire to "atake" (catch up with, overtake) the Canterbury pilgrims.[2] The other sense of moving forward is epistemic and about the acquisition of alchemical and deceptive abilities with questionable effectivity.

That the proximity of characters to each other or to objects is significant in Middle English literature may not come as a surprise, but the perspective that the physical sciences offer further opens up literary works to insights about propinquity for all kinds of objects and not just characters, and it uncovers an array of material, emotional, and other effects. Bringing the scientific developments to bear on the literature is not a matter of the direct citation of natural philosophy in the literature but more a sense that scientific ideas about local area, motion, and so on had changed apprehensions of space in the culture, including an awareness that proximate entities of all kinds could affect each other in distinct ways given particular conditions. This chapter on the topic of propinquity focuses on Chaucer's *Legend of Good Women*. It especially investigates moments in which lovers near each other but also takes up other occasions in the *Legend* and other pertinent writings in which a person or object approaches another entity so that distinctions between beings threaten to collapse. Part of my argument is that proximities develop throughout the poem across the different legends; they contribute to its structure. The first section of the following discussion examines moments in which medieval science's attention to proximate objects finds certain parallels in the Prologue to the *Legend of Good Women* and the first Legend of Cleopatra. Nearness is arguably the spur that begins the collection of legends, and it has ambiguous yet potentially positive or ameliorative qualities in the beginning of Chaucer's poem. The second part of the chapter continues a consideration of the *Legend* in light of natural philosophy's interest in the distance between objects, including the distance between nonhuman things, not just people, and especially the science's new interest in objects that approach each other rather than heading toward a goal. The section discusses perhaps *the* story of propinquity the Middle Ages adapted from classical literature, namely Pyramus and Thisbe. The Legenda Tesbe pushes questions about proximity to an extreme in that it focuses on the acute distance between characters and objects that near each other so much that their identities threaten to collapse into one another. The third section of the chapter looks at the remaining legends, especially the Legend of Ariadne, for how proximate distances devolve in tenor and become instrumental in deaths and betrayals as the poem develops.

In his influential essay on heterotopias, "Of Other Spaces," originally given as a lecture in 1967, Michel Foucault makes the following claim:

> The present epoch will perhaps be above all the epoch of space. We are in the epoch of simultaneity: we are in the epoch of juxtaposition, the

epoch of the near and far, of the side-by-side, of the dispersed. We are at a moment, I believe, when our experience of the world is less that of a long life developing through time than that of a network that connects points and intersects with its own skein.

L'époque actuelle serait peut-être plutôt l'époque de l'espace. Nous sommes à l'époque du simultané, nous sommes à l'époque de la juxtaposition, à l'époque du proche et du lointain, du côte à côte, du dispersé. Nous sommes à un moment où le monde s'éprouve, je crois, moins comme une grande vie qui se développerait à travers le temps que comme un réseau qui relie des points et qui entrecroise son écheveau.

"Of Other Spaces" has had a large effect on postmodernist, urban, literary, and other scholarship. It has also influenced medieval studies despite the fact that Foucault draws historical distinctions between the "epoch" of the present that he describes and earlier times in history. "Of Other Spaces," in fact, starts with a schematic genealogy that draws distinctions between medieval ideas about space and Galileo's innovations in the late-sixteenth and seventeenth centuries, both of which are further distinguishable, Foucault sketches out, from twentieth-century ideas about space. While he notes some continuities between the medieval era and today—namely, traces of sanctified spaces—his argument is that proximity is a twentieth-century problem, especially "knowing what relations of propinquity, what type of storage, circulation, marking, and classification of human elements should be adopted in a given situation in order to achieve a given end" (*de savoir quelles relations de voisinage, quel type de stockage, de circulation, de repérage, de classement des éléments humains doivent être retenus de préférence dans telle ou telle situation pour venir à telle ou telle fin.*).[3]

The following discussion proposes that Foucault's observations pertain to the late medieval era at least as much as the modern one because proximity was one of the important organizing features of medieval spaces. Medieval science and literature, rather than being principally teleological (whether that end has to do with humans or is less sociological and more directly spatially focused), concentrate on intensifications of proximate relations that are directionally varied. These intense proximities—motile, internally graduated, and affective—do not determine a situation or character so much as express dynamic possibilities. The features of proximity that survive from the Middle Ages into the early modern period or indeed today are thus larger in number and scale than Foucault suggests.

Just as the previous chapter examined the late medieval mechanics of propinquity in light of theories about affect, so does this chapter in part consider the topic of affection. It does so not to collapse historical distinctions between medieval and modern ideas about feelings but instead to notice that where Brian Massumi and others have suggested "an intrinsic connection between movement and sensation," fourteenth-century natural science and literature delve thoroughly into that "connection."[4] Modern theorists of affect coincidentally draw attention to ideas that Oresme and others had in fact explored in depth some 650 years earlier, namely that an extensive change such as motion implicates an "intensive," or qualitative, modification.[5] Fourteenth-century culture had available a mechanical theory of emotions in terms of intensities that went well beyond their relationships within the "intellectual soul" or among "preconscious, non-discursive, non-narrated" feelings.[6] Given sufficient proximity, affective phenomena of all kinds could interact with various emotions within a person, and they could intentionally or unintentionally bring about change in other people. Intensities could also interact among nonhuman animals, indeed, among all objects in the world, including physical things, thoughts, and imagined entities. Nothing was inert when it came to intensities. Attention to affective configurations, with the kind of complexity that Oresme's and other scholastic philosophers' theories of the world demands, is an important element in a study of propinquity.

Brief examples can serve to give an initial sense of the diverse kinds of nearness in Middle English literature and to indicate the complexities involved when two (or more) entities come together. Romance is a genre that obviously lends itself to the topic. The bedroom scenes in *Sir Gawain and the Green Knight* and *Troilus and Criseyde* may be said to generate their tensions through the exquisite calibrations of social and physical distance, the two often coming into conflict with each other. Other moments in these two stories also depend in large part on the proximity of people to other people, people to locations, and people to objects. In *Sir Gawain*, one could consider the Green Knight's entry on horseback into the crowded hall to stand before Arthur and Guinevere, the head rolling around under the feet of Camelot's courtiers, the reciprocity of physical intimacy between Gawain and the host at the castle, the odd "besideness" of the threatening chapel near that castle, and the ax blade's nearness to, and touch ("snyrt") on, Gawain's neck.[7] In Chaucer's historical tragedy, Troilus is drawn closer and closer to Criseyde over the course of the narrative, and Criseyde herself is in a very complicated position, caught among the forces of his approach (and later Diomede's), the propinquity of the Greek army, and her genealogy. These scenes and narratives are at least partially explainable without an appeal to natural philosophy, but in some cases scientific ideas offer a strong parallel in terms of the complexities of propinquity's effects on characters and situations.

It can be noted, at least, that natural philosophers on the subject of proximity would caution against placing too much emphasis on moments in which characters appear to affect each other across large distances. Such moments in literature tend to involve prayer, mystical experiences, miracles, dreams, and so on, and thus are arguably special cases of cause-effect relationships. The poignancy of romances and other genres associated with *fin amor* in which lovers are separated conversely arises from the fact that lovers cannot communicate with each other over a great distance. They may appear to influence each other through their thoughts, but they tend more so to affect themselves—for instance, one aspect of their imagination becoming dominant at the expense of one's own bodily health. For example, in the *Book of the Duchess*, Alcione refuses to eat because of the suspicion that Ceyx is lost at sea, and the narrator of the *Book* similarly worries about her character. She is stuck, as it were, until she can see or learn of her husband. She is released from her grief only when Juno summons Morpheus, who sends Ceyx's "dreynte body" to the foot of her bed and promises that she can also see it afterward in person on the shore while she is awake.[8] Physical proximities alleviate her sorrow. In other instances, sometimes a character thinking of another can coincide, often ironically, with what the object of thought is doing, but these moments are examples of coincidence rather than cause-effect situations. The much more typical relationship of one person or object affecting another involves physical closeness.

Moments when physical nearness is important for one character to affect another may seem too ubiquitous in literature to have significant meaning, but some writings go into extensive detail and signal direct engagement with the subject of propinquity itself. Consider again *Troilus and Criseyde* as in part a poem that registers parallels between the hero and the heroine in terms of their qualities. First, the famous opening scene in *Troilus and Criseyde* when Troilus falls in love in the temple depends on the chance propinquity of Criseyde. Troilus is happy to scorn lovers until he sights Criseyde, and when he does, the first idea he raises is a question of location. He asks "wher hastow woned?"[9] *Woned* is the past participle of the Old English verb *wunian*, meaning "to dwell or live habitually."[10] Troilus asks where she has been, so the place reference is to a certain extent metaphorical, as in the expression "Where have you been all my life?" But the topic of locational relations continues when Troilus thereafter admires Criseyde's appearance and movements in ways that suggest she is proportional to herself, to her gender, and to her estate:

> . . . alle hire lymes so wel answerynge
> Weren to wommanhod, that creature

Was nevere lasse mannyssh in semynge;
And ek the pure wise of hire mevynge
Shewed wel that men myght in hire gesse
Honour, estat, and wommanly noblesse.[11]

As the scene continues, the intricate mechanics of sight confirm the suitability of the two lovers-to-be because their qualities correspond in a physical manner. Most oddly and complexly, Chaucer's narrator describes a moment in which Criseyde's suitability transfers to Troilus, and Troilus's own apperception comes to match her appearance. As Sarah Stanbury says succinctly, Chaucer makes use of "the elision of the transitive with the intransitive" in this scene.[12] Troilus's eyes see through the crowd in the temple until they land on Criseyde and stay there, the motion coming to a stop: "ther it stente."[13] He admires her physical form, and then he notices her own looking about, which, we are informed, was "somdel deignous" (somewhat disdainful) because she appears to be challenging those around her in the temple who might question her right to be there. Troilus continues to observe Criseyde as she

                      let falle
Hire look a lite aside in swich manere,
Ascaunces [As if to say], "What, may I nat stonden here?"
And after that hir lokynge gan she lighte, [she began to brighten her
    manner of looking]
That nevere thoughte hym seen so good a syghte.

And of hire look in him ther gan to quyken
So gret desir and such affeccioun,
That in his herte botme gan to stiken
Of hir his fixe and depe impressioun.[14]

The scene appears to have analogues with passages, if not sources, in Aristotle and elsewhere on medieval optics as Stanbury and other critics have noted, but Oresme also singles out the eyes in his theory of how intensities can affect each other. Eyes, he says, are most effective in communicating affects between people, and between objects and people, "For eyes, being of rare complexion, are like certain holes in the head, and are joined or connected to the organs of the interior senses. Therefore, they are notably altered as the result of accidents of the soul; and most of all the signs of thoughts and affections appear in them [the eyes]" (oculi namque sunt quasi quedam foramina capitis et sunt complexionis rare et coniuncti sive connexi organis interiorum sensuum; ideo notabilius immutantur propter accidentia anime et in eis apparent maxime signa cogitationum vel affectionum).[15] In the scene Criseyde changes her way of looking about her by "lightening" it, and Troilus thereafter likes her own way of looking so much that where before his gaze was stuck on her, now her "looks" awaken ("quyken") his "desir and affeccioun." Readers may be forgiven for mistaking Troilus's admiration of Criseyde as a kind of objectification when they read that he likes "hire look," but what he is really noticing and what affects him is her own manner of looking. Chaucer makes this clear with the syntactical symmetry of the phrases "of hire look in him ther gan to quyken" and "That in his herte botme gan to stiken / Of hir." Troilus sees her seeing; he looks at her looking. And he likes the way she looks about her so much that where before his look stuck in her, now her looks "stiken" in the bottom of his heart.

The amity between them, beginning as it does with a sharply focused symmetry between Troilus's and Criseyde's attention, results in a "latent," or hidden, effect, just as Oresme theorizes.[16] Troilus, we learn following the transfer of looks, at this stage is "unwar that Love hadde his dwellynge / Withinne the subtile [furtive] stremes of hire yen." We are told "That sodeynly hym thoughte he felte dyen, / Right with hire look, the spirit in his herte."[17] Again, Chaucer plays with the ambiguity of "Right with hire look" between the way she is seen and the way she looks about her, but the passage again tilts toward the latter. Criseyde's powers of observation and self-awareness in the temple are what stir Troilus's desire. Chaucer employs ideas about proximity and proportional affect to capture the complex dynamics of the situation in a manner that Oresme would have recognized. Critics have long understood the imagery of Troilus's eye piercing through the crowd and stopping on Criseyde, but contemporary physics would notice the reciprocity of the mechanics of propinquity in the scene. The space between the two characters becomes the crucial efficient cause rather than there being an end point as in a *locus naturalis* and the looks being directed to only one place. It is a concentrated narrative scene in which emotional configurations are altered by the configurations of a proximate object without the awareness of the viewer.

The dark mirror image of this moment in *Troilus and Criseyde* occurs in Robert Henryson's continuation of Chaucer's poem, *The Testament of Cresseid*. The *Testament* evokes the same instant and similar dynamics in its climax, and it explores in still greater depth the terrible ironies of the proximity between the two characters. In the *Testament*, Criseyde, reduced to poverty and disfigured beyond conscious recognition by leprosy, is begging with her fellow lepers on the side of a road, and Troilus is traveling back to Troy after a triumphant defeat of the Greeks that day. Troilus hears the lepers, and "neirby the place can pas / Quhair Cresseid sat, not witting quhat scho was [not knowing what she was]."[18] He has been thinking of her, and she of him, in the intervening time since they separated, but it takes physical proximity for the *Testament* to reach its climax even though he never knowingly recognizes her.

The imprinting process of memory and of optics plays a role in Henryson's scene, as in the temple scene in *Troilus and Criseyde*, but optical laws and romance conventions do not adequately explain the actions and interactions. The moment in the *Testament* relies on the physical nearness not only of the two former lovers but also on the close propinquity of the physical Criseyde with the image of Criseyde in Troilus's memory, which the hero struggles to register. Without his conscious recognition, he nevertheless comes to juxtapose and compare two Criseydes: one the remembered image of her, and the other the image of her in the form of the woman before him. In the scene, Criseyde first looks at Troilus, and he reacts when he sees her "with ane blenk," thinking he has seen the woman beside him somewhere before. He cannot quite place her, but seeing her reminds him, ironically of course, of Criseyde and her own "amorous blenking" in the past; and Henryson's narrator goes on to explain how when an image is imprinted deeply enough in memory, it can reappear.[19] The insight echoes not only medieval mnemonics but also proximity because Troilus, while not recognizing Criseyde, projects a Criseyde from his memory onto the woman before him. The remembered impression of Criseyde "deludis" Troilus's "wittis outwardly" (his outward senses), and the force of the disturbing impression physically superimposes the remembered Criseyde from the past onto the leper in the present. Henryson's narrator comments that

> Na wonder was, suppois in mynd that he
> Tuik hir figure sa sone, and lo now quhy:
> The idole of ane thing in cace may be [The image of a thing in a
>   certain case may be]
> Sa deip imprentit in the fantasy
> That it deludis the wittis outwardly
> And sa appeiris in forme and lyke estait
> Within the mynd as it was figurait.[20]

Troilus is affected physically as well as mentally. He becomes feverish, sweating and trembling, until out of pity and in memory of Criseyde he casts down jewels for the lepers. In the reciprocating scene, the discussion in the *Testament* moves from the physical presence of Criseyde on the ground beside Troilus on horseback to the psychological realm of one image from the past repeating and reappearing before Troilus in his "mynd," then back to the physical realm, with a fluidity that reminds readers of Chaucer's own ability to move so easily among these worlds in *Troilus and Criseyde*. The simple point is that none of it can occur without physical proximity, whether of the actual person or of her image, but the scene also more complexly suggests a certain intensification of proximate significance that affects Troilus's configurations. The physical and the mental-imagistic, and the present and the past, affect each other given the odd echoes that proximity causes.

Nearness in the Prologue and the Legend of Cleopatra

Chaucer's *Legend of Good Women* brings men and women, people and objects, and things and other entities together in ways that Chaucer found in his sources, that are at least partly explainable in terms of romance conventions of secretive and illicit lovers, and that also may be accounted for simply in terms of the necessities of the narrative. It is telling, nevertheless, to look a little more closely at the poem in light of the subject of nearness, the significance of which is first signaled in the Prologue through descriptions of the narrator's physical actions. These moments in the Prologue are oddly circular and repetitive before they to a certain extent resolve themselves through pressure coming to bear on the narrator's physical position, making it tempting to read the narrator's movements in the Prologue as foreshadowing the whole of the *Legend*'s unsatisfying repetitions and inconclusiveness, as critics used to describe the *Legend*.[21] But the following analysis of the poem does not read it as carelessly structured. Its sustained attention to proximity instead provides another key to unlocking the *Legend*'s interests. In the poem, physical nearness initially generates ambiguity. This uncertainty thereafter takes a dark turn with the Legend of Thisbe, a story about the propinquity of lovers to which Chaucer adds his own elements. Finally, the proximities in the narrative become darker and darker as the poem progresses.

The Legend of Good Women begins as a return "to bokes," "to the doctrine of . . . olde," and to "olde appreved stories," all set against the inability to know with certainty about heaven and hell, and also against an incredulity on the part of those who believe only the things they can see with their own eyes.[22] The poem sets up a competition, as it were, between returning to *auctoritee* and personal empiricism of a particular kind. The *Legend* as a text also has a kind of recursive skepticism because of Chaucer returning to the *Legend* to revise his Prologue. The narrator's own actions in terms of physical space and movement at the beginning of the poem are similarly iterative, even comically so in ways that seem to echo the *House of Fame*, *Troilus and Criseyde*, and the *Canterbury Tales*, in which Chaucer casts himself as inadequate to his task. The action in the *Legend* starts as apparently innocent when the dreamer describes himself as getting up every May dawn to observe a daisy's flowering, but the cyclical nature of the aubade soon becomes stronger when he offers an image of himself, upon nightfall and the closing of the daisy, hurrying back to his house in order to get to bed early so that he can go out again in the morning. But instead of returning home as he says he does, he in fact returns to an arbor in his garden next to his house, and there he sleeps and dreams of going back out into nature and looking on the daisy once more.[23]

This somewhat obsessively looping motion would be understood in medieval mechanics as causing increasing intensity, and the action indeed becomes most concentrated when the dreamer is confronted by the God of Love and his queen Alceste, who is dressed as a bejeweled daisy, because the narrator is out of place in their meadow, a kind of natural space often strongly associated with aristocratic owners.[24] Love looks on the dreamer "sternely . . . / So that his loking dooth myn herte colde" before the company of ladies sings a ballad and more in honor of the lady, but it is not long before the dreamer again becomes the object of Love's stern attention.[25] He has been observing the procession of ladies and the seating of the god and his lady all by "estaat."[26] Silence ensues, and the narrator-dreamer remains still until the God of Love looks on him again and asks who he is. Where the G version has Love ask simply what the dreamer is doing there in his presence, the F version sharpens the focus on women and proximity when the god asks what he is doing there "so nygh myn oune floure, so boldely." Before the dreamer can reply, Love continues by warning "Yt were better worthy, trewely, / A worm to neghen ner my flour than thow."[27] The narrator's presence in the G manuscript has to do with rank since so much of the dress, procession, seating, and even the silence has been according to station. In the F version, the dreamer is lower than a worm in the god's estimation, and the issues are of rank plus gender in the spirit of warnings about physical associations between men and women in conduct literature as in "The Babees Book," "How the Good Wiff Tauȝte Hir Douȝtir," and *The Book of the Knight of La Tour-Landry*.[28] As the previous chapter pointed out, contemporary natural history on the science of mechanics, like texts on medieval sociality, also concentrated on moments in which one entity near another one has the ability to affect it, no matter what kind of entity, so it is tempting to read the interspecies threat from the God of Love to the dreamer in this light, but there is not a strong suggestion of science in this scene. The proximity is simply cast as too much in social senses of rank and gender, and the implications and results are colored by an undefined danger.

The remainder of the Prologue moves away from such direct challenges to the dreamer in part because the daisy—Queen Alceste—intercedes, but not before the dreamer attempts to answer Love's question about why he is there with another question that prompts further explanation of the god's anger and weariness at the narrator for being an enemy of love in his writings.[29] The proximities are again not particularly philosophical, but suffice it to note at this stage that the narrator's problem is in part caused by his proximity to the God of Love and the daisy. The topic of closeness to a person or object, along with accompanying emotional reactions, then comes up again in the first of the legends, about Marc Anthony and Cleopatra, which the dreamer is set to tell in penance for his transgressions against the God and Goddess of Love. The physical and textual transgressions against love beget another recursive reaction in that they give rise to a tale that is also about propinquity.

In the Legenda Cleopatrie, love is unambiguously true for both Anthony and Cleopatra, as it will be for Pyramus and Thisbe, unlike other tales that become progressively more about men feigning to love women and then outright deceiving and violating them. Anthony kills himself first, and we are told that the lesson of the tale is that women can be as true as, if not truer than, men who say they will die if their ladies do not return love. Cleopatra makes an elaborate shrine for Anthony's body, and in her final speech tells of how she swore to him that she would not only be his but also feel exactly as he did. She says that "in myself this covenaunt made I tho [then]," referring to the moment she swore "frely" to be his. She proclaims:

> "That ryght swich as ye felten, wel or wo,
> As fer forth as it in my power lay,
> Unreprovable unto my wyfhod ay [always],
> The same wolde I fele, lyf or deth."

Her cry is that just as before when she promised to feel what he feels, now she will feel the same even though he is dead. She therefore casts herself into a "pit" of serpents.[30] Whether the moment is histrionic or not, ironic or not (given that the poem ends and she does not appear to meet him again), their feelings are intended to match. The topic of "wyfhod" is the crucial addition to the subject so that she is another representative of good women.

It is true that so far there is only suggestive evidence of the influence of contemporary scientific ideas about motion and proximity on Chaucer's *Legend of Good Women*. The discourse of the poem is high and courtly, and there is scarce indication of the systematic theory seen in a work such as Oresme's treatise on qualities and motions. The dream vision works according to its own logic of "curtesye" and appears not to require specific intensities and configurations.[31] Even the scientific influence of Boethian ideas about dreams, let alone his ideas about motion in the universe, is muted here. Yet physical proximity is an issue that sets the narrator on his path of recuperating his reputation for the God and Goddess of Love, and the Legend of Cleopatra depicts "fredom," noble generosity, as the exact match not just of rank but also of love and emotions: "ryght swich as ye felten . . . The same wolde I fele." Cleopatra and Anthony find a symmetry in their relationship that extends to the equivalence of their emotions whether alive or dead. Neither character in the opening legend, moreover, is dominant or moving, as it were, toward a final destination. Instead, although Anthony dies first, the two have sworn to be like each other, so Cleopatra's death meets and matches his.

True Departure in the Legend of Thisbe

Following the Legend of Cleopatra, Chaucer's Legenda Tesbe Babilonie squarely addresses proximal spaces and the dynamics of beings close to each other. Its central issue is not of a moving object heading toward one final terminus, an idea Chaucer evokes and problematizes in the discussion of sound in the *House of Fame* as well as in the rush of people at the end of that poem and elsewhere. Rather, the Legend of Thisbe is about the tensile distance between objects and characters in which entities move toward each other or are otherwise so near each other as to invoke strain. As in the physics, the legend's central issue concerns the measure between objects, between objects and things, between things in themselves, and between locations. It examines the qualities of the dividing wall and what a distinction between two proximate entities can enable. It comes in the end of the poem to equate a wall with death in a way that is quite fitting given the poem's previous legend, of Cleopatra, where she tried to make her feelings and her own demise match that of her lover's own death.

The critical reception of the *Legend of Good Women* has moved away from readings of the collection of stories as consistently ironic and instead perceives ambiguities and ambivalences. The special issue of the *Chaucer Review* on the poem, *Looking Forward, Looking Back on the Legend of Good Women*, explicitly rejects irony in its approaches to the poem that include several affective readings.[32] The Legend of Thisbe is most commonly read for its complexity, critics emphasizing Thisbe's agency in the poem in particular and also perceiving the intricacies of the symmetries in the story. Sheila Delany describes the Legend of Thisbe as "subversive of sentiment" and notes in particular the equality of the two protagonists. Kara Doyle claims that Chaucer depicts the relationship and actions of the two lovers in such a way that "egalitarianism operates without irony."[33] Steven Kruger's "Passion and Order in Chaucer's *Legend of Good Women*" remains one of the most discerning readings of the Thisbe story in that it sees a structural tension in the narrative that is emblematic of the *Legend* as a whole. The Legend of Thisbe explores the "conflict between passionate emotion and social structure" in which the two forces are "repeatedly opposed to each other . . . but they do not become the two perfectly antithetical terms of a perfectly symmetrical opposition. . . . Structures may oppose passion, but they cannot do so perfectly. At times, obstruction itself provides an opening for love. At times, attempts to control emotions only make them grow in strength. . . . Hidden and inert within the structures by which a society orders experience are the seeds of disorder. The flaws in societal structure . . . provide passion with a means of expression."[34]

Chaucer's interest also lies in proximate relations that are as structural and complex as these conflicts. Propinquity is presented in the *Legend of Good Women* especially in terms of affective associations between people. In the *Chaucer Review* special issue, Glenn Burger's study of affect and marital relations in the poem describes how Chaucer "radically destabilizes the structures of feeling—classically expressed in terms of male love service to the lady and female pity for the lover—through which courtly love makes erotic desire legible as an ennobling emotion."[35] Lynn Shutters also argues that the *Legend* as a whole contains stories that "do not dictate or fix social forms so much as they facilitate contact between them, often to unexpected effect." Along with other literary works, the poem "allows us to detect the messiness of marital affection, to recover marital affection not as a stable, coherent emotion, but as an emotion in progress."[36] The relations between people, especially in marriage, are one of the poem's central concerns, but the poem also engages in some depth with relations between people and objects, and between objects themselves, an interest that similarly occupied the natural scientists.

As told in English by Chaucer, John Gower, and John Metham (in an adaptation based on Christine de Pizan), the Legend of Thisbe addresses the topic of proximity particularly in terms of space, distance coming into particularly sharp focus in Chaucer's rendering of the story of the tragic lovers. At the end of Gower's version in the *Confessio amantis*, Pyramus has killed himself in "folhaste," and Thisbe likewise impales herself with the same sword, and "thus bothe on o swerd bledende [bleeding] / Thei weren founde ded liggende [lying]."[37] Gower offers an ethical critique of the characters' hurried suicides, Allen Shoaf pointing out that the legend appears under the heading of *ira* in book 3 of the *Confessio*, and under a subject heading of "contek." Shoaf traces the etymology of *contek* as having possible origins in a sense of contact and touch so that the tale comes to be about "the extraordinary fact that Pyramus and Thisbe never actually make contact until Pyramus is in his death throes." The poem suggests the somewhat paradoxical idea that "'contek' prevents contact."[38] Indeed, the Latin gloss at the opening of the poem seems to make a point of connecting motion with contact in scientific language with its emphasis on *accelerantes ex impetuositate* when the gloss says, "Here the Confessor sets out an exemplum against those who, in the cause of love, many times stumble into their own destruction because of too much rushing due to impetuosity" (Hic in amoris causa ponit Confessor exemplum contra illos qui in sua dampna nimis accelerantes ex impetuositate seipsos multociens offendunt).[39]

Chaucer renders the tale somewhat differently. He announces the topic of proximity and barriers at the beginning of the Legend of Thisbe by describing how Queen Semiramus built Babylon, surrounding it with ditches and "walles make [made] / Ful hye, of hard tiles [bricks] wel ybake." The fortifications separate the city from the area around it, and Pyramus and Thisbe's error later in the poem will in part be to venture out from the town into the sundered and alien countryside. In the meantime, the next wall in the poem, a "ston-wal" that sits between two great lords of the city, separates the lovers. In Chaucer's addition to the tale, diverging from his sources, words can initially circumvent the divide in that descriptions of the two young people reach around the wall between the two houses due to intermediaries: "The name of everych [each one] gan to other sprynge / By women that were neighebores aboute."[40] The tale thereby introduces the issue of proximity via the additional topic of speech, with the alacrity of "sprynge" already suggesting the lovers' desire to know each other.

The Legend of Thisbe soon closes off the possibility of circumventing the wall or of otherwise crossing its division, and instead the focus of the tale narrows to the object between the two houses and the two lovers, the wall that paradoxically separates and joins them. The wall is an unusual object in itself in that it is also self-divided, separated, as it were, from itself. Unlike Gower's version of the Ovid where the lovers make a hole in the wall, the barrier in Chaucer's poem has a history that means it is already "clove a-two, ryght from the cop [top] adoun, / Of olde tyme of his fundacioun." The wall's age, therefore, echoes the opening of the Prologue with its laying out of a conflict between empirical observation and the learning from "bokes," "doctrine of . . . olde," and "olde appreved stories." It too is an object that occupies a divide between seeing things with one's own eyes and authority, the latter here presented by the fathers who forbid the two young people to meet but also by the wall itself. The wall has a role as divider and as an exponential causal factor in increasing the love between Pyramus and Thisbe. It is likened to fire that gets hotter if the embers are covered: "Forbede a love, and it is ten so wod [ten times as mad]."[41]

The cleft in the wall is the tightest it can be, Chaucer also differing from Christine de Pizan's and John Metham's later versions in which Thisbe puts her belt buckle through the wall or appears to throw a piece of glass over the wall to get Pyramus's attention, and where the lovers can see light through the crack or even pass gold rings through it. In Chaucer it is "so narw and lyte" that it cannot be discovered by anyone except the two.[42] The effect is to concentrate again on the affective conundrums that proximate relations cause and also to recall the subject of propinquity together with speech. The setup in Chaucer's version is one in which the hidden proximity between the two pieces of the wall enables only words to move between Pyramus and Thisbe, where before language had to reach the other by means of the neighbors around them. Their words are like confession when, "with a soun as softe as any shryfte, / They lete here wordes thourgh the clifte pace."[43] The simile is appropriate given that medieval confession was often in public and not in a private confessional box. The sound is, like in the *House of Fame*, also made physical in the way it "paces" through the wall's cleft.

The wall in the Legend of Thisbe offers Chaucer a way to consider proximity's role in making two entities. He appears interested in how one entity potentially becomes two, and two things might remerge into one via a necessarily quantitative distinction that is nevertheless due to qualitative causes. Just as some fourteenth-century science measured change as a quantity that could be represented by numbers and plot points on a graph even though it was admitted that many species of change were continuous, so too Chaucer meditates on what makes one entity divisible owing to qualitative characteristics. The wall is barely two parts, so close to itself that it is as though it were one entity; even though it is cracked, its cleft is so narrow that it exists as if it were not fissured. The lovers protest about it, addressing the wall as one thing and complaining that even though it is split, they want it to be even more divided: "Why nylt thow cleve or fallen al a-two?" And an alternative degree of proximity soon enters when they go on to praise the wall as though they were indebted to its oneness, complaining in the second-person singular that "yit be we to thee holde," because the wall "sufferest for to gon / Oure wordes thourgh thy lym and ek thy ston."[44] The two lovers will also come to be described in a similar fashion.

The Legend of Thisbe thus does not establish the grounds of the poem on the basis of simple separation that might be resolved by one usually male lover achieving his aim; rather, the problem is the proximal distance between people and between objects, especially ones that are ambiguously self-divided such as the wall. It is as though Chaucer were examining the Aristotelian idea of two things that join together becoming one or coming to occupy the same place. The poem centers on the difference between two things, threatening to collapse the distinction between them by having a wall that has two parts that are so close together as to almost be one, or a narrative in which the wall is desired to be far enough apart that the lovers can unite. In either case, Pyramus and Thisbe soon express hopes that the claustrophobic oscillations of distinctions and distances will find release in the broad spaces outside the walls of Babylon, but even there the potential opening out of the poem's spatial problems does not resolve the tensions but instead introduces the opposite issue of too large a distance. When the lovers plan to leave the city entirely in order to meet, they acknowledge that the "feldes ben so brode and wide" that they could get lost, so they have to agree to convene at one "mark," under a particular tree.[45]

They go to the meeting place, but Pyramus is too slow; the lion has bloodied Thisbe's token, and he commits suicide. Now death comes to stand metonymically for the wall. Thisbe's words remain focused on death as the joining factor with potentially as much ambiguity and complexity as the wall formerly embodied. Thisbe's final speech after discovering Pyramus's dead body is a compact series of phrases and lines that offers a study in the kinds of complexities that may be found in late medieval mechanics and that the legend has already established as part of its theme. Even the temporality seems to wobble at the end when she speaks. In a play on "folwe" and "felawe," and with compressed syntax, Thisbe indicates that she will cause Pyramus's death even though he has already expired, and she also claims that she has already been the cause of his death. She says to her lover's body before her that she "wol thee folwe ded, and I wol be / Felawe and cause ek of thy deth."[46]

Where the wall's lime and stone confuse desire, which is beholden to the wall's unity while also hoping to cleave it, Thisbe concludes her speech with another complex plaintive meditation on what is joined with, and separated by, what:

> "And thogh that nothing, save the deth only,
> Mighte thee fro me departe trewely,
> Thow shalt no more departe now fro me
> Than fro the deth, for I wol go with thee."[47]

Here at the end of the legend, a play on "thee" and "the deth" repeats the proximal conflations. Death is the only thing that can "trewely" part the couple, but Pyramus is said not to have any more ability to leave Thisbe than he is able to leave death, so Thisbe, Pyramus, and death can all join together. In Gower's version, the confessor draws a lesson from the tale that the lover should understand his heart's ability to wait, but it is Metham who makes the most surprising attempt to circumvent any of these proximal difficulties in the conclusion of the Legend of Thisbe by having a holy hermit revive the dead lovers, who convert to Christianity and begin to spread faith throughout the city.[48] Chaucer, on the other hand, remains intent on concentrating the force of the tale on the puzzle of distinguishing between entities.

The topic of the last speech in the Legend of Thisbe comes to be actual or genuine separation, to "departe trewely," as Thisbe says, and Chaucer in the narrative continues to explore what might actually be able to uncouple the lovers. On one side is the force of love that brings the two together, and on the other are the fathers who did not let them join. The item that now stands in the place between them is the grave, just as it was in the story of Cleopatra. Thisbe continues her complaint:

> "And now, ye wrechede jelos fadres oure,
> We that whilom [formerly] were children youre,
> We preyen yow, withouten more envye,
> That in o grave yfere we moten lye,
> Sith love hath brought us to this pitous ende."[49]

"Yfere" (also "in-feere") is an adverb meaning "together" and is derived from the Old English noun *geféra/gefér*, meaning "companion." In his writings, Chaucer uses it principally to describe lovers and other people coming together and also qualities that people admire "all together" in each other.[50] Thisbe hopes that they might finally unite by coming together in death.

Mutually proceeding toward each other is a subject that is related to the larger theme of "trouthe" in the *Legend of Good Women*, a word that appears in some form approximately sixty times in the poem. Chaucer concludes the Legend of Thisbe with a lesson that describes Pyramus's suicide as that of a "trewe and kynde" man.[51] He also depicts Thisbe as equal to Pyramus in daring and ability as though the legend's attempted uniformities can resolve the inequality in his narrator's treatment of men and women, the problem that he is supposed to address by composing the *Legend of Good Women*. But the death duplicates the wall in a way that suggests that the poem does not plot out such a clear trajectory toward truth. The proximities of these objects are too intense, and, instead of enabling couples and themes to coincide, they confound the lovers in that they establish, frustrate, and part their spatial desire.

Remitting Propinquity in the *Legend of Good Women*

Fourteenth-century ideas about mechanics that reduced the significance of a final end point in motion and instead drew attention to all locations in an object's period or distance of motion offer an interpretive lens through which to see Middle English literature differently. It is possible, for instance, to consider moments in which characters' emotions intensify as the distance decreases, such as in the significant moment of the Knight's closing appeal to the Pardoner to "drawe thee neer" to the Host, a request that suggests that propinquity can be effective as a cause and a sign of attempted reconciliation.[52] Indeed, the *Canterbury Tales* as a whole concerns proximal distances between pilgrims and characters more than it does the end point of the journey at Canterbury. Even John Lydgate's *Siege of Thebes* barely registers the Canterbury it has left and the Southwark to which the pilgrims are returning, turning attention instead to the relationships among the pilgrims and the spaces they occupy. In the *Legend of Good Women*, Chaucer employs ideas about proximity and proportional affect to capture the complex dynamics of situations in a manner that Oresme would have recognized. In fact, a pattern becomes discernible in the poem when one consider proximity. The legends devolve from themes of mutuality between men and women, as in the stories of Anthony and Cleopatra, and Pyramus and Thisbe; through narratives of inequality in which men dissemble and women are the unfortunate recipients of their misfortune, as in Dido, Hypsipyle, and Medea; to the rapes of Lucrece and Philomela, and the genealogy of infidelity in the Legend of Hypermnestra. This is neither a consistent decline nor a constant increase in the proximate violence of men against women, but a pattern is discernible. Matters have already taken a dark turn in the stories of Cleopatra and Thisbe, of course, if not in the Prologue itself, but at least a remnant of the fraught mutualities that we saw there continues to a lesser extent in the following legends of Dido and of Hypsipyle and Medea. For instance, audiences are warned at the beginning of the Legenda Didonis martiris that Aeneas will betray Dido, but the two meet up together because of something beyond both their control, namely a sudden storm that frightens everyone when they are out hunting. A cave is the site in which Dido and Aeneas "began the depe affeccioun / Betwixe hem two."[53] In the Legenda Ysiphile et Medee, Jason marries Hypsipyle, albeit under false pretenses, and after Medea tells him how he can gain the golden fleece, they meet together in a manner that is again characterized by being "in-feere," together.[54]

These moments of mutuality find more extended expression in the Legend of Ariadne. Where the Legend of Thisbe involved a wall and death, the narrative of Theseus, Phaedra, and Ariadne also involves walls but in the end comes to center on a bed and an island as meeting—and sundering—objects and locations. It is a more complicated story than that of Pyramus and Thisbe because it involves more characters and locations, but, because it tries to compress the characters, events, and locations, its effect is paradoxically more thematically diffuse. Walls are introduced early in the legend when Scylla, the daughter of King Nisus of Megara ("Alcathoe" in the legend), betrays her city to Minos, king of Crete, after she falls in love with Minos one day when she spies him while she "stod upon the wal, / And of the sege saw the maner al."[55] Minos is seeking revenge for his son's death in nearby Athens, and the fall of Megara means the people of Athens must give a child every year to be savagely consumed by the Minotaur. The wall thus is the site of the beginning of the terror until the noble son of Aegeus, Theseus, is called on as sacrifice. He is initially imprisoned in the base of a tower, and here another wall enters the story because the side of his jail is also the wall of Phaedra and Ariadne's house. It is now their turn to stand on a third wall, and there one night they "herden al" Theseus's "compleynynge as they stode on the wal / And lokeden upon the bryghte mone."[56] They take pity on him and agree to help him escape. Theseus promises, it is not clear whether to Phaedra or Ariadne, never to leave them but always to remain as their servant and to forsake his own home. He will of course betray this promise and more after he expresses a love he claims he has always had for Ariadne alone.

The movements and locations in the Legenda Adriane de Athenes are complicated in that the story involves leaving and returning to Athens. Theseus will go back to Athens with Phaedra, but the story does not end in that one destination. Indeed, the legend is more about a geographical midpoint, the place between their two homelands: the island where Ariadne dies, which is described as "an yle amyd the wilde se."[57] Like the wall in the Legend of Thisbe, the island, and Ariadne and Theseus's shared bed there, are the objects of complaint whose qualities come to be particularly cruel for the abandoned heroine. On waking to find that Theseus is gone, she searches in her bed, runs to the beach, and sees his vessel sailing away. Even though she tries to signal him with her headscarf and cries out to him to turn back, she is left alone on the island. She returns to their place, kissing the steps where he walked away from her as she returns, then addresses their bed in ways that echo Pyramus's and Thisbe's complaints to the wall. The bed is numerically singular but also in some ways dual: "Thow bed . . . that hast receyved two, / Thow shalt answere of two, and nat of oon!" As with that earlier legend, her complaint is futile, and she immediately also realizes that she is caught between Athens, where she cannot go, and a home to which she cannot return should another boat come, because she and Theseus stole away from her father. The island is like the bed in that it is between her former life and her future one. Her closing question is about her location, "Where shal I . . . become?"[58]

The last in the *Legend of Good Women*, of Hypermnestra and Lynceus, is one of the darkest, historically pressured as it is with a lineage of men's infidelity leading up to the moment of Hypermnestra's father asking her to cut her new husband's throat on their wedding night because he has had dreams that Lynceus will murder him. The naturalness and unnaturalness of human behavior is taken up as a topic again, and the subject of relative rates of motion is also explicitly addressed. Hypermnestra debates within herself whether to kill her husband, considers her nature and appearance as a "mayde," and goes against her father's command, thus breaking the cycle of betrayal, at least for a moment.[59] But she is abandoned by Lynceus because he is a swifter runner than she, and the audience gets a glimpse of the impending moment when her father will ominously "hente," or capture and imprison her:

> This Lyno swift was, and lyght of fote,
> And from his wif he ran a ful good pas [at a fast pace].
> This sely [wretched] woman is so weik—Allas!—
> And helples, so that or [ere] that she fer wente,
> Hire crewel fader dide hire for to hente.

Lynceus leaves her behind, and the narrator complains that the "unkynde" Lynceus has abandoned her.[60] The closing image of the Legenda Ypermystre is of the heroine sitting down until she is captured and put in prison, and the audience is told plainly and abruptly at the end of the whole *Legend of Good Women* that the "tale is seyd for this conclusioun."[61] While there has been debate about whether the *Legend* is finished, it seems fitting that it ends with an image of stasis. It is not as though Chaucer has, as critics have posited, become "bored" with a project that he returned to at least once, nor is the tone of this ending particularly ironic.[62] The end is a point where the tale posits, albeit briefly, a moment of stillness that the *Legend of Good Women* has so far denied, one in which Hypermnestra's attributes sink to stillness under the weight of untrue men. This is not to say that truth is locatable in the unmoving bodies of women, but instead to suggest the relative and proximate motions of qualities raise the question that Ariadne asks. As the woman with a thread that has come to stand also for narrative progress, she inquires "Where shal I . . . become?"

Chaucer's *Legend of Good Women*, particularly the Legend of Thisbe but other narratives as well, engages with the same topic that fascinated late medieval scholastic science, namely the proximate relationships between entities.

Chaucer's retelling of these famous stories obviously centers on affective interactions between characters, but it also includes propinquity between other things: arbors and home, daisies and dreamers, loves, words and walls, cities and islands, beds and past lives. A major interest of the poem lies in the ways that people and things might move toward each other, almost to merge, before being parted only to return to potential oneness. The *Legend* is structured such that proximal relations move farther and farther away from balanced relationships between men and women to situations in which the men exploit distance for their own ends, abandoning the women who then turn to questions of what will happen to them, what will come to them, and where they will go.

Afterword

Ubiquitous Being in the Pardoner's Prologue and Tale

Geoffrey Chaucer's Pardoner pauses by a tavern's sign to drink and eat. He describes how he begins each sermon by telling where he is from and that he preaches in many churches. He says that when he gives his lesson, he stretches his neck out and wags it toward the east and west. He admits to the Canterbury pilgrims that he will go anywhere to earn a living and will have a mistress in every town. Gluttony, he expounds, has corrupted all the world, making people work in the East, West, North, and South to satisfy greed. He describes drunks who have consumed adulterated wine from Lepe and do not know whether they are in Spain or at home in Cheapside. The tale that he tells about three degenerates who frequent stews and taverns is set in Flanders. Death is understood to live in a town just over a mile away. At a stile in a fence, the three rioters meet an old man, who describes his journey through cities and villages all the way to India in the farthest East. In the Pardoner's Prologue and Tale, individual and local places (a particular town or fence stile) and large spaces and peripatetic characters (from England to the cardinal points of the globe) are massed together. In bringing spaces that are opposite in scale together, Chaucer explores human ubiquity in such a way that he critiques a limited understanding of space and offers instead ubiquity as a model for living. Ubiquity, or being everywhere, however, appears to contradict fundamental tenets of medieval natural science and modern thinking about space and its relation to being. From ancient times to modern ones, philosophy has argued for a close tie between place and identity. Chaucer's Pardoner's Prologue and Tale present a challenge to those ideas because ubiquity suggests no attachment to a place.

This book has explored the fundamental and common apprehensions of space in science, mechanical arts, and literature of the late Middle Ages, a dynamic period of change in essential notions of place, motion, proximity, and more. The aim of this study has been to proceed deliberately to discern how the culture of the time understood space, first in scientific writings and images on space, and then to see how literature takes up the philosophical ideas and complicates them in its narratives. One guide throughout has been to avoid as much as possible deducing the historical characteristics of space from unrepresentative or specific kinds of space, such as protonationalist discourses, sacred sites, cities, entryways, and so on. The intention instead has been to look at the most direct evidence about how people saw the spaces around them. I have tried to maintain a tight focus on space itself by beginning with an examination of it as local and horizontal, and only then moving out to other key aspects of areal extent, namely motion and proximity.[1]

In this afterword, I offer a brief excursus into the polar opposite of that tight focus on the inherent characteristics of space—namely, ubiquity (being everywhere)—by analyzing Chaucer's Pardoner's Prologue and Tale. The topic of ubiquity leads back to philosophical ideas about the relationship between place and being. Aristotle asks rhetorically in his *Physics*: Where are the unreal creatures, the hybrid goat-stag and the mythological sphinx? In the thirteenth century, when Aquinas examines the *Physics* in his *Expositio in VIII libros Physicorum Aristotelis*, he suggests that the goat-stag and the sphinx do not exist anywhere because they are *fictitia*. Aristotle's argument is that imagined creatures are not anywhere because they do not exist, and Aquinas's point is the same except he adds the suggestion that fictional things also do not exist and are therefore nowhere.[2] In both philosophers' theories, existence and location in space are bound up with each other. To be is to be somewhere.

Heidegger, Merleau-Ponty, Jeff Malpas, and others return to Aristotle's and Aquinas's perceptions to reiterate that to be is to be in a specific place, that individual location and identity come together. In "Building Dwelling Thinking," Heidegger contends that "to say that mortals *are* is to say that in *dwelling* they persist through spaces by virtue of their stay among things and locales."[3] Merleau-Ponty in the *Phenomenology of Perception* reasons that "every conceivable being is related either directly or indirectly to the perceived world, and since the perceived world is grasped only in terms of direction, we cannot dissociate being from orientated being. . . . We cast anchor in some 'setting' which is offered to us."[4] Malpas affirms that "the very possibility of the appearance of things—of objects, of self, and of others—is possible only within the all-embracing compass of place. It is, indeed, in and through place that the world presents itself."[5] The Pardoner's Prologue and Tale strain against these senses of place and identity because they seem to posit that ubiquity, in which a person does not occupy any specific location, can enable being. So can one indeed be ubiquitous, be everywhere? If so, how is that being different from a singular emplacedness?

First, what if one starts to address the topic of place's relation to being from the point of view of ubiquity instead of the emplacedness of Aristotle and Aquinas? If ubiquity is being everywhere, what is the opposite of ubiquity? The answer is not being nowhere but instead haecceity, uniquity, locicity, let us say *ubity*, or whereness. But ubiquity is really a special condition of space whose opposite only appears to be ubity. Ubiquity is not detachment but more properly attachment to many and dispersed places. It is *everywhereness*. Ubity and ubiquity are in fact two sides of the same coin that both involve emplacedness. It remains to be seen at what points the whereness of ubity and everywhereness coincide, or whether ubiquity is a special case of ubity, or if there is still a strong sense of difference between being everywhere and being one place in particular. It also remains to be discovered how a being, such as a human, can be a ubiquitarian or ubiquitist.[6]

Philosophy has generally considered ubiquity as a special restricted case, but history reveals broader ideas about everywhereness. Ubiquity appears in Aristotle's *Physics*, but it is only time and dimensions (or directions) that are "equally everywhere and with everything."[7] The philosopher of science Max Jammer reviews the idea of God's ubiquity in the Talmudic tradition.[8] Aquinas interprets Aristotle to say that God is in all things everywhere in the sense that he is true being and therefore the agent of being in all.[9] Nevertheless, there are other historical senses of ubiquity that are distinguishable from Talmudic theology, Aristotelian physics, and Thomistic philosophy in which the idea of ubiquity applies to entities other than time, dimensions, and the divine. The word *ubiquity* does not appear in English until the sixteenth century, but senses of everywhereness appear quite frequently in Middle English. Chaucer uses the word *everywhere* to refer to every place in the most general sense and to all sites across a region and throughout a city. For example, he employs it in reference to the presence of pagans in Britain in "that contree everywhere" in the Man of Law's Tale, when Griselda must "worshipe" Walter's new bride "In word and werk, bothe heere and everywheere" in the Clerk's Tale, as a paraphrase of Ephesians 4.5–6 about the Divine "Aboven alle and over alle everywhere" in the Second Nun's Tale and to the news that prisoners will be exchanged at the beginning of book 4 of *Troilus and Criseyde*, "anon was couth in every strete, / Bothe in th'assege, in town, and everywhere."[10] Other Middle English words commonly connote everywhereness, such as the adverb *thikke*, which refers to people, including their vices, appearing profusely across an area.[11]

Although ubiquity is related in some ways to migrancy and exile, globalism and cosmopolitanism, it is nevertheless also distinguishable from them. In the lineage of Hugh of St. Victor, Erich Auerbach, and Edward Said, exile emphasizes the role of home. I am referring to Hugh's well-known saying, "The tender beginner is he whose homeland is sweet; he is already strong for whom every place is like his native one; but he is truly perfect for whom all the world is a place of exile" (Delicatus ille est adhuc cui patria dulcis est; fortis autem iam, cui omne solum patria est; perfectus vero, cui mundus totus exsilium est. ille mundo amorem fixit, iste sparsit, hic exstinxit).[12] Maire Said and Edward Said translated Auerbach's "Philology and *Weltliteratur*," and Auerbach in turn drew inspiration from Hugh.[13] Said writes that Hugh "makes it clear that the 'strong' or 'perfect' man achieves independence and detachment by working through attachments, not by rejecting them. Exile is predicated on the existence of, love for, and bond with, one's native place; what is true of all exile is not that home and love of home are lost, but that loss is inherent in the very existence of both."[14] However, in contrast, ubiquity does not especially concern itself with home or a home in a significant manner, nor does ubiquity necessarily entail the Freudian sense of "working through" loss that Said expresses. In addition, although critics have differentiated between globalism and cosmopolitanism, ubiquity is distinct from both.[15] It does not necessarily involve the global transmission of capital, diseases, or information, nor is it bound up with distantial or scalar jumping that cosmopolitanism often implies (again, see Hugh of St. Victor). I might add in the current context that ubiquity is also distinct from a form of cosmopolitanism identified with city refuges, a feature of cosmopolitanism that Jacques Derrida identified in his late work.[16]

Chaucer offers the Pardoner's Prologue and Tale as a case study in ubiquity in relation to being by putting constricted senses of place side by side with ubiquitous existence. A number of other Canterbury tales also address individual places and large areas in the same narrative. The Man of Law's Tale has remarkable changes in scale from the relatively local sense of Rome, a city in Syria, and King Alla's realm in the north of England to the expansive sense of "lond and see / Bothe north and south, and also west and est" where Constance spends years afloat.[17] But the tale is episodic and structured, and the land spaces are internally differentiated. Patricia Ingham is astute in observing of the tale that its "view of the differences within the 'West'—after we have already witnessed difference within Syria—troubles any stable opposition of 'East' to 'West'" as generalizable entities.[18] In ways that are very different from the exposed senses of separation in the Man of Law's Tale, the Canon's Yeoman's Tale begins with a palpable sense of homelessness that is nevertheless not dissimilar to Constance's exiles, but it settles down to a hopeful sense of new beginnings when the Yeoman joins the company of Canterbury pilgrims. His restlessness finds potential relief in a restorative *communitas* even if the group he joins is fractured in multiple ways and directions.

The Pardoner's Prologue and Tale, however, emphasize singular senses of place alongside large spaces and ubiquitous characters. On the one hand, Chaucer is interested in exploring what it means to have the spatial perceptions that the taverner and three rioters possess in the tale—namely, a more constricted understanding of the locations they frequent. On the other hand, he pushes the idea of spatial experience to its limit via the old man, the Pardoner, and the personified figure of greed. These three figures suggest an interest in the idea of spatial ubiquity in that the Prologue and Tale engage with the implications of a person's ability to travel and perhaps to exist everywhere.

First, the Pardoner's Prologue and Tale present, and are critical of, a restricted and often local sense of space. As noted, the Prologue and Tale are full of references to specific places. The Pardoner insists that the pilgrims pause by a particular "alestake" to drink and eat before he will tell his tale. He describes how he begins each sermon by telling where he is from. The tale he tells is set in Flanders. The three rioters meet the old man at a particular stile in a fence. Perhaps the most telling example of a restricted sense of space occurs early in the narrative of the three debauchees. Already drinking in a tavern in the morning, they hear a bell tolling before a corpse being taken to its grave. One of them asks his boy to find out who has died, but the boy says he already knows. It was an "old felawe" of theirs, he informs them, and he adds that "Deeth . . . in this contree al the peple sleeth." The taverner who is also there precipitates the three men's misinterpretation of Death's realm when he immediately misconstrues the boy's report. He interprets the sense of "contree" as local, saying:

> The child seith sooth [truth], for he [Death] hath slayn this yeer,
> Henne [Hence] over a mile, withinne a greet village,
> Bothe man and womman, child, and hyne [servant], and page;
> I trowe [believe] his habitacioun be there.[19]

The rioters follow the taverner's lead in this narrow understanding of "contree," and they bond together to go and kill Death where they think he will be with ironically fitting consequences. The Prologue and Tale are critical of the spatial myopia at play here.

Second, the Pardoner's Prologue and Tale present large areas through three figures, each of whose sense of a vast space is different. One is the Pardoner himself, who describes how he preaches in many churches, during which he stretches his neck out to the east and west. He says he is sure to authorize his preaching with his bulls and relics, but his problem is not his insistence on authority even though they are fakes. The issue is rather of profusion: too many places, too many bulls, too many relics, too many languages (to "saffron" his sermons), too many stories ("an hundred false japes moore"), too many gestures with his head nodding and tongue and hands moving fast, too many *exempla*, too much gold, wool, cheese, and bread. He will "have a joly wenche in every toun" and insists that while he is able, he "wol preche and begge in sondry londes."[20]

The Pardoner's moving around is similar to a second figure who has an even more expansive sense of space—namely, the glutton the Pardoner describes at length at the beginning of his tale. The glutton is someone who, like the Pardoner himself, runs the risk of affecting other people, but where the Pardoner might actually save souls even if he is in sin, the glutton is a figure who does not have any, even incidental, positive effect. To the glutton, the whole world exists so that people can satisfy him or attempt to sate his desires. His throat and mouth put people to "swynke" in "est and west and north and south," so that everywhere (and everyone) is alike to him; all work "To gete a glotoun deyntee mete and drynke!"[21] The Pardoner continues to describe a species of glutton—the drunkard—a type of person who, as it were, unthinkingly takes the Pardoner's and the glutton's approach to places to an extreme. Drunks who have consumed adulterated wine from Lepe do not know whether they are in Spain or at home in Cheapside; the drunkard does not know where he is. His state of drunkenness is a delirium, and this mania means he is lost in more than one sense.[22] He has lost his soul through his sin, but he has also lost his sense of space and therefore has misplaced his very identity.

The third figure who also appears everywhere is the old, poor man that the rioters encounter on their search for Death. He too moves around, but his peripatetic actions are cast in empathetic terms. Later in the tale, when the proudest of the three men challenges the old man about why he continues to live so long, he answers that even if he were to walk to the farthest reaches of the earth, to India, he cannot find anyone in a city or a village who would change his youth for his age.[23] A sense of this geographical span is fairly common in Chaucer as analyses have pointed out. We hear of it in the Wife of Bath's Prologue in her remark on "any wyf from Denmark unto Ynde," and the *Boece* contains an admonition to those who cannot put aside "thi foule dirke desires," whose "lordschipe[s]" "strecche so fer that the contre of Ynde quaketh at thy comaundementz . . . and that the laste ile that highte Tyle [Thule] be thral to the."[24] The old man, of course, embodies the temporarily realistic understanding of *contemptus mundi*, but there is also a sense of barren homogeneity to the earth over which he treads that I posit tips the space from globalism to ubity and so ubiquity. Here I diverge again from Foucault's reading "Of Other Spaces" where he reiterates a medieval–early modern divide. Foucault says that the medieval space is "a hierarchic ensemble" (*un ensemble hiérarchisé*) and the early modern, with Galileo, is radically different, one of "extension" (*l'étendue*) within "an infinite and infinitely open space" (*un espace infini, et infiniment ouvert*).[25] Already here, for the old man, everywhere is alike; no place offers him respite, so he must always travel. Chaucer's superbly economic language is evident when his character states flatly "I moot [must] go thider as I have to go." The acclaimed moment is when the man knocks on the ground and appeals to the earth, his "leeve mooder," to take him in. Here his staff serves as a kind of pointer to indicate the place where he would like to go and remain, a sharp perpendicularity to contrast with his constant horizonality.[26] He has to continue moving across the earth, which keeps him from its depth and a singularity of place that he dearly desires.

The Pardoner's Prologue and Tale therefore cast the Pardoner's own opportunistic ubiquity in a negative light and the glutton's hunger as leading to a lack of awareness of where one is, yet the narrative also offers the old man's perpetual wandering as a positive, if pitiable, model. I am not so interested in the contrasts between what these two groups obviously represent morally or theologically so much as the fact that the Pardoner-glutton and old man are two sides of the same spatial coin. Though very different in their attitudes toward everywhere, they embody a sense of ubiquity, of everywhere being available to them. Chaucer, deliberately I would argue, puts the restricted senses of space side by side with the negative and positive models of everywhereness. It is easy, for instance, to discern parallels between the three rioters with their fatal drunkenness and the glutton and the drunkard at the beginning of the Pardoner's sermonizing.

The very last remaining references to space in the Pardoner's Tale turn on an intimate proximity. They are the Pardoner's offer to Harry Bailey that he come and kiss the relics along with the Knight's attempt at reconciliation when he asks the host to kiss the Pardoner and also for the Pardoner to "drawe . . . neer."[27] Such moments have been read in light of queer theory and in terms of the touch and affective responses more generally. Queer theories interpret Harry Bailey's violent reaction against the Pardoner as a kind of panic that reveals the Host's coercive character, and they have interpreted the Pardoner's own performance of his trickery as calling the *communitas* of the pilgrim and the entire *Canterbury Tales*'s storytelling competition into question.[28] Affective criticism on this topic comes in the wake of Carolyn Dinshaw's discussion of touching the past in *Getting Medieval* and *How Soon Is Now?*, which extend queer theory into the realm of the temporal in its advocacy for touching the past.[29]

As the chapter on proximity suggested in the case of Pyramus and Thisbe and the *Legend of Good Women* more generally, drawing near can threaten to collapse separate distinct beings into one, and the intimacy of the kiss and the cross-spatial and cross-temporal touch in the Pardoner's Tale seem to have this kind of potential ability. These are moments that vibrate with the tensile distances of proximate bodies and, in addition to queering social cohesion, threaten to violate one of the principles of Aristotelian physics that two objects cannot occupy the same place or else they would be one. The situations also suggest a further investigation on Chaucer's part of the potential limitations of being together in a limited space. The *communitas* that the Knight encourages at the end brings bodies together in a small area, curtailing for the moment the Pardoner's ubiquity and the pilgrimage travel that Harry Bailey has also transformed into a competition. The obverse to this competitive conviviality is the taverner's and the three rioters' overliteralization of their understandings of "contree" spaces and where Death has recently been. Their apprehension of space is too close up, too uninformed by a broader understanding, and the end of the tale evokes their mistakes again.

So what does this mean? Is Chaucer saying anything coherent in putting the spatially narrow and the everywhere next to each other in one prologue and tale? Note, first, that purely on the level of space, he is not suggesting that a local particularity is better than ubiquity or, vice versa, that constant wandering and ubiquity are better than remaining in one locale. Both senses and experiences of space have the potential to be misleading. The issue at hand here is whether Chaucer is doing more than merely juxtaposing the nearsighted with being everywhere. In an essay titled "Thinking Topographically," Jeff Malpas meditates on boundaries as the spatial philosopher Edward Casey did before him. Boundaries cause, or mean, separation but also junction; they divide and join. Malpas, following Heidegger and others, writes that because of this duality, "the identity of the place that the boundary defines is also indeterminate, with every place having enfolded in it, and being enfolding within, other places." But he is sure to clarify that "[t]his does not mean that individual places lack any character that belongs to them." "[R]ather . . . their character is such as always to admit of other possibilities, other descriptions—is always such as to implicate other places."³⁰ Malpas insists again that a being does not give up a sense of a particular place, but his point is that the relation of one space to another or others changes the character of the individual space and therefore the being. The nature of the changed space and the changed being he does not explore further.

This "enfolding" of other places into one site and the "implication" of one place as it folds out to other places may be true as far as it goes. But what if one location enfolds and implicates the largest space, one even larger and different in nature than the global, namely a ubiquitous one? For this is what Chaucer does. He does not have one character appreciate or consider both kinds of places at once, but he implies an extreme sense of the interconnectedness of local space with being everywhere. It is possible, he suggests, for a local space to "implicate" ubiquity, for everywhereness to be enfolded in the local. The Pardoner's Prologue and Tale, therefore, go further than the much later philosophers in answering the question of how ubiquity relates to, can be "enfolded" in, being. Chaucer suggests that the inclusion of ubiquity in a locale also admits deaths into a present space. The three rioters die, but that admittance of death into a space is not true for the old man, whose desire for termination remains unsatisfied even though he is as close to ubiquitous as possible. For him, every location is imbued with God's will that may be encountered at any time, but also, for him, instead of death, the presence of everywhere in a local space empties out the local space. Or if not emptying it completely, which Malpas has suggested is impossible, the enfolding of everywhere in somewhere is a process that makes a space less physical and more difficult to perceive. It has transformed the old man's body, after all. He remarks for his audience to look, "Lo how I vanysshe, flessh, and blood, and skyn!" His face is pale and withered.³¹

When the pilgrims at the start of Chaucer's *Canterbury Tales* come from every shire of England, pilgrims including the Knight lately come from a voyage from even farther away, and they go to the Tabard in Southwark and pause in particular places along the way to Canterbury—by an alestake, at Deptford, Rochester, and Sittingbourne—each of the locations is infused with their presence, their bodies and clothes, their horses and weapons, their dialects, and their narratives of themselves and their tales of others, meaning that Chaucer's places are typically enfolded with the layered implications of other particular spaces. Occasionally, however, when the other space is of an even more extreme nature than the space of the migrant, the cosmopolitan, and the global traveler, when it is ubiquity, there is the chance of the opposite effect. The Pardoner's Prologue and Tale show how a particular place can be deracinated, turned inside out into a spectral thinness.

NOTES

Introduction

1. Historians of the scientific texts I discuss use the words *science* and *scientific* to denote medieval natural philosophy and other writings, and I follow them. David Lindberg and Michael Shank define the medieval sciences broadly as "attempts to acquire, to evaluate, or to create systematic knowledge of the natural world." Introduction, *The Cambridge History of Science*, ed. David Lindberg and Michael Shank, vol. 2 (Cambridge: Cambridge University Press, 2013), 7.

2. Brian Harley and David Woodward, "Concluding Remarks," in *The History of Cartography*, vol. 1, *Cartography in Prehistoric, Ancient, and Medieval Europe and the Mediterranean*, ed. Brian Harley and David Woodward (Chicago: University of Chicago Press, 1987), 504.

3. Aristotle, *Physics*, bks. 3–4, trans. Edward Hussey, Clarendon Aristotle Series (Oxford: Oxford University Press, 1983), 4.1.208^{a}27–31. Italics in original.

4. Thomas Aquinas, *Commentary on Aristotle's Physics*, bks. 3–8, trans. Pierre H. Conway (Columbus, OH: College of St. Mary of the Springs, 1958–1962), 4.1.407, html edition by Joseph Kenny, http://www.dhspriory.org/thomas/Physics4.htm.

5. Joel Kaye, *A History of Balance, 1250–1375: The Emergence of a New Model of Equilibrium and Its Impact on Thought* (Cambridge: Cambridge University Press, 2014), 15. Kaye discusses Thomas Bradwardine, Nicole Oresme, and the Merton School at Oxford in chapter 8, all of whom are important in my study.

6. Ernest Moody, "Empiricism and Metaphysics in Medieval Philosophy," *Philosophical Review* 67 (1958): 147; Edward Grant, *A History of Natural Philosophy: From the Ancient World to the Nineteenth Century* (Cambridge: Cambridge University Press, 2007), 200, 231–32. See also Antonia Gransden, "Realistic Observation in Twelfth-Century England," *Speculum* 47 (1972): 29–51; Peter King, "Mediaeval Thought-Experiments: The Metamethodology of Mediæval Science," in *Thought Experiments in Science and Philosophy*, ed. Tamara Horowitz and Gerald J. Massey (New York: Rowman and Littlefield, 1991), 43–64.

I am indebted to the following histories of science used throughout this work: Anneliese Maier, *On the Threshold of Exact Science: Selected Writings of Anneliese Maier on Late Medieval Natural Philosophy*, trans. Steven D. Sargent (Philadelphia: University of Pennsylvania Press, 1982); Marshall Clagett, ed., *The Science of Mechanics in the Middle Ages* (Madison: University of Wisconsin Press, 1959); and *Nicole Oresme and the Medieval Geometry of Qualities and Motions: A Treatise on the Uniformity and Difformity of Intensities Known as Tractatus de configurationibus qualitatum et motuum*, trans. Marshall

Clagett (Madison: University of Wisconsin Press, 1968); Ernest A. Moody, *Studies in Medieval Philosophy, Science, and Logic: Collected Papers, 1933–1969* (Los Angeles: University of California Press, 1975); Edward Grant, ed., *A Source Book in Medieval Science* (Cambridge, MA: Harvard University Press, 1974); and Grant, *A History of Natural Philosophy*; and David C. Lindberg, *The Beginnings of Western Science: The European Scientific Tradition in Philosophical, Religious, and Institutional Context, 600 B.C. to A.D. 1450* (Chicago: University of Chicago Press, 1992).

7. In the *Summa theologiae*, Aquinas states that "[t]he proper object of the human intellect . . . since it is joined to a body, is a nature or 'whatness' found in corporeal matter," and "the proper object proportional to our intellect is the nature of sensible things" (Intellectus . . . humani, qui est conjunctus corpori, proprium objectum est quidditas sive natura in materia corporali existens and proprium objectum intellectui nostro proportionatum est natura rei sensibilis). Thomas Aquinas, *Summa theologiae, Opera Omnia* (1888; Pamplona: University of Navarra), 1, question 84, art. 7–8, http://www.corpusthomisticum.org/sth1084.html. See also question 88 in the same work. (This and all subsequent uncited translations are my own.) In Aquinas's commentary on Aristotle's *Physics*, he further describes how essential the *quidditates* of the world are for study: "[N]atural science, which is called physics, deals with those things which depend upon matter not only for their existence, but also for their definition" (de his vero quae dependent a materia non solum secundum esse sed etiam secundum rationem, est naturalis, quae physica dicitur). Aquinas, *Commentaria in octo libros Physicorum* (Turin: Textum Leoninum, 1954), http://www.corpusthomisticum.org/cpy011.html; Aquinas, *Commentary on Aristotle's Physics*, bks. 1–2, trans. Richard J. Blackwell, Richard J. Spath, and W. Edmund Thirlkel (New Haven, CT: Yale University Press, 1963), lectio. 1, chap. 1, para. 3, http://www.dhspriory.org/thomas/Physics1.htm.

8. Thomas Hoccleve, *The Regiment of Princes*, ed. Charles R. Blyth, TEAMS (Kalamazoo, MI: Medieval Institute Publications, 1999), 2087–88.

9. Bert Hansen, "An Overview," in *Nicole Oresme and the Marvels of Nature: A Study of His De Causis Mirabilium with Critical Edition, Translation, and Commentary*, trans. Bert Hansen (Toronto: Pontifical Institute of Mediaeval Studies, 1985), 70.

10. Hugh of St. Victor, *Hugonis de Sancto Victore Didascalicon de Studio Legendi: A Critical Text*, ed. Charles Henry Buttimer, Studies in Medieval and Renaissance Latin 10 (Washington, DC: Catholic University Press, 1939), 1.8, 1.9, 2.1; *The Didascalicon of Hugh of St. Victor: A Medieval Guide to the Arts*, trans. Jerome Taylor (New York: Columbia University Press, 1991), 55–56, 62.

11. Hugh of St. Victor, *Didascalicon*, 2.20; *Didascalicon*, trans. Taylor, 75.

12. Robert Kilwardby, *De ortu scientiarum*, ed. Albert G. Judy (Oxford: Clarendon Press, 1976), 42.393. On the significance of the mechanical arts at the time, see George Ovitt Jr., "The Status of the Mechanical Arts in Medieval Classifications of Learning," *Viator* 14 (1983): 89–106; Elspeth Whitney, "Paradise Restored: The Mechanical Arts from Antiquity through the Thirteenth Century," *Transactions of the American Philosophical Society* 80 (1990): 1–169 (who discusses Kilwardby on 118–23); J. D. North, "Astronomy and Mathematics," in *The History of the University of Oxford*, vol. 2, *Late Medieval Oxford*, ed. J. I. Catto and Ralph Evans (Oxford: Oxford University Press, 1992), 167–68; Joan Cadden, "The Organization of Knowledge," in Lindberg and Shank, *The Cambridge History of Science*, 2:240–67, especially 2:245.

13. Taylor, introduction, *Didascalicon*, 38–39. See also Lisa H. Cooper, *Artisans and Narrative Craft in Late Medieval England* (Cambridge: Cambridge University Press, 2011).

14. Jeff Malpas, "Place and Singularity," in *The Intelligence of Place: Topographies and Poetics*, ed. Jeff Malpas (London: Bloomsbury, 2015), 79.

15. The sense of objectivity I gather from medieval mechanics and literature is not as strong as ideas about object-oriented ontology. Also, one might be tempted to think that medieval optics is a pertinent field to consider on the topic of objects in space. However, optics was a subdiscipline of physics, and it is different from the more central topic of space and motion I take up. Studies of optics frequently address tensions between authoritative accounts of space and individual viewers' observations whereas ideas about space in medieval mechanics is not interested in what it would likely consider mere error. For discussions of optics, see for example, Peter Brown's *Chaucer and the Making of Optical Space*, which follows Linda Holley, Sarah Stanbury, A. C. Spearing, and others who have considered the role of optics in science and Middle English literature. The emphasis on optics in these analyses points away from objectivity and toward subjectivity and misperceptions—or as Brown calls it, how "space was mediated through an individual's eye." Brown, *Chaucer and the Making of Optical Space* (Bern: Peter Lang, 2007), 316. Brown cites Linda Holley, *Chaucer's Measuring Eye* (Houston, TX: Rice University Press, 1990); Sarah Stanbury, *Seeing the Gawain-Poet: Description and the Act of Perception* (Philadelphia: University of Pennsylvania Press, 1991); and A. C. Spearing, *The Medieval Poet as Voyeur: Looking and Listening in Medieval Love-Narratives* (Cambridge: Cambridge University Press, 1993).

16. Geography as a discipline and the subfield of historical geography, along with forms of geocriticism, which are important in this study, struggle with taking into account art and literature in relation to the more scientific discussions of spatial phenomena. Marc Brosseau, in a review essay, critiques how geographers have used poetry and prose with a "will to transform fictional literature into a reservoir of positive geographical data." Geographers, he says, "remain within a generally unexamined mimetic conception of literature: we go from viewing literature as the reflection of reality to considering it as the reflection of the soul contemplating or experiencing this same reality." Marc Brosseau, "Geography's Literature," *Progress in Human Geography* 18 (1994): 337–38. Another scholar writing from the perspective of geography, Joanne Sharp, describes a situation in which "[h]umanist geographers have referred to literature on the whole without comment on genre, mode of production, or range of consumption. Critical geographies on the other hand, have regarded literature as a material artefact that fulfils a role designated by its position in various social and economic processes." Sharp criticizes both approaches because they "offer a limited vision of the relationship between geography and literature." Joanne P. Sharp, "Towards a Critical Analysis of Fictive Geographies," *Area* 32 (2000): 327–34. Other overviews of the relationship between geography and literature include Douglas C. D. Pocock, "Geography and Literature," *Progress in Human Geography* 12 (1988): 87–102; and Fabio Lando, "Fact and Fiction: Geography and Literature," *GeoJournal* 38 (1996): 3–18.

17. Alexander N. Gabrovsky, *Chaucer the Alchemist: Physics, Mutability, and the Medieval Imagination* (New York: Palgrave Macmillan, 2015), 25.

18. Study of other scales—universal, global, and national—is well established. A helpful introduction to universal space may be found in Edward Grant, *Much Ado about*

Nothing: Theories of Space and Vacuum from the Middle Ages to the Scientific Revolution (Cambridge: Cambridge University Press, 2008). Even though the focus is on cartography, *The History of Cartography* volumes, edited by Harley and Woodward, contain probably the best introductions to ideas about the whole earth, with volume 1 on prehistoric, ancient, and medieval Europe and the Mediterranean, and volume 2 on Islam and South Asia. They are available online at http://www.press.uchicago.edu/books/HOC/index.html. Late medieval British and English nationalism has been addressed in many works, including the following: Thorlac Turville-Petre, *England the Nation: Language, Literature, and National Identity, 1290–1340* (Oxford: Clarendon Press, 1996); Derek Pearsall, "The Idea of Englishness in the Fifteenth Century," in *Nation, Court, and Culture: New Essays on Fifteenth-Century English Poetry*, ed. Helen Cooney (Dublin: Four Courts, 2000), 15–27; Michelle R. Warren, *History on the Edge: Excalibur and the Borders of Britain, 1100–1300* (Minneapolis: University of Minnesota Press, 2000); Patricia Clare Ingham, *Sovereign Fantasies: Arthurian Romance and the Making of Britain* (Philadelphia: University of Pennsylvania Press, 2001); Geraldine Heng, *Empire of Magic: Medieval Romance and the Politics of Cultural Fantasy* (New York: Columbia University Press, 2003); Daniel Birkholz, *The King's Two Maps: Cartography and Culture in Thirteenth-Century England* (New York: Routledge, 2004); Kathy Lavezzo, ed., *Imagining a Medieval English Nation* (Minneapolis: University of Minnesota Press, 2004); Ardis Butterfield, "Nationhood," in *Chaucer: An Oxford Guide* (Oxford: Oxford University Press, 2005), 50–65; Kathy Lavezzo, *Angels on the Edge of the World: Geography, Literature, and English Community, 1000–1534* (Ithaca, NY: Cornell University Press, 2006); David Matthews, "Laurence Minot, Edward III, and Nationalism," *Viator* 38 (2007): 269–88; Andrea Ruddick, *English Identity and Political Culture in the Fourteenth Century* (Cambridge: Cambridge University Press, 2013); and Susan Nakley, *Living in the Future: Sovereignty and Internationalism in the Canterbury Tales* (Ann Arbor: University of Michigan Press, 2017).

19. Studies focusing on particular spaces in Middle English and French literature include Barbara A. Hanawalt and Michal Kobialka, eds., *Medieval Practices of Space* (Minneapolis: University of Minnesota Press, 2000); Kathryn L. Lynch, ed., *Chaucer's Cultural Geography* (New York: Routledge, 2002); Clare A. Lees and Gillian R. Overing, eds., *A Place to Believe In: Locating Medieval Landscapes* (University Park: Pennsylvania State University Press, 2006); Sarah Kay, *The Place of Thought: The Complexity of One in Late Medieval French Didactic Poetry* (Philadelphia: University of Pennsylvania Press, 2007); William Woods, *Chaucerian Spaces: Spatial Poetics in Chaucer's Opening Tales* (Albany: State University of New York Press, 2008); S. A. Mileson, *Parks in Medieval England* (Oxford: Oxford University Press, 2009); Gillian Rudd, *Greenery: Ecocritical Readings of Late Medieval English Literature* (Manchester: Manchester University Press, 2010); Julian Weiss and Sarah Salih, eds., *Locating the Middle Ages: The Spaces and Places of Medieval Culture* (London: King's College, Centre for Late Antique and Medieval Studies, 2012); Dorsey Armstrong and Kenneth Hodges, *Mapping Malory: Regional Identities and National Geographies in Le Morte Darthur* (New York: Palgrave Macmillan, 2014); Mary C. Flannery and Carrie Griffin, eds., *Spaces for Reading in Later Medieval England* (New York: Palgrave Macmillan, 2016); and Laura Varnam, *The Church as Sacred Space in Middle English Literature and Culture* (Manchester: Manchester University Press, 2018).

20. Stuart Elden, *The Birth of Territory* (Chicago: University of Chicago Press, 2013), 144–45, who cites Pietro Janni, *La mappa e il periplo: Cartografia antica e spazio odologico* (Rome: Giorgio Bretschneider, 1984), 58–65, and A. D. Lee, *Information and Frontiers: Roman Foreign Relations in Late Antiquity* (Cambridge: Cambridge University Press, 1993), 86–87.

21. Nicholas Howe, *Writing the Map of Anglo-Saxon England: Essays in Cultural Geography* (New Haven, CT: Yale University Press, 2008), 5–7.

22. Helen Cooper has suggested evidence of "linearity" in fourteenth- and fifteenth-century romance in *The English Romance in Time: Transforming Motifs from Geoffrey of Monmouth to the Death of Shakespeare* (Oxford: Oxford University Press, 2004), 67–69. See also Robert Allen Rouse, "What Lies Between? Thinking through Medieval Narrative Spatiality," in *Literary Cartographies: Spatiality, Representation, and Narrative*, ed. Robert T. Talley Jr. (New York: Palgrave Macmillan, 2014), 13–29.

23. Cosmological space was one of the important concerns of the late Middle Ages, and the issues occupied many of the same philosophers that I discuss here. Distinct interest arises in the question of a vacuum that may exist within this cosmos and/or outside it; see Grant, *Much Ado about Nothing*. Ideas about the cosmos, like ways of thinking about more local spaces, progressed beyond Aristotle and Aquinas. I would add that *Scribes of Space* also does not take up the larger, often noted debates surrounding William of Ockham's nominalist controversies. Although many of the philosophers overlap, the philosophical investigations that are most relevant here are about the specific mechanics of velocities, resistance, and so on, a discussion that also extends to the practical arts.

24. For one example of usage of the term *spaceship earth*, see Denis Cosgrove, *Geography and Vision: Seeing, Imagining and Representing the World* (London: Tauris, 2008), 19.

25. See Shayne Aaron Legassie, *The Medieval Invention of Travel* (Chicago: University of Chicago Press, 2017), especially vii–x, 16–20. Legassie mentions Margery and discusses Mandeville at some length.

26. John Murdoch, "From Social into Intellectual Factors: An Aspect of the Unitary Character of Late Medieval Learning," in *The Cultural Context of Medieval Learning: Proceedings of the First International Colloquium on Philosophy, Science, and Theology in the Middle Ages, September 1973*, ed. John Murdoch (Dordrecht, Holland: Reidel, 1975), 287, 340. "Measurement" here is not to be understood in a modern sense of exact standards but remained within the context of philosophical discussion. On measurement, see Anneliese Maier, "The Achievements of Late Scholastic Natural Philosophy," in Maier, *On the Threshold of Exact Science*, 143–70; Maier's essay provides a good introduction to the innovations in science in the fourteenth century. For twelfth-century science prior to the rediscovery of Aristotle, see Nadja Germann, "Natural Philosophy in Earlier Latin Thought," in *The Cambridge History of Medieval Philosophy*, ed. Robert Pasnau and Christina van Dyke, 2 vols. (Cambridge: Cambridge University Press, 2010), 1:219–31.

Aristotle's *libri naturales* and the *Metaphysics* were banned from the faculty of arts at the University of Paris in the early years of the thirteenth century after first being reintroduced in the West in the twelfth century. But by 1262 William of Moerbeke was translating (or retransmitting earlier translations of) Aristotle, and Thomas Aquinas would soon thereafter write his *Expositio in VIII libros Physicorum Aristotelis*. The bishop of Paris condemned a further set of scientific ideas in 1277, and although the condemnations changed the course of medieval science, they did not stop its study, and the censures were annulled in 1325. On the history of Aristotelian texts in curricula, see the introduction to Thomas Aquinas, *Commentary on Aristotle's Physics*, bks. 1–2, html edition by Joseph Kenny, xxi, http://www.dhspriory.org/thomas/Physics.htm; D. A. Callus, "The Introduction of Aristotelian Learning in Oxford," *Proceedings of the British Academy* 29 (1943): 229–81; Grant, *Source Book*, 42–52; James A. Weisheipl, "The Interpretation of Aristotle's *Physics* and the Science of Motion," in *The Cambridge History of Later Medieval Philosophy: From the Rediscovery of Aristotle to the Disintegration of Scholasticism, 1100–1600*, ed. Norman Kretzmann, Anthony Kenny, and Jan Pinborg (Cambridge: Cambridge University Press, 1982), 521–23; John Marenbon, ed., *Aristotle in Britain during the Middle Ages: Proceedings of the International Conference at Cambridge, 8–11 April 1994*, Société Internationale pour l'Etude de la Philosophie Médiévale 5 (Turnhout: Brepols, 1996), especially Francesco del Punta, Silvia Donati, and Cecilia Trifogli, "Commentaries on Aristotle's *Physics* in Britain, ca. 1250–1270," 265–83; Cecilia Trifogli, "Change, Time, and Place," in Pasnau and van Dyke, *Cambridge History of Medieval Philosophy*, 1:267–78.

On the reorganization of universities due to Aristotle's works, see Ernest Moody, preface to *Quaestiones Super Libris Quattuor de Caelo et Mundo*, by Jean Buridan, ed. Ernest Moody (Cambridge, MA: Medieval Academy of America, 1942), xiv; and Cadden, "The Organization of Knowledge," 2:240–67, who also emphasizes continuities with mechanical arts. For the further influence of universities outside their walls, also taken up in chapter 6, see William Courtenay, *Schools and Scholars in Fourteenth-Century England* (Princeton, NJ: Princeton University Press, 1987); and Michael Shank, "Schools and Universities in Medieval Latin Science," in Lindberg and Shank, *Cambridge History of Science*, 2:207–39. For a useful bibliography of Aristotelian and pseudo-Aristotelian works translated into Latin, Arabic, Hebrew, and Greek, see Appendix B: Medieval Translations, in Pasnau and van Dyke, *Cambridge History of Medieval Philosophy*, 2:793–832.

27. *Nicole Oresme and the Medieval Geometry of Qualities and Motions* (hereafter cited as *Qualities and Motions* with translations by Clagett unless otherwise noted).

28. Lefebvre's aim was to counter this type of "endless division" (which he viewed as analogous to the division of labor in society) with a "unitary theory" that sought to reassemble the fragmentary disciplines of physical/sensory, mental/imaginary, and social/ideological space. Henri Lefebvre, *The Production of Space* (Oxford: Blackwell, 1991), 8, 11. One summary of place in relation to the "spatial turn" may be found in Charles W. J. Withers, "Place and the 'Spatial Turn' in Geography and in History," *Journal of the History of Ideas* 70 (2009): 637–58.

29. Edward Relph, "Disclosing the Ontological Depth of Place: *Heidegger's Topology* by Jeff Malpas," *Environmental and Architectural Phenomenology Newsletter* 19 (2008): 5.

30. Michel Foucault, "Des espaces autres," in *Dits et écrits, 1954–1988*, tome 4, 1980–1988 (Paris: Gallimard, 1994), 752–62; Foucault, "Of Other Spaces," trans. Jay Miskowiec, *Diacritics* 16 (1986): 22–27; Pierre Bourdieu, *The Field of Cultural Production: Essays on Art and Literature*, ed. Randal Johnson (New York: Columbia University Press, 1993); Michel de Certeau, *The Practice of Everyday Life*, trans. Steven F. Rendall (Berkeley: University of California Press, 1984).

31. For Walter, "Modern 'space' is universal and abstract, whereas a 'place' is concrete and particular. People do not experience abstract space; they experience places. . . . Abstract space is infinite; in modern thinking it means a framework of possibilities. A place is immediate, concrete, particular, bounded, finite, unique. Abstract space is repetitive and uniform. Abstraction moves away from the fullness of experience." Eugene Victor Walter, *Placeways: A Theory of the Human Environment* (Chapel Hill: University of North Carolina Press, 1988), 142. See also his discussion of a distinction between space and place on 120–21, 125–26. Edward Casey's works exploring distinctions between place and space are often cited on this topic: *Getting Back Into Place: Toward a Renewed Understanding of the Place-World* (Bloomington: Indiana University Press, 1993); "How to Get from Space to Place in a Fairly Short Stretch of Time: Phenomenological Prolegomena," in *Senses of Place*, ed. Steven Feld and Keith H. Basso (Santa Fe, NM: School of American Research Press, 1996), 13–52; *The Fate of Place: A Philosophical History* (Berkeley: University of California Press, 2013). However, for discussions of Aristotle's term *topos*, see *Physics*, trans. Hussey, note to 4.1.208^{a}27; see also Aristotle, *Physics*, bks. 3–4, vol. 1, trans. Philip H. Wicksteed and Francis M. Cornford (Cambridge, MA: Harvard University Press, 1963), 267.

32. Doreen Massey, "Introduction: Geography Matters," in *Geography Matters! A Reader*, ed. Doreen Massey and John Allen (Cambridge: Cambridge University Press, 1984), 4.

33. Doreen Massey, *For Space* (London: Sage, 2005), 18.

34. Massey, "Introduction: Geography Matters," 5. See also several works by Jeff Malpas on the relation of space to the social or "intersubjective": introduction to *The Intelligence of Place*, ed. Jeff Malpas, 1–10; *Place and Experience: A Philosophical Topography* (Cambridge: Cambridge University Press, 1999), 35–36; "Thinking Topographically: Place, Space, and Geography," http://jeffmalpas.com/downloadable-essays/, accessed June 10, 2018; and *Heidegger and the Thinking of Place: Explorations in the Topology of Being* (Cambridge, MA: MIT Press, 2012), 65–66. Another relevant discussion is Tim Cresswell, *Place: An Introduction*, 2nd ed. (Malden, MA: Wiley-Blackwell, 2014), 55–56.

35. James J. Gibson, *The Senses Considered as Perceptual Systems* (Boston: Houghton Mifflin, 1966), 59.

36. Bernard Cohen, for example, despite his acknowledgment that Galileo based his work on predecessors, emphasizes the "great Scientific Revolution" of the sixteenth and seventeenth centuries and Kepler, Galileo, and Newton. *The Birth of a New Physics*, rev. ed. (New York: W. W. Norton, 1985), xi.

37. Samuel Y. Edgerton, *The Renaissance Rediscovery of Linear Perspective* (New York: Basic Books, 1975), 9. Art historians, for example, Margaret Goehring, diverge from Edgerton by examining the "general cognitive context for understanding space and place" that existed in the late medieval period, such as looking to the revival of Aristotelian learning in twelfth-century philosophy. Goehring examines the particular ways that landscape was represented in the Middle Ages due to "how artists . . . visualized the organization of certain environments, and why this organization was deemed necessary." *Space, Place, and Ornament: The Function of Landscape in Medieval Manuscript Illumination* (Turnhout: Brepols, 2014), 15, 27.

38. Pierre Duhem, *Le système du monde: Histoire des doctrines cosmologiques de Platon à Copernic*, vol. 7 (Paris: Hermann, 1956), 3–4.

39. Linda Ehrsam Voigts, "Scientific and Medical Books," in *Book Production and Publishing in Britain 1375–1475*, ed. Jeremy Griffiths and Derek Pearsall (Cambridge: Cambridge University Press, 1989), 351. Voigts cites Lucien Febvre and Henri-Jean Martin for this idea in *The Coming of the Book: The Impact of Printing 1450–1800*, trans. David Gerard (London: Verso, 1976), 258–60, 276–77.

40. Patrick Gautier Dalché, "The Reception of Ptolemy's *Geography* (End of the Fourteenth to Beginning of the Sixteenth Century)," in *The History of Cartography*, vol. 3, *Cartography in the European Renaissance*, ed. David Woodward (Chicago: University of Chicago Press, 2007), 296. Gautier Dalché cites Paul Lawrence Rose, "Humanist Culture and Renaissance Mathematics: The Italian Libraries of the *Quattrocento*," *Studies in the Renaissance* 20 (1973): 46–105, esp. 56.

41. Patrick Gautier Dalché, *La Géographie de Ptolémée en Occident (IVe–XVIe siècle)* (Turnhout: Brepols, 2009), 8. Misapprehensions about medieval space and the Middle Ages more generally enfold and reinforce each other, from Jacob Burckhardt on Italian Renaissance superiority in science and geography to Hans Blumenberg, and they continue to the present day. Additional analyses and critiques of misrepresentations of medieval geography include essays in Keith Lilley, ed., *Mapping Medieval Geographies* (Cambridge: Cambridge University Press, 2013). See especially Lilley's introduction; Jesse Simon, "Chorography Reconsidered: An Alternative Approach to the Ptolemaic Definition," 23–44; and Meg Roland, "'After poyetes and astronomyers': English Geographical Thought and Early English Print," 127–51, esp. 128.

42. Martin Willis provides an overview of studies on mainly Enlightenment and post-Enlightenment relations between literature and natural philosophy in *Literature and Science* (New York: Palgrave, 2014). See also Gillian Beer, *Darwin's Plots: Evolutionary Narrative in Darwin, George Eliot and Nineteenth-Century Fiction*, rev. ed. (1983; Cambridge: Cambridge University Press, 2000); Alice Jenkins, *Space and the 'March of Mind': Literature and the Physical Sciences in Britain 1815–1850* (Oxford: Oxford University Press, 2007); Charlotte Sleigh, *Literature and Science* (New York: Palgrave Macmillan, 2010); Margareth Hagen, Randi Koppen, and Margery Vibe Skagen, eds., *The Art of Discovery: Encounters in Literature and Science* (Aarhus, Denmark: Aarhus University Press, 2010); Margareth Hagen and Margery Vibe Skagen, eds., *Literature and Chemistry: Elective Affinities* (Aarhus, Denmark: Aarhus University Press, 2014); Janine Rogers, *Unified Fields: Science and Literary Form* (Montreal, QC: McGill-Queen's University Press, 2015). Kellie Robertson describes an "overlap in subject matter between . . . 'popular' poetry and the language of natural philosophy—what today we call 'physics'" in the Middle Ages in "Medieval Materialism: A Manifesto," *Exemplaria* 22 (2010): 111; see also her *Nature Speaks: Medieval Literature and Aristotelian Philosophy* (Philadelphia: University of Pennsylvania Press, 2017).

1. Local Space, Edges, and Contents

1. R. A. Skelton and P. D. A. Harvey, eds., *Local Maps and Plans from Medieval England* (Oxford: Clarendon Press, 1986), 3.

2. All from Skelton and Harvey, *Local Maps and Plans*: Judith A. Cripps, "Barholm, Greatford, and Stowe, Lincolnshire," 263–88; P. D. A. Harvey, "Boarstall, Buckingham-

shire," 211–19; M. W. Barley, "Sherwood Forest, Nottinghamshire," 131–39; F. Hull, "Isle of Thanet, Kent," 119–26.

3. Denis Cosgrove, "Introduction: Mapping Meaning," in *Mappings*, ed. Denis Cosgrove (London: Reaktion, 1999), 9. See also Julio Escalona, "The Early Middle Ages: A Scale-Based Approach," in *Scale and Scale Change in the Early Middle Ages: Exploring Landscape, Local Society, and the World Beyond*, ed. Julio Escalona and Andrew Reynolds (Turnhout: Brepols, 2011), 1–22.

4. Hugh of St. Victor, *Didascalicon*, 2.9, 2.13; *Didascalicon*, trans. Taylor, 68, 70.

5. Hugh of St. Victor, *Didascalicon*, 1.8, 1.11; *Didascalicon*, trans. Taylor, 55, 60.

6. Dick Harrison, *Medieval Space: The Extent of Microspatial Knowledge in Western Europe during the Middle Ages* (Lund: Lund University Press, 1996), 1. Harrison's study is one of the few that directly examine local areas.

7. Harrison, *Medieval Space*, 2, 17. On the distances of royal travel, which could vary from eleven to thirty-six miles in a day, see Michael Prestwich, "The Royal Itinerary and Roads in England under Edward I," in *Roadworks: Medieval Britain, Medieval Roads*, ed. Valerie Allen and Ruth Evans (Manchester: Manchester University Press, 2016), 177–97.

8. Lilley, introduction, *Mapping Medieval Geographies*, 5.

9. Simon, "Chorography Reconsidered," 44. Evidence for medieval geographical knowledge derived from the *Geography* is also discussed in John Moffitt, "Medieval *Mappaemundi* and Ptolemy's *Chorographia*," *Gesta* 32 (1993): 59–68; Gautier Dalché, "The Reception of Ptolemy's *Geography*, 285–364; Gautier Dalché, *La Géographie de Ptolémée en Occident*, 87–142; Keith D. Lilley, "Geography's Medieval History: A Neglected Enterprise?" *Dialogues in Human Geography* 1 (2011): 147–62.

10. Introduction, *Ptolemy's Almagest*, trans. G. J. Toomer (London: Gerald Duckworth, 1984), 1–2.

11. O. A. W. Dilke, "The Culmination of Greek Cartography in Ptolemy," in Harley and Woodward, *History of Cartography*, 177; J. M. Fletcher, "Developments in the Faculty of Arts, 1370–1520," in *The History of the University of Oxford*, vol. 2, *Late Medieval Oxford*, ed. J. I. Catto and Ralph Evans (Oxford: Oxford University Press, 1992), 323; Linne R. Mooney, "A Middle English Text on the Seven Liberal Arts," *Speculum* 68 (1993): 1029. On Chaucer and Ptolemy, see Karl Young, "Chaucer's Aphorisms from Ptolemy," *Studies in Philology* 34 (1937): 1–7; and *Sources and Analogues of the Canterbury Tales*, ed. Robert M. Correale and Mary Hamel, vol. 2 (Cambridge: D. S. Brewer, 2005), 381n1.

12. *Ptolemy's Geography: An Annotated Translation of the Theoretical Chapters*, trans. J. Lennart Berggren and Alexander Jones (Princeton, NJ: Princeton University Press, 2000), 1.1. I use this translation even though Berggren and Jones inadvertently emphasize the Kantian distinction when they translate *chōrographia* as "regional cartography," thus suggesting its subordinate role. The following discussion draws on Matthew Boyd Goldie, "An Early English Rutter: The Sea and Spatial Hermeneutics in the Fourteenth and Fifteenth Centuries," *Speculum* 90 (2015): 722–23.

13. χῶρος and χώρα, *A Greek-English Lexicon*, ed. Henry George Liddell and Robert Scott (Oxford: Clarendon Press, 1940), Perseus Digital Library, http://www.perseus.tufts.edu/hopper/text?doc=Perseus:text:1999.04.0057:entry=xw%3dros1. It is well known that the lineage of the term extends from origins in Parmenides to Plato's *Timaeus* and Aristotle's works (the latter usage of which has a great influence on medieval

physics), and on to Julia Kristeva and Jacques Derrida among others. Other works consulted on the nature of the *khôros* include *Plato's Cosmology: The Timaeus of Plato*, trans. and commentary Francis M. Cornford (New York: Routledge, 1935); A. E. Taylor, *A Commentary on Plato's Timaeus* (Oxford: Clarendon Press, 1962); F. Lukerman, "The Concept of Location in Classical Geography," *Annals of the Association of American Geographers* 51 (1961): 200–201; Päivi Kymäläinen and Ari A. Lehtinen, "Chora in Current Geographical Thought: Places of Co-Design and Re-Membering," *Geografiska Annaler, Series B, Human Geography* 92 (2010): 251–61; and Simon, "Chorography Reconsidered," 23–44.

14. Lukerman, "Concept of Location," 195.
15. *Ptolemy's Geography*, 1.1.
16. *Ptolemy's Geography*, 8.1.
17. Simon, "Chorography Reconsidered," 36.
18. *Ptolemy's Geography*, 1.1.
19. *Ptolemy's Geography*, 1.1.
20. For example, see Neal Smith, "Contours of a Spatialized Politics: Homeless Vehicles and the Production of Geographical Scale," *Social Text* 33 (1992): 54–81.
21. *Ptolemy's Geography*, 1.1.
22. Daniel K. Connolly, *The Maps of Matthew Paris: Medieval Journeys through Space, Time, and Liturgy* (Woodbridge, Suffolk: Boydell and Brewer, 2009), 2–3, 28–29. See also Katharine Breen, *Imagining an English Reading Public, 1150–1400* (Cambridge: Cambridge University Press, 2010), 138–44, 167–71; and Kathryn M. Rudy, *Virtual Pilgrimages in the Convent: Imagining Jerusalem in the Late Middle Ages* (Turnhout: Brepols, 2011).
23. *Ptolemy's Geography*, 1.1.
24. Skelton and Harvey, *Local Maps and Plans*, 3.
25. *Ptolemy's Geography*, 1.1.
26. Moffitt, "Medieval *Mappaemundi* and Ptolemy's *Chorographia*," 61.
27. *Ptolemy's Geography*, 1.1.
28. Michael Curry, "Toward a Geography of a World without Maps: Lessons from Ptolemy and Postal Codes," *Annals of the Association of American Geographers* 95 (2005): 682; Simon, "Chorography Reconsidered," 23–44. See also Kenneth Olwig, "*Choros*, *Chora*, and the Question of Landscape," in *Envisioning Landscapes, Making Worlds: Geography and the Humanities*, ed. Stephen Daniels, Dydia DeLyser, J. Nicholas Entrikin, and Douglas Richardson (Milton Park, Abingdon, Oxon: Routledge, 2011), and the works cited therein in notes 8, 9, 10, and 12. For a discussion of chorography in relation to spaces in Anglo-Saxon literature, see Alfred Hiatt, "Beowulf Off the Map," *Anglo-Saxon England* 38 (2009): 11–40.
29. Simon, "Chorography Reconsidered," 31. John Agnew's distinctions among "location," "locale," and "sense of place" seem helpful here. Agnew defines a "location" as "the representation in local social interaction of ideas and practices derived from the relationship between places. In other words, location represents the impact of the 'micro order' in a place (uneven economic development, the uneven effects of government policy, segregation of social groups etc.)." He distinguishes that from a "locale," which "refers to the structured 'micro sociological' content of place, the settings of everyday, routine special interaction provided in a place," and a "sense of place," which "refers to the subjective orientation that can be engendered by living in a place."

Agnew, *Place and Politics: The Geographical Mediation of State and Society* (Boston: Allen and Unwin, 1987), 5.

30. Sir Isaac Newton, *Philosophiae Naturalis Principia Mathematica* (1686; Project Gutenberg, 2009), definition 8, http://www.gutenberg.org/files/28233/28233-h/28233-h.htm; Newton, *The Principia: Mathematical Principles of Natural Philosophy*, trans. I. Bernard Cohen and Ann Whitman (Berkeley: University of California Press, 2014), 408–9. A useful discussion of the Newtonian distinction appears in Max Jammer, *Concepts of Space: The History of Theories of Space in Physics* (Cambridge, MA: Harvard University Press, 1954), 98–99.

31. Harley and Woodward, "Concluding Remarks," 1:505.

32. Maurice Merleau-Ponty, *Phenomenology of Perception*, trans. Colin Smith (London: Routledge and Kegan Paul, 1962), 101, 102.

33. Merleau-Ponty, *Phenomenology of Perception*, 253, 304.

34. See Malpas's ideas throughout this work and, for a helpful beginning, the introduction to his book *Place and Experience*, 1–18.

35. Edward S. Casey, "Do Places Have Edges? A Geo-Philosophical Inquiry," in *Envisioning Landscapes, Making Worlds: Geography and the Humanities*, ed. Stephen Daniels, Dydia DeLyser, J. Nicholas Entrikin, and Doug Richardson (Abingdon, Oxon: Routledge, 2011), 67–69. See also Massimo Cacciari, "Place and Limit," and Casey, "Place and Edge," both in Malpas, *Intelligence of Place*, 13–22 and 23–38.

36. T. A. Heslop, "Eadwine and His Portrait," in *The Eadwine Psalter: Text, Image, and Monastic Culture in Twelfth-Century Canterbury*, ed. Margaret Gibson, T. A. Heslop, and Richard W. Pfaff (University Park: Pennsylvania State University Press, 1992), 184.

37. Nicholas Pickwoad and Francis Woodman describe the original binding and orientation. Nicholas Pickwoad, "Codicology and Palaeography," in Gibson, Heslop, and Pfaff, *The Eadwine Psalter*, 5–6; Francis Woodman, "The Waterworks Drawings of the Eadwine Psalter," in Gibson, Heslop, and Pfaff, *The Eadwine Psalter*, 171.

38. William Urry, "Canterbury, Kent, circa 1153 × 1161" in Skelton and Harvey, *Local Maps and Plans*, 43–58. See also Roberta Magnusson, *Water Technology in the Middle Ages: Cities, Monasteries, and Waterworks after the Roman Empire* (Baltimore: Johns Hopkins University Press, 2003), 53–115; and Woodman, "Waterworks Drawings," 175–76.

39. Robert Willis, "The Architectural History of the Conventual Buildings of the Monastery of Christ Church in Canterbury," *Archaeologia Cantiana* 7 (1868): 11.

40. The real exception to the containment and equilibrium of the local space in both manuscript images is a boat at the top of the folio 286r. Its lines are less elegantly rendered than the other lines on the page, and they are not shaded. It is difficult to tell, but it looks like it may be drawn in the same hand as the fainter "heshesh" below the orchard's trees. It may be that the vessel is a later addition, a whimsical annotation, but like marginal comments, diagrammatic arrows, and other images on the edges and elsewhere in other manuscripts, it is possibly indicative of an early reaction. What could the "heshesh" writer have been thinking as he or she looked at the page? The vessel has a shallow draft and curiously forked bowsprit and bumpkin at the stern, but the River Stour is (and was) some distance from the cathedral, and a masted boat would have been impractical on the river given the stone bridges that crossed it and still cross it. The boat is impossible to interpret with any confidence, yet its doodler seems to be picking

up on the nonrepresentational aspects of the manuscript's diagram and the larger image, signaling not a real boat in a real place but instead a sign of a vessel such as might ply the shallow waters of a river or another waterway. Another point is that the boat and person in it are not out of scale with the rest of the plan, and the image is oriented like the other structures, meaning that the annotator shares a basic sense of space similar to that of the original artist, and he or she seems to understand that objects in a chorographic space are categorizable as signs. The boat, like the buildings and the pipes—indeed everything on these maps in Trinity College, MS R.17.1—corresponds in some way with the observable phenomena of Canterbury Cathedral, but the objects are also abstract signs to depict the structures.

41. William St John Hope, *The History of the London Charterhouse from Its Foundation until the Suppression of the Monastery* (London: Society for Promoting Christian Knowledge, 1925), 107–44, with the plan facing page 107. Hope dates the first version of the map to shortly after 1422, whereas David Knowles and W. F. Grimes date it to about 1500. Knowles and Grimes, *Charterhouse, the Medieval Foundation in the Light of Recent Discoveries* (London: Longmans, Green, 1954), 36.

42. F. Hull, "Cliffe, Kent," in Skelton and Harvey, *Local Maps and Plans*, 99–105.

43. Hull, "Cliffe, Kent," 104.

44. Casey, "Place and Edge," 30.

45. Roland Barthes, *Empire of Signs*, trans. Richard Howard (New York: Hill and Wang, 1982), 43. The miniature has an "exaggeration of interiority," according to Susan Stewart, or what Edward Casey calls a certain "nonexchangability" between one place and another. Stewart, *On Longing: Narratives of the Miniature, the Gigantic, the Souvenir, the Collection* (Durham, NC: Duke University Press, 1993), 44; Casey, *Representing Place: Landscape Painting and Maps* (Minneapolis: University of Minnesota Press, 2002), 157.

46. H. S. A. Fox, "Exeter, Devonshire," in Skelton and Harvey, *Local Maps and Plans*, 163–69.

47. M. G. Snape, "Durham," in Skelton and Harvey, *Local Maps and Plans*, 189–94. Other local maps are similar in their presentation of parcels of property. See Skelton and Harvey, *Local Maps and Plans*, map numbers 16, 18, 20, 22, 23, 27, and 29.

48. See, for example, Skelton and Harvey on "Influences and Traditions," introduction, *Local Maps and Plans* 33–39.

49. M. W. Beresford, "Inclesmoor, West Riding of Yorkshire," in Skelton and Harvey, *Local Maps and Plans*, 161.

50. Beresford, "Inclesmoor, West Riding of Yorkshire," 153–61; Al Oswald, John Goodall, Andrew Payne, and Tara-Jane Sutcliffe, *Thornton Abbey, North Lincolnshire: Historical, Archaeological and Architectural Investigations*, English Heritage Research Department Report 100–2010 (Portsmouth: English Heritage, 2010), 105–6.

51. Quoted in Beresford, "Inclesmoor, West Riding of Yorkshire," 159.

52. There has been speculation about which map came first: whether DL 42/12 was a smaller, more schematic map that precedes the larger embellished MPC 1/56, whether the large map came first and was simplified in the smaller map, or whether the two are preceded by a lost earlier archetype. This problem has not been resolved, and indeed such speculation might reveal more about the expectations we bring to maps than anything else; the large map, for instance, has been described as a later "skillful elaboration" of the small one, and the small one a later "distillation" of the

larger. Beresford, "Inclesmoor, West Riding of Yorkshire," 160–61. Harvey hypothesizes further about a lost original in *The History of Topographical Maps: Symbols, Pictures and Surveys* (London: Thames and Hudson, 1980), 95.

53. Harvey has suggested the influence of these graphics. See his examples of Sluis in the Low Countries from the early fourteenth century or the rental houses of Gloucester street by street with images from 1455 in *History of Topographical Maps*, 88–90.

54. Derek J. Price, "Medieval Land Surveying and Topographical Maps," *Geographical Journal* 121 (1955): 1–7.

55. William L. Howarth, "Imagined Territory: The Writing of Wetlands," *New Literary History* 30 (1999): 520.

56. Beresford, "Inclesmoor, West Riding of Yorkshire," 159.

57. Curry, "Toward a Geography," 682.

58. On thirteenth-century items on the Hereford map, for example, see Valerie Flint, "The Hereford Map: Its Author(s), Two Scenes and a Border," *Transactions of the Royal Historical Society*, 6th ser., 8 (1998): 19–44.

59. On the altar, see Urry, "Canterbury, Kent, Late 14th Century x 1414," in Skelton and Harvey, *Local Maps and Plans*, 107–17.

60. Hull, "Isle of Thanet, Kent," 122; Antonia Gransden, *Historical Writing in England*, vol. 2, *1307 to the Early Sixteenth Century* (1974; New York: Routledge, 1996), 354–55.

61. Thomas of Elmham, *Historia Monasterii S. Augustini Cantuariensis*, ed. Charles Hardwick, Rolls Series 8 (London: Longman, 1858), 207.

62. John Gower, *Confessio amantis*, *The Complete Works*, ed. G. C. Macauley, vol. 3 (Oxford: Clarendon Press, 1901), 7.530; Geoffrey Chaucer, "To Rosemounde," in *Riverside Chaucer*, ed. Larry D. Benson, 3rd ed. (Boston: Houghton Mifflin, 1987), 2; Robert Henryson, *Orpheus and Eurydice*, in *The Complete Works*, ed. David J. Parkinson, TEAMS (Kalamazoo, MI; Medieval Institute Publications, 2010), 223, http://d.lib.rochester.edu/teams/text/parkinson-henryson-complete-works-orpheus-and-eurydice. All quotations from Chaucer are from the *Riverside Chaucer*; prologues and tales from the *Canterbury Tales* are cited by title.

63. For one discussion of the terminology involved, see Harley and Woodward in the preface to *History of Cartography*, 1:xvi; and David Woodward, "Medieval *Mappaemundi*," in Harley and Woodward, *History of Cartography*, 1:287–88.

64. Hull, "Isle of Thanet, Kent," 123.

65. Hull, "Isle of Thanet, Kent," 123.

66. Catherine Delano-Smith and Roger J. P. Kain, *English Maps: A History* (Toronto: University of Toronto Press, 1999), 29.

67. Thomas of Elmham, *Historia Monasterii S. Augustini Cantuariensis*, 207.

68. Thomas of Elmham, *Historia Monasterii S. Augustini Cantuariensis*, 207–8.

69. John Matthews Manly, "Chaucer and the Rhetoricians," in *Chaucer Criticism*, ed. Richard J. Schoeck and Jerome Taylor, vol. 1 (1926; Notre Dame, IN: University of Notre Dame Press, 1960), 271. Manly to an extent follows John Dryden, and both concentrate on Chaucer's depictions of characters rather than places. On Chaucer and realism, see Elizabeth Robertson, "Modern Chaucer Criticism," in *Chaucer: An Oxford Guide*, ed. Steve Ellis (Oxford: Oxford University Press, 2005), 358–62; Geoffrey W. Gust, *Constructing Chaucer: Author and Autofiction in the Critical Tradition* (New York: Palgrave MacMillan, 2009), 81–84.

70. Gransden, for example, describes the realism of *L'Histoire de Guillaume le Maréchal* (1225–1226) and Jean Le Bel's *Vrayes Chroniques* (1352–1361) in *Historical Writing in England*, vol. 1, *c. 500 to c. 1307* (1974; New York: Routledge, 1996), 305–14 and *Historical Writing in England*, 2:83–89, and she explores the feature further in "Realistic Observation in Twelfth-Century England," 29–51. The Kendrick quotation appears in that work; T. D. Kendrick, *British Antiquity* (London: Methuen, 1950), 134.

2. Local Literature

1. Aristotle, *Physics*, 4.1.208^{a}29–30.
2. Stephanie Trigg, introduction to *Congenial Souls: Reading Chaucer from Medieval to Postmodern* (Minneapolis: University of Minnesota Press, 2002), xv–xvi.
3. The term *setting*, incidentally, is a literary device and concomitant set of approaches that were not defined until the nineteenth century. *Oxford English Dictionary*, s.v. "setting" (n., sense 6b), http://www.oed.com.
4. Prologue of the Prioress's Tale, 7.454–59, 7.468–70, 7.485.
5. Prioress's Tale, 7.488–96.
6. Louise O. Fradenburg, "Criticism, Anti-Semitism, and the Prioress's Tale," *Exemplaria* 1 (1989): 98.
7. Prioress's Tale, 7.568. On space in the tale, see Kathy Lavezzo, *The Accommodated Jew: English Antisemitism from Bede to Milton* (Ithaca, NY: Cornell University Press, 2016), 106–34. A summary of earlier analyses of privy spaces in relation to anti-Semitism, including in the Prioress's Tale, appears in Heather Blurton and Hannah Johnson, *The Critics and the Prioress: Antisemitism, Criticism, and Chaucer's Prioress's Tale* (Ann Arbor: University of Michigan Press, 2017), 87–90.
8. Prioress's Tale, 7.594–96.
9. Prioress's Tale, 7.600–606.
10. The *Middle English Dictionary* provides several definitions of "inwith" and categorizes this example in the tale as a preposition meaning "Within the terminal limit of (a period of time), before the expiration of" (definition 3a). But other senses of "inwith" seem more appropriate. It is a difficult word to read in this context because it could refer to where Jesus reveals the space to be, or it could be where "in hir thoght" the space is revealed. The *Dictionary* defines "inwith" as an adverb and a preposition spatially in both of these ways. *Middle English Dictionary*, s.v. "inwith" (adv.) and "inwith" (prep.), ed. Hans Kurath, Sherman M. Kuhn, and Robert E. Lewis (Ann Arbor: University of Michigan Press, 1952–2001), http://quod.lib.umich.edu/m/med/.
11. Prioress's Tale, 7.613–20.
12. Prioress's Tale, 7.568, 7.571, 7.572, 7.604, 7.606.
13. Prioress's Tale, 7.683–90.
14. *The Middle English Translation of Guy de Chauliac's Treatise on "Apostemes": Book II of the Great Surgery*, ed. Björn Wallner, Publications of the New Society of Letters at Lund 80 (Lund: Lund University Press, 1988), 87.
15. *Lyf of the Noble and Crysten Prynce, Charles the Grete*, ed. Sidney J. H. Herrtage, EETS, e.s., 37 (London: N. Trübner, 1881), 1.
16. *Middle English Dictionary*, s.v. "locale" (adj.). Other dictionaries consulted include *A Dictionary of English Etymology*, ed. Hensleigh Wedgwood, vol. 2 (London: Trübner,

1862), the *Oxford English Dictionary*, and the *Anglo-Norman Dictionary*, The Anglo-Norman On-Line Hub, Universities of Aberystwyth and Swansea (http://www.anglo-norman.net/).

17. Guillaume de Deguileville, *The Pilgrimage of the Lyfe of the Manhode: From the French*, ed. William Addis Wright, Roxburghe Club (London: J. B. Nichols, 1869), 49.

18. Guillaume de Deguileville, *Le pélerinage de vie humaine*, ed. J. J. Stürzinger, Roxburghe Club (London: J. B. Nichols, 1893), 3221.

19. Ptolemy's Geography, 1.1.

20. *Middle English Dictionary*, s.v. "place," "space."

21. Aristotle, *Physics*, bks. 3–4, 4.4.212^{a}5–6, 4.4.212^{a}20–21. For a helpful analysis of Aristotle's ideas about place, see Benjamin Morison, *On Location* (Oxford: Oxford University Press, 2002), especially 2 and 11–53; and for a further critique of Morison's interpretation, see David Bostock, *Space, Time, Matter, and Form* (Oxford: Oxford University Press, 2006), 128–34.

22. Edward Grant, "Place and Space in Medieval Physical Thought," in *Motion and Time, Space and Matter: Interrelations in the History of Science and Philosophy*, ed. Peter K. Machamer and Robert G. Turnbull (Columbus: Ohio State University Press, 1976), 137. See also Jammer, *Concepts of Space*, 15–16; and Edward Grant, "The Medieval Doctrine of Place: Some Fundamental Problems and Solutions," in *Studi sul xiv secolo in memoria di Anneliese Maier*, ed. Maierù and Paravicini Bagliani (Rome: Storia e Letteratura, 1981), 57–79.

23. Cecilia Trifogli, *Oxford Physics in the Thirteenth Century (ca. 1250–1270): Motion, Infinity, Place, and Time* (Leiden: Brill, 2000), 16–17, 177–83.

24. Edward Grant, "The Concept of *Ubi* in Medieval and Renaissance Discussions of Place," *Manuscripta: A Journal for Manuscript Research* 20 (1976): 71–80. See also Pierre Duhem's three chapters on place and cosmology, "Theory of Place before the Condemnations of 1277," "Theory of Place from the Condemnations of 1277 to the End of the Fourteenth Century," and "Place in Fifteenth-Century Cosmology," in *Medieval Cosmology: Theories of Infinity, Place, Time, Void, and the Plurality of Worlds*, ed. and trans. Roger Ariew (Chicago: University of Chicago Press, 1985), 139–78, 179–268, 269–94.

25. Knight's Tale, 1.1895–96.

26. Louise O. Fradenburg, "Sacrificial Desire in Chaucer's *Knight's Tale*," *Journal of Medieval and Early Modern Studies* 27 (1997): 67.

27. Reeve's Tale, 1.4120–24.

28. Woods, *Chaucerian Spaces*, 78. See also Peter Brown, "The Containment of Symkyn: The Function of Space in the 'Reeve's Tale,'" *Chaucer Review* 14 (1980): 225–36.

29. W. Rothwell, "Anglo-French and English Society in Chaucer's 'The Reeve's Tale,'" *English Studies* 87 (2006): 521–23. Dictionaries consulted include the *Anglo-Norman Dictionary*; *Old French Online*, Linguistics Research Center, University of Texas at Austin (https://lrc.la.utexas.edu/eieol_base_form_dictionary/ofrol/15); the *Middle English Dictionary*; the *Oxford English Dictionary*; and *A Dictionary of English Etymology*.

30. Gower, *Confession amantis*, 2.3370.

31. Robert Manning of Brunne, *The Story of England*, pt. 1, ed. Frederick J. Furnivall, Rolls Series 87 (London: Longman, 1887), University of Michigan, Corpus of Middle English Prose and Verse, http://name.umdl.umich.edu/AHB1379.0001.001; *Peter Langtoft's Chronicle (as Illustrated and Improv'd by Robert of Brunne)*, pt. 2, ed. Thomas

Hearne (Oxford: Printed at the Theater, 1725), 5398, 5504, 5587, 14620, 15169, University of Michigan, Corpus of Middle English Prose and Verse, http://name.umdl.umich.edu/ABA2096.0001.001. Exeter appears in the Petyt manuscript of Mannyng; see note to line 5281.

32. *Romaunt of the Rose*, 1448.

33. Knight's Tale, 1.1971–72; Reeve's Tale, 1.4295.

34. Trigg, introduction, *Congenial Souls*, xiii–xvii. See also Phillipa Hardman, "Lydgate's Uneasy Syntax," and Robert Myer-Lee, "Lydgate's Laureate Prose," both in *John Lydgate: Poetry, Culture, and Lancastrian England*, ed. Larry Scanlon and James Simpson (Notre Dame, IN: University of Notre Dame Press, 2006), 12–35, 36–60.

35. John Lydgate, *The Siege of Thebes*, ed. Robert R. Edwards, TEAMS (Kalamazoo, MI: Western Michigan University, 2001), Pro.48.

36. John Lydgate, *Fall of Princes*, ed. Henry Bergen, 4 vols. (Washington, DC: Carnegie Institute of Washington, 1923–1927), 4.39. See also 8.2835.

37. *Siege of Thebes*, Pro.19.

38. *Siege of Thebes*, Pro.130–57. The "space of a bowe draught" is a familiar expression of a limited spatial distance in Middle English.

39. *Siege of Thebes*, 1.301–2, 1.322–24. Michelle Warren mentions these passages in "Lydgate, Lovelich, and London Letters," in *Lydgate Matters: Poetry and Material Culture in the Fifteenth Century*, ed. Lisa H. Cooper and Andrea Denny-Brown (New York: Palgrave Macmillan, 2007), 131–32.

40. *Siege of Thebes*, Pro.90.

41. An additional possible interpretation is that the large set of buildings on the left is another version of the cathedral itself pulled out from any realistic space, a kind of blowup of the structure we see in the town. This is a technique of highlighting and adding detail within the same picture frame that was not unusual in medieval illuminations and other visual arts. The buildings, though they might look like a gothic cathedral, however, do not resemble Canterbury. The artistic technique of a multipart illumination containing a magnification of detail furthermore typically has clear borders around each part of an image, and this one does not. If it were a blowup, then it also should not occupy the same area as the pilgrims in the foreground.

42. Jean A. Givens, *Observation and Image-Making in Gothic Art* (Cambridge: Cambridge University Press, 2005), 101–4. Contrast this image with images from Lydgate's *Lives of St. Edmund and St. Fremund* in British Library, Harley 2278, whose presentation, according to Sarah Salih, "does not constitute pure or mimetic landscape art or description, for the landscape never appears alone, and verisimilitude is not a main concern." "Lydgate's Landscape History," in Weiss and Salih, *Locating the Middle Ages*, 83.

43. *Siege of Thebes*, 1.187.

44. *Siege of Thebes*, 1.240.

3. Horizonal Space

1. James J. Gibson, *The Perception of the Visual World* (Cambridge, MA: Riverside, 1950), 6–7.

2. I proposed the term *horizonal* in "An Early English Rutter," 702–3, 726–27. Thank

you to Steven F. Kruger for suggesting the term, which has a legacy that can be traced back to Martin Heidegger and beyond. See the discussion and notes on *waagerecht* and *Gegend / Gegnet* in Heidegger's *Country Path Conversations*, trans. Bret W. Davis (Bloomington: Indiana University Press, 2010), 52–53, 73–74, and *auf* in Heidegger, *Ontology: The Hermeneutics of Facticity*, trans. John van Buren (Bloomington: Indiana University Press, 1999), 103–4.

3. Bartholomaeus Anglicus, *On the Properties of Things, John Trevisa's Translation of Bartholomaeus Anglicus De Proprietatibus Rerum, a Critical Text*, ed. M. C. Seymour, 3 vols. (Oxford: Clarendon Press, 1975–1988), 1.457. Elsewhere, Trevisa adds a comment that "'orisoun' is a straunge terme and moche i-vsed in astronomye; and to wite and knowe what 'orisoun' is to mene, takeþ hede þat if a man stondiþ on a gret hille oþir in a greet playn and large, so þat he se þe wolkin [sky] alle aboute withoute eny lette, hym schal seme þat þe wolkin touchith þe erþe al aboute [him] in euery seyde, and þat a cercle of þe wolkyn biclippith þe erþe al aboute as fer as his siȝt strecchith; þat cercle hatte [is called] 'orisoun', and is þe myddel cercle betwene þe partie þat we seeþ of heuen and þe partie þat we seeþ noȝt. And þis cercle is bitwene þe forseid parties, and is þe nethemest of þat on and ouemest [topmost] of þat oþir." Bartholomaeus, *On the Properties of Things*, 1.83. See also the narrator's apostrophizing about January in the Merchant's Tale: "O Januarie, what myghte it thee availle, / Thogh thou myghtest se as fer as shippes saille?" 4.2107–8.

4. James J. Gibson, *The Ecological Approach to Visual Perception* (New York: Houghton Mifflin, 1979), 127–28.

5. Gibson, *Ecological Approach*, 127.

6. In his philosophical discussion of spatiality, Malpas is careful to distinguish between a creature that experiences space via "the possession merely of certain behavioural capacities or dispositions" and a being that has "a grasp of the concept of space," which "requires . . . a spatial framework." He clarifies that a "grasp of the concept of space need not involve a grasp of any particular *theory* of space or of any specific term used to refer to space" but instead "the possession of complex spatial and topographic frameworks" that are "necessary for the sort of experience and understanding of the world that is characteristic of human experience and thought." Malpas, *Place and Experience*, 46–49.

7. Aristotle, *Physics*, bks. 3–4, 4.4.212^{a}5–6, 4.4.212^{a}20–21.

8. Astrolabes are explained with many illustrations in Robert T. Günther, *The Astrolabes of the World*, 2 vols. (Oxford: Oxford University Press, 1932). Fourteen English medieval astrolabes and Chaucer's *Treatise on the Astrolabe* are described on 463–83. Further summary and analysis of astrolabes and Chaucer's *Treatise* appear in *A Treatise on the Astrolabe*, ed. Sigmund Eisner, Variorum Edition, vol. 6, pt. 1 (Norman: University of Oklahoma Press, 2002) and the *Riverside Chaucer*, 1092–102. Paragraph numbers refer to the *Riverside* edition. J. A. W. Bennett's discussion of Merton College covers astrolabes and astronomical treatises: *Chaucer at Oxford and at Cambridge* (Oxford: Oxford University Press, 1974), 33, 58–85.

9. Bertrand Westphal, *The Plausible World: A Geocritical Approach to Space, Place, and Maps*, trans. Amy D. Wells (New York: Palgrave Macmillan, 2013), 46, 48. Westphal bases some of his ideas on Paul Zumthor, *La mèsure du monde: Représentation de l'espace au Moyen Âge* (Paris: Seuil, 1993), 61–62.

10. David Woodward, "Maps and the Rationalization of Geographic Space," in *Circa 1492: Art in the Age of Exploration*, ed. Jay A. Levenson (Washington, DC: National Gallery of Art, 1991), 83, 85.

11. *Middle English Dictionary*, s.v. "abstract."

12. *Paradiso*, *The Divine Comedy*, ed. and trans. Charles S. Singleton, 3 vols. (Princeton, NJ: Princeton University Press, 1975), 22.133–38.

13. *Boece*, 4.pr.6.217–19.

14. *Troilus and Criseyde*, 5.1815–16. Compare *Boece*, 2.pr.7.23–49.

15. *The Writings of Julian of Norwich: A Vision Showed to a Devout Woman and a Revelation of Love*, ed. Nicholas Watson and Jacqueline Jenkins (University Park: Pennsylvania State University Press, 2006), 139.

16. On "homely," see note to line 15 on page 136 (line 15 page 137) in *The Writings of Julian of Norwich*.

17. John Metham, *Amoryus and Cleopes*, ed. Stephen F. Page, TEAMS (Kalamazoo, MI: Medieval Institute Publications, 1999), 304–5.

18. *House of Fame*, 907.

19. Analysis of the image may be found in Joyce Coleman, "Illuminations in Gower's Manuscripts," in *The Routledge Research Companion to John Gower*, ed. Ana Saez-Hidalgo, Brian Gastle, and R. F. Yeager (New York: Routledge, 2017), 118–21. Ways that audiences may have interacted with *mappaemundi* more generally are discussed in Ruth Evans, "Getting There: Wayfinding in the Middle Ages," in Allen and Evans, *Roadworks*, 138–39.

20. David Woodward, "Medieval *Mappaemundi*," 1:334–58.

21. Tony Campbell, "Portolan Charts from the Late Thirteenth Century to 1500," in Harley and Woodward, *History of Cartography*, 1:377–78.

22. P. D. A. Harvey, "Local and Regional Cartography in Medieval Europe," in Harley and Woodward, *History of Cartography*, fig. 20.8 and 1:473. See also Harvey, *History of Topographical Maps*, fig. 41 and 79–81.

23. The terminology, as with most terminology used with maps, is partly or wholly anachronistic. A true rhumb line is a loxodrome, a line of constant compass bearing. See the discussion of rhumb lines and terminology in general in Campbell, "Portolan Charts," 1:375–81.

24. See J. B. Harley and David Woodward, "Greek Cartography in the Early Roman World," in Harley and Woodward, *History of Cartography*, 1:165, 1:174; Dilke, "The Culmination of Greek Cartography in Ptolemy," esp. 1:179–80, 1:184–85, 1:187.

25. Woodward, "Medieval *Mappaemundi*," 1:286; Harvey, "Local and Regional Cartography in Medieval Europe," 1:496.

26. Harvey, "Local and Regional Cartography," fig. 20.11, 1:475–76, 1:496–97. A twelfth-century Eastern geography, Abu ʿabd-Allah Muhammad al-Idrīsī's *Nuzhat al-mushtaq fi-ikhtirāq al-āfāq*, describes areas by (latitudinal) climes and (longitudinal) sections to form a grid, and his descriptions would therefore key to a map, but no medieval example with an illumination survives.

27. Bibliothèque Nationale de France, ms. Latin 4939, http://gallica.bnf.fr/ark:/12148/btv1b55002483j.

28. Riccardo Manzotti, *The Spread Mind: Why Consciousness and the World Are One* (New York: OR Books, 2018), vii.

29. The development of the two types of quadrants—the *quadrans vetus* and the *quadrans novus*—from the astrolabe is summarized in Elly Dekker, "'With his sharp lok perseth the sonne': A New Quadrant from Canterbury," *Annals of Science* 65 (2008): 201–4.

30. For a summary of surviving works of practical geometry, see H. L. L. Busard, introduction to *De arte mensurandi*, by Johannes de Muris, trans. H. L. L. Busard (Stuttgart: Franz Steiner Verlag, 1998), 7–12.

31. Moody, "Empiricism and Metaphysics," 147.

32. Hugh of St. Victor, *Practica geometriae, Hugonis de Sancto Victore Opera Propaedeutica*, ed. Roger Baron (Notre Dame, IN: University of Notre Dame Press, 1966), prologue; *Practical Geometry (Practica Geometriae), Attributed to Hugh of St. Victor*, trans. Frederick A. Homann (Milwaukee, WI: Marquette University Press, 1991), 33.

33. Hugh of St. Victor, *Didascalicon*, 2.6; *Didascalicon*, trans. Taylor, 67.

34. Hugh of St. Victor, *Didascalicon*, 2.13; *Didascalicon*, trans. Taylor, 70.

35. Hugh of St. Victor, *De arca Noe morali*, *Practical Geometry*, trans. Homann, 85–86.

36. On the astrolabe, see, for example, Hugh of St. Victor, *Practica geometriae*, 1.10–1.18; *Practical Geometry*, trans. Homann, 40–47.

37. Hugh of St. Victor, *Practica geometriae*, 1.18; *Practical Geometry*, trans. Homann, 46.

38. Hugh of St. Victor, *Practica geometriae*, 2.36; *Practical Geometry*, trans. Homann, 55.

39. *Artis cuiuslibet consummatio*, *Practical Geometry in the High Middle Ages: Artis Cuiuslibet Consummatio and the Pratike De Geometrie*, trans. Stephen K. Victor (Philadelphia: American Philosophical Society, 1979), 1.34, 1.36, 2.30–31, 2.33.

40. Authorship, dates, and manuscripts are discussed in Nan Hahn, "Medieval Mensuration: 'Quadrans Vetus' and 'Geometrie Due Sunt Partes Principales,'" *Transactions of the American Philosophical Society* 72 (1982): xv–xx; and Wilbur Knorr, "The Latin Sources of the *Quadrans vetus*, and What They Imply for Its Authorship and Date," in *Texts and Contexts in Ancient and Medieval Science: Studies on the Occasion of John E. Murdoch's Seventieth Birthday*, ed. Edith Sylla and Michael McVaugh (Leiden: Brill, 1997), 23–67.

41. The *Quadrans vetus* discusses heights, widths, and depths in paragraphs 48–52, 65–76, and 77–78, in Hahn, "Medieval Mensuration," 65–72 and 90–106.

42. Chaucer, *Treatise on the Astrolabe*, paragraph 41.

43. *The Practica Geometriae of Dominicus de Clavasio*, ed. H. L. L. Busard, *Archive for History of Exact Sciences* 2 (1965): 520–21.

44. The words appear throughout the *Practica Geometriae*, pages 524–75 in Busard's edition.

45. Hugh of St. Victor, *Practica geometriae*, 1.18; *Practical Geometry*, trans. Homann, 46.

46. *Artis cuiuslibet consummatio*, *Practical Geometry in the High Middle Ages*, 2.32.

47. Hugh of St. Victor, *Practica geometriae*,1.22; *Practical Geometry*, trans. Homann, 48.

48. Hahn, "Medieval Mensuration," paragraphs 60–63, 66–68, pages 83–88, 92–96.

49. "A Method Used in England in the Fifteenth Century for Taking the Altitude of a Steeple or Inaccessible Object," in *Rara Mathematica; or, A Collection of Tretises on the Mathematics and Subjects Connected with Them*, ed. James Orchard Halliwell (London: J. W. Parker, 1839), 27–28.

50. Robertson, "Medieval Materialism," 108.

51. Certeau describes the medieval map (there is only one type of medieval map in his analysis) as one that "included only the rectilinear marking out of itineraries (performative indications chiefly concerning pilgrimages), along with the stops one was to make (cities which one was to pass through, spend the night in, pray at, etc.) and distances calculated in hours or in days, that is, in terms of the time it would take to cover them on foot. Each of these maps is a memorandum prescribing actions." In contrast, the early modern map "erases" the itineraries and pushes the signs of travel to its margins so that it "became more autonomous . . . a formal ensemble of abstract places." While Certeau is in fact more interested in premodern maps, his reliance on the itinerary map as evidence for all medieval spatial understanding, and his sharp contrasts with early modern cartography are not particularly sustainable. It is at least possible to propose that the late medieval era is when we see both kinds of spatial tendencies, and more, in scientific, literary, and mapmaking practices. Certeau, *Practice of Everyday Life*, 120–21. Certeau bases his argument on George Kimble's 1938 *Geography in the Middle Ages* and other works. Kimble, *Geography in the Middle Ages* (London: Methuen, 1938).

52. These assertions about limited medieval knowledge, particularly of mathematics, coincide too easily with prejudices about the Middle Ages and at other times are more carefully articulated. See, for example, H. C. Darby, "The Agrarian Contribution to Surveying in England," *Geographical Journal* 82 (1933): 529–35; E. G. R. Taylor, "The Surveyor," *Economic History Review* 17 (1947): 121–33; E. G. R. Taylor, *The Haven-Finding Art: A History of Navigation from Odysseus to Captain Cook* (London: Hollis and Carter, 1956), 158–59; Lon Shelby, "The Geometrical Knowledge of Mediaeval Master Masons," *Speculum* 47 (1972): 404–405; Hahn, "Medieval Mensuration," xi–xiv.

53. Bennett, *Chaucer at Oxford and at Cambridge*, 65–75.

54. These last two arguments about cartography are made by Skelton and Harvey, who nevertheless offer some evidence for land measurement, as below, and conclude that "research is still needed to elucidate the chronology of the union of map-making and surveying in England." R. A. Skelton and P. D. A. Harvey, "Surveying in Medieval England," in Skelton and Harvey, *Local Maps and Plans*, 19.

55. Skelton and Harvey, "Surveying in Medieval England," 14.

56. *Riverside Chaucer*, 1195; Chaucer, *Treatise on the Astrolabe*, 103.

57. "The Richard II Quadrant," 1860,0519.1, accessed June 23, 2018, www.britishmuseum.org/collection, British Museum. http://www.britishmuseum.org/research/collection_online/collection_object_details.aspx?objectId=55091&partId=1; Silke Ackermann and John Cherry, "Richard II, John Holland and Three Medieval Quadrants," *Annals of Science* 56 (1999): 3–23.

58. Introduction, *Practica Geometriae of Dominicus de Clavasio*, 520; Voigts, "Scientific and Medical Books," 382–84.

59. Hugh of St. Victor, *Practica geometriae*, prologue, prenotanda 2; *Practical Geometry*, trans. Homann, 33–34.

60. Homann, introduction, *Practical Geometry*, 24. On the practical application of "practical geometry," see *Practical Geometry in the High Middle Ages*, trans. Victor, 53–

73. Victor writes that "the contents of the practical geometries hint strongly at some connections with practical concerns" (54).

61. *Artis cuiuslibet consummatio, Practical Geometry in the High Middle Ages*, 1, prologue.

62. Ptolemy's *Geography*, 1.1.

63. Quintilian, *Insitutio oratoria*, trans. H. E. Butler, Loeb Classical Library (Cambridge, MA: Harvard University Press, 1920–1922), 8.3.61–62, 9.2.40–44.

64. Ernst Robert Curtius, *European Literature and the Latin Middle Ages* (Princeton, NJ: Princeton University Press, 1952), 202n37.

65. Matthew of Vendôme, *Ars Versificatoria, Mathei Vindocinensis: Opera*, ed. Franco Munari, vol. 3 (Rome: Edizioni di Storia e Letteratura, 1988); *Ars Versificatoria: The Art of the Versemaker*, trans. Roger Parr (Milwaukee, WI: Marquette University Press, 1981), 1.109–11, 1.113. Further discussion of Matthew in relation to Middle English poetry may be found in Derek A. Pearsall, "Rhetorical 'Descriptio' in 'Sir Gawain and the Green Knight,'" *Modern Language Review* 50 (1955): 129–34.

66. Gerald of Wales refers to the *Topographia* with this name in his *De rebus a se Gestus* and elsewhere. *Giraldi Cambrensis opera*, ed. J. S. Brewer, vol. 1, Rerum Britannicarum medii aevi scriptores (London: Longman, Green, Longman, and Roberts, 1861), 80.

67. On the *Cosmographia* and Gerald's approach to the natural world, Robert Bartlett describes his "detailed, individualized observation, alive to the natural world as a motley multitude of particulars rather than a cosmic pattern." *Gerald of Wales, 1146–1223* (Oxford: Clarendon Press, 1982), 133.

68. Gerald of Wales, *Topographia Hibernica, Giraldi Cambrensis Opera*, vol. 5, ed. James F. Dimock, Rerum Britannicarum medii aevi scriptores (London: Longman, Green, Longman, and Roberts, 1861), 7; Gerald of Wales, *The Historical Works of Giraldus Cambrensis Containing the Topography of Ireland, and the History of the Conquest of Ireland*, ed. Thomas Wright, trans. Thomas Forester (London: H. G. Bohn, 1863), 7.

69. Jeffrey J. Cohen, *Hybridity, Identity, and Monstrosity in Medieval Britain: On Difficult Middles* (New York: Palgrave Macmillan, 2006), 85.

70. Gerald of Wales, *Topographia Hibernica*, 5.7–8; Gerald of Wales, *Historical Works of Giraldus Cambrensis*, trans. Thomas Forester, 6–8.

71. Lavezzo, *Angels on the Edge of the World*, 61–62. Lavezzo cites Pliny's observations about environmental determinism in which creatures not only match but also are a consequence of their loci. Pliny the Elder, *Natural History*, trans. H. Rackham (Cambridge, MA: Harvard University Press, 1938), 1.320–23. On Gerald's connections with cartographic activities at Hereford Cathedral, see Daniel Birkholz, "Hereford Maps, Hereford Lives: Biography and Cartography in an English Cathedral City," in Lilley, *Mapping Medieval Geographies*, 225–49.

72. *The Anonimalle Chronicle, 1333 to 1381*, ed. V. H. Galbraith (Manchester: Manchester University Press, 1927), 143.

73. *Anonimalle Chronicle*, 143.

74. *Anonimalle Chronicle*, 144.

75. The climax of the revolt has been discussed, for example, in Charles Oman, *The Great Revolt of 1381* (Oxford: Clarendon Press, 1906); Rodney Hilton, *Bond Men Made Free: Medieval Peasant Movements and the English Rising of 1381* (New York: Viking, 1973);

R. B. Dobson, ed., *The Peasants' Revolt of 1381* (London: Macmillan, 1983); R. H. Hilton and T. H. Aston, eds., *The English Rising of 1381* (Cambridge: Cambridge University Press, 1984); and Steven Justice, *Writing and Rebellion: England in 1381* (Berkeley: University of California Press, 1994).

76. *Anonimalle Chronicle*, 141.

77. Westphal, The Plausible World, 46; Woodward, "Maps and the Rationalization of Geographic Space," 85.

78. *Middle English Dictionary*, s.v. "frounter"; *Anglo-Norman Dictionary*, s.v. "frunter" (n., sense 2).

79. Bartholomaeus, On the Properties of Things, 1.457.

4. Horizontal and Abstracted Spaces

1. Gibson, *Ecological Approach*, 127.

2. Woodward, "Maps and the Rationalization of Geographic Space," 84.

3. George Puttenham, *Arte of English Poesie*, ed. Gladys Dodge Willcock and Alice Walker (Cambridge: Cambridge University Press, 1936), 239.

4. *Proslepsis* is elongated *paralipsis* or *occultatio*. "Paralipsis occurs when we say that we are passing by, or do not know, or refuse to say that which precisely now we are saying" (Occultatio est cum dicimus nos praeterire aut non scire aut nolle dicere id quod nunc maxime dicimus). *Ad C. Herennium*, trans. Harry Caplan, Loeb Classical Library (Cambridge, MA: Harvard University Press, 1954), 4.26.36.

5. Clerk's Prologue and Tale, 4.39–56.

6. Petrarch, *Historia Griseldis*, 17.3.61–63; *Le Livre Griseldis*, lines 9–12; both in Thomas J. Farrell and Amy W. Goodwin, "The Clerk's Tale," in *Sources and Analogues of the Canterbury Tales*, vol. 1, ed. Robert M. Correale and Mary Hamel (Cambridge: D. S. Brewer, 2002), 101–67. On Chaucer's use of Giovanni Boccaccio's *Decameron*, Petrarch's *Historia*, *Le Livre Griseldis*, and other sources, see the groundbreaking work of J. Burke Severs, "The Clerk's Tale," in *Sources and Analogues of Chaucer's Canterbury Tales*, ed. W. F. Bryan and Germaine Dempster (1941; New York: Humanities Press, 1958), 288–331; and Severs, *The Literary Relationships of Chaucer's Clerkes Tale* (New Haven, CT: Yale University Press, 1942). For Boccaccio, see *Decameron*, ed. Vittore Branca (Torino: Einaudi, 1980), 1232–48.

7. Clerk's Prologue and Tale, 4.57–65.

8. Severs, "Clerk's Tale," 288; William Ellen Bettridge and Francis Lee Utley, "New Light on the Origin of the Griselda Story," *Texas Studies in Literature and Language* 13 (1971): 153–208; Farrell and Goodwin, "Clerk's Tale," 101; Richard Firth Green, "Griselda in Siena," *Studies in the Age of Chaucer* 33 (2011): 4.

9. Clerk's Prologue and Tale, 4.1135–36.

10. Studies of the *Book*'s composition include Barry Windeatt's edition, *The Book of Margery Kempe*, ed. Barry Windeatt (Cambridge: D. S. Brewer, 2000), 5–9; Lynn Staley, *Margery Kempe's Dissenting Fictions* (University Park: Pennsylvania State University Press, 1994), 1–38; Nicholas Watson, "The Making of *The Book of Margery Kempe*," in *Voices in Dialogue: Reading Women in the Middle Ages*, ed. Linda Olson and Kathryn Kerby-Fulton (Notre Dame, IN: University of Notre Dame Press, 2005), 395–434; and the works cited below.

11. For a description of *dispositio*, see Rita Copeland and Ineke Sluiter, eds., *Medieval Grammar and Rhetoric: Language Arts and Literary Theory, AD 300–1475* (Oxford: Oxford University Press, 2009), 39–42. Thomas of Chobham, among others who follow Cicero's *De inventione*, links *dispositio* with *memoria* in his thirteenth-century *Summa de arte praedicandi*, 2.2, in *Medieval Grammar and Rhetoric*, 632. Mary Carruthers also shows the connection between the two in *The Book of Memory: A Study of Memory in Medieval Culture* (Cambridge: Cambridge University Press, 1990), 82–86. On book 14 of Isidore of Seville's *Etymologies* and its exploration of relations between geography and memorial arts, see Andy Merrills, "Geography and Memory in Isidore's *Etymologies*," in Lilley, *Mapping Medieval Geographies*, 45–64. On Kempe and the memorial arts, see Staley, *Margery Kempe's Dissenting Fictions*, 86–88.

12. The word *qarantal* appears to derive from Latin, meaning "forty."

13. On the word *Sarazyn* and Saracens more generally, see Jeffrey J. Cohen, *Medieval Identity Machines* (Minneapolis: University of Minnesota Press, 2003), 188–221; and Anthony Bale, *Feeling Persecuted: Christians, Jews and Images of Violence in the Middle Ages* (London: Reaktion, 2010), 165.

14. *Book of Margery Kempe*, 173.

15. Analyses of the spaces of the scene include Ruth Summar McIntyre, "Margery's 'Mixed Life': Place Pilgrimage and the Problem of Genre in *The Book of Margery Kempe*," *English Studies* 89 (2008): 655–56; and Albrecht Classen, *East Meets West in the Middle Ages and Early Modern Times: Transcultural Experiences in the Premodern World* (Berlin: Walter de Gruyter, 2013), 57–58.

16. Gibson, *Ecological Approach*, 127.

17. Anthony Goodman, *Margery Kempe and Her World* (Harlow, Essex: Pearson, 2002), 186; Diane Watt, "Faith in the Landscape: Overseas Pilgrimages in *The Book of Margery Kempe*," in Lees and Overing, *A Place to Believe In*, 170–87; Bale, *Feeling Persecuted*, 161–65. Bale draws on Rosalyn Voaden, "Travels with Margery: Pilgrimage in Context," in *Eastward Bound: Travel and Travellers, 1050–1550*, ed. Rosamund Allen (Manchester: Manchester University Press, 2004), 177–95.

18. *Book of Margery Kempe*, 163.

19. *The Writings of Julian of Norwich*, 279, 281, 289, 299, 301, 335, 337, 373.

20. Virginia Raguin, "Real and Imagined Bodies in Architectural Space: The Setting for Margery Kempe's *Book*," in *Women's Space: Patronage, Place, and Gender in the Medieval Church*, ed. Virginia Raguin and Sarah Stanbury (Albany: State University of New York Press, 2005), 105. Windeatt's notes on Kempe cite two uses of the city metaphor (though he incorrectly says "syte" means "city" in chapter 67 [2794]; it means "sight").

21. *Book of Margery Kempe*, 372.

22. *Book of Margery Kempe*, 333–34.

23. *Book of Margery Kempe*, 75–78.

24. Yi-Fu Tuan, *Space and Place: The Perspective of Experience* (Minneapolis: University of Minnesota Press, 1977), 46.

25. Though by no means all of them. Voaden notes that Lynn was a major international port, yet its activities and buildings, mere steps from the church, are not described. "Travels with Margery," 179–80.

26. *Book of Margery Kempe*, 328. Raguin points out that the *Book* "is rife with specific notations of place," but Sarah Stanbury writes that "as readers have noted, there is an absence of descriptive detail and attention to place in Kempe's *Book*, and although she tells us a great deal about where she goes and with whom she talks, she gives us little information about the surface texture of either people or places." Raguin, "Real and Imagined Bodies in Architectural Space," 105; Stanbury, "Margery Kempe and the Arts of Self-Patronage," in Raguin and Stanbury, *Women's Space*, 78.

27. *Book of Margery Kempe*, 66–69.

28. *Book of Margery Kempe*, 330–31.

29. Dee Dyas, *Pilgrimage in Medieval English Literature, 700–1500* (Cambridge: D. S. Brewer, 2001), 130.

30. *Book of Margery Kempe*, 360.

31. *Book of Margery Kempe*, 145. In the second book, when Kempe travels with her daughter-in-law to the Low Countries, they are described as stopping at a church five or six miles outside Lynn. *Book of Margery Kempe*, 392 and 394.

32. *Book of Margery Kempe*, 213–14.

33. Bale discusses the replicas in *Feeling Persecuted*, 149–59, before going on to discuss Kempe. Samuel Fanous notes similarities between the Franciscans and Kempe in "Measuring the Pilgrim's Progress: Internal Emphases in *The Book of Margery Kempe*," in *Writing Religious Women: Female Spiritual and Religious Practices in Late Medieval England*, ed. Denis Renevey and Christiania Whitehead (Toronto: University of Toronto Press, 2000), 159.

34. *Book of Margery Kempe*, 420.

35. *Book of Margery Kempe*, 176, 221. On medieval guides, see Evans, "Getting There," 134–37.

36. Kate Parker, "Lynn and the Making of a Mystic," in *A Companion to The Book of Margery Kempe*, ed. John H. Arnold and Katherine J. Lewis (Cambridge: D. S. Brewer, 2004), 55–73.

37. *Book of Margery Kempe*, 123.

38. Pardoner's Prologue, 6.417–19.

39. *Book of Margery Kempe*, 292.

40. *Book of Margery Kempe*, 267.

41. *Book of Margery Kempe*, 104, 225.

42. *Book of Margery Kempe*, 205–6.

43. *Book of Margery Kempe*, 172, 176.

44. *Book of Margery Kempe*, 415–17.

45. Sebastian Sobecki, "'The writyng of this tretys': Margery Kempe's Son and the Authorship of Her Book," *Studies in the Age of Chaucer* 37 (2015): 271; see also 275–78.

46. *Book of Margery Kempe*, 41.

47. *Middle English Dictionary*, s.v. "conformen" (v., sense 3).

48. *Book of Margery Kempe*, 146.

49. *Middle English Dictionary*, s.v. "convenyens."

50. "Yf the namys of the placys be not ryth wretyn, late no man merveylyn, for sche stodyid mor abowte [concentrated more on] contemplacyon than the namys of the placys, and he that wrot hem had nevyr seyn hem, and therfor have hym excusyd." *Book of Margery Kempe*, 401.

51. Margery's own lack of measurement appears on 169, 214, 285, 323, and 355. In contrast, Christ's measurement of her tears and so on occurs on 299, 301, 376, and 418. Both priests' inability to measure is on 294. See also 311 for another example of a priest not being able to "mesuryn" himself in the presence of a weeping "holy woman."

52. The literature on Mandeville is extensive. Work on authorship and the composition may be found in Michael J. Bennett, "Mandeville's Travels and the Anglo-French Moment," *Medium Aevum* 75 (2006): 273–92; Iain Macleod Higgins, Introduction, *The Book of John Mandeville: With Related Texts*, edited and translated by Higgins (Indianapolis, IN: Hackett, 2011), ix–xii; W. Mark Ormrod, "John Mandeville, Edward III, and the King of Inde," *Chaucer Review* 46 (2012): 314–39; Legassie, *Medieval Invention*, 78–82; and the works cited below. On the representation of Others, see, for example, Shirin A. Khanmohamadi, *In Light of Another's Word: European Ethnography in the Middle Ages* (Philadelphia: University of Pennsylvania Press, 2013), 113–44.

53. Anthony Bale, "'ut legi': Sir John Mandeville's Audience and Three Late Medieval English Travelers to Italy and Jerusalem," *Studies in the Age of Chaucer* 38 (2016): 203–4.

54. *Mandeville's Travels*, ed. M. C. Seymour (Oxford: Clarendon Press, 1967), 3.

55. Mary B. Campbell, *The Witness and the Other World: Exotic European Writing, 400–1600* (Ithaca, NY: Cornell University Press, 1988), 129–30.

56. *Mandeville's Travels*, 5.
57. *Mandeville's Travels*, 106.
58. *Mandeville's Travels*, 1.
59. *Mandeville's Travels*, 39.
60. *Mandeville's Travels*, 11.
61. *Mandeville's Travels*, 15.
62. *Mandeville's Travels*, 32.
63. *Mandeville's Travels*, 119.
64. *Mandeville's Travels*, 54–55, 72.
65. *Mandeville's Travels*, 228.
66. *Mandeville's Travels*, 222.
67. *Mandeville's Travels*, 131–32. There is no single Antarctic star in the Southern Hemisphere.

68. On this episode, see my *The Idea of the Antipodes: Place, People, and Voices* (New York: Routledge, 2010), 67–69.

69. *Mandeville's Travels*, 132–34.

70. Where Iain Higgins reads this episode as suggesting all the world is relative to the West (and vice versa), it seems that Mandeville is being more critical and radical. Higgins, *Writing East: The "Travels" of Sir John Mandeville* (Philadelphia: University of Pennsylvania Press, 1997), 137.

71. *Mandeville's Travels*, 136–37.
72. Westphal, *Plausible World*, 46.

5. The Science of Motion

1. Anneliese Maier, "Causes, Forces, and Resistance," in Maier, *On the Threshold of the Exact Science*, 51.

2. Robertson, *Nature Speaks*, 9.

3. Maier, "Causes, Forces, and Resistance," 48–49.

4. Aristotle, *Physics*, bks. 1–2, trans. W. Charlton, Clarendon Aristotle Series (Oxford: Oxford University Press, 1970), 2.1.192^{b}10–22.

5. Aristotle, *Physics*, 3.1.200^{b}14. On Aristotle's ideas about nature and kinēsis, see Helen S. Lang, *Aristotle's Physics and Its Medieval Varieties* (Albany: State University of New York Press, 1992), 23–24; and Morison, *On Location*, 11–15.

6. Grant, *History of Natural Philosophy*, 174–75.

7. Bartholomaeus, *On the Properties of Things*, 516.

8. British Library, Royal MS 12 G II, fol. 1v, and Oxford, Merton College MS 272, fol. 176, both in Robert Bartlett, *The Natural and the Supernatural in the Middle Ages* (Cambridge: Cambridge University Press, 2008), 37n8. The Royal quotation also appears in Charles Burnett, "The Introduction of Aristotle's Natural Philosophy into Great Britain," in Marenbon, *Aristotle in Britain during the Middle Ages*, 46. Roger French attributes the first text to Henry of Rainham "noting down what he heard in lectures in Oxford before 1300" in *Canonical Medicine: Gentile Da Foligno and Scholasticism* (Leiden: Brill, 2001), 30n43. The second quotation is from Geoffrey of Aspall from the thirteenth century, in translation only in John E. Murdoch and Edith D. Sylla, "The Science of Motion," in *Science in the Middle Ages*, ed. David C. Lindberg (Chicago: University of Chicago Press, 1978), 206. On Aspell, see Enya Macrae, "Geoffrey of Aspall's Commentaries on Aristotle," *Mediaeval and Renaissance Studies* (Warburg Institute) 6 (1968): 94–134; James Weisheipl, "Science in the Thirteenth Century," in *The History of the University of Oxford*, vol. 1, *The Early Oxford Schools*, ed. J. I. Catto (Oxford: Clarendon Press, 1984), 464; and Trifogli, *Oxford Physics in the Thirteenth Century (ca. 1250–1270)*, 25.

9. Anneliese Maier, "The Nature of Motion," in Maier, *On the Threshold of Exact Science*, 22; Murdoch and Sylla, "Science of Motion," 206–7; Clagett, *Science of Mechanics*, 422. Medieval philosophers were not so interested in Aristotle's ontological discussion of motion, although it had some bearing on their discussions. On motion's ontology, see Johannes Thijssen, "The Nature of Change," in Pasnau and van Dyke, *Cambridge History of Medieval Philosophy*, 1:279–90.

10. On Albertus Magnus, see Grant, *History of Natural Philosophy*, 164–65. On *ens mobili*, see Lang, *Aristotle's Physics*, 167–68; and Edith D. Sylla, "A Guide to the Text" of Buridan, in *Quaestiones super octo libros Physicorum Aristotelis (secundum ultimam lecturam)* bks. 1–2, ed. Michiel Streijger and Paul J. J. M. Bakker, History of Science and Medicine Library 50 (Leiden: Brill, 2015), lxxxv–lxxxvi.

11. Aquinas, *Commentaria in octo libros Physicorum*, http://www.corpusthomisticum.org/cpy011.html; Aquinas, *Commentary on Aristotle's Physics*, 1, lectio 1, chap. 1, par. 3, http://www.dhspriory.org/thomas/Physics1.htm.

12. Ernest A. Moody, "Buridan, Jean," in *Complete Dictionary of Scientific Biography*, vol. 2 (Detroit: Charles Scribner's Sons, 2008), 603; see also Jack Zupko, "John Buridan," in *The Stanford Encyclopedia of Philosophy*, Spring 2014, http://plato.stanford.edu/archives/spr2014/entries/buridan/.

13. Buridan, *Quaestiones super octo libros Physicorum Aristotelis*, 287$^{9-13}$; Sylla, "A Guide to the Text," cl7.

14. *Boece*, 3.pr.12.219. The location of this realm is incidentally not as clear as is often assumed, for while the fixed stars are in some schemes the truly unmoving

objects, in other conceptions the earth at the center of the universe is another fixed object, while in still other descriptions the earth under the moon is the least fixed of all.

15. *Boece*, 4.pr.6.71.

16. *Boece*, 4.pr.2.97–114.

17. Aristotle, *Physics*, bk. 8, trans. Daniel W. Graham, Clarendon Aristotle Series (Oxford: Oxford University Press, 1999), 8.6.258^{b}10–259^{a}7.

18. Aquinas, *Commentaria in octo libros Physicorum*, http://www.corpusthomisticum.org/cpy07.html#72553; Aquinas, *Commentary on Aristotle's Physics*, 8, http://www.dhspriory.org/thomas/Physics8.htm. The concluding identification with God is in 8, lectio 22, par. 1172.

19. Aristotle, *Physics*, 8.4.255^{a}3, 8.255^{b}14–15, 8.5.256^{a}19.

20. Aquinas, *Commentaria in octo libros Physicorum*, http://www.corpusthomisticum.org/cpy03.html#72022; Aquinas, *Commentary on Aristotle's Physics*, 3, lectio 9, chap. 5, par. 359 and par. 365, http://www.dhspriory.org/thomas/Physics3.htm.

21. Aquinas, *Commentaria in octo libros Physicorum*, http://www.corpusthomisticum.org/cpy03.html#72051; Aquinas, *Commentary on Aristotle's Physics*, 4, lectio 11, chap. 8, par. 522, http://www.dhspriory.org/thomas/Physics4.htm. See also where Aquinas comments on Aristotle as saying, "Natural motion is prior to compulsory since compulsory motion is only a departure from natural motion" (Motus naturalis est prior violento, cum motus violentus non sit nisi quaedam declinatio a motu naturali). Aquinas, *Commentaria in octo libros Physicorum*, http://www.corpusthomisticum.org/cpy03.html#72051; Aquinas, *Commentary on Aristotle's Physics*, 4, lectio 11, chap. 8, par. 524, http://www.dhspriory.org/thomas/Physics4.htm.

22. Aquinas, *Commentaria in octo libros Physicorum*, http://www.corpusthomisticum.org/cpy07.html#72553; Aquinas, *Commentary on Aristotle's Physics*, 8, lectio 7, chap. 4, par. 1023, http://www.dhspriory.org/thomas/Physics8.htm.

23. Aquinas, *Commentaria in octo libros Physicorum*, http://www.corpusthomisticum.org/cpy03.html#72022; Aquinas, *Commentary on Aristotle's Physics*, 4, lectio 8, chap. 5, par. 492–93, http://www.dhspriory.org/thomas/Physics4.htm. On Aristotle's idea of natural place, see Peter Machamer, "Aristotle on Natural Place and Natural Motion," *Isis* 69 (1978): 377–87.

24. Aristotle, *Physics*, 8.4.254^{b}8–255^{a}6.

25. Aristotle, *Physics*, 8.4.255^{a}2–3; Aristotle, *Physics*, 4.8.215^{a}2–3.

26. Aristotle, *Physics*, note to 4.8.215^{a}1–14 (218).

27. Aristotle, *De caelo*, trans. J. L. Stocks (Oxford: Clarendon Press, 1922), 1.8.277^{a}29, http://classics.mit.edu/Aristotle/heavens.1.i.html.

28. Clagett, *Science of Mechanics*, 542.

29. Aquinas, *In libros Aristotelis De caelo et mundo expositio: The Heavens*, trans. Fabian R. Larcher and Pierre H. Conway, 2 vols. (Columbus, OH: College of St. Mary of the Springs, 1964), bk. 1, lectio. 17, 173–74, http://dhspriory.org/thomas/DeCoelo.htm#1-17.

30. Maier, "Causes, Forces, and Resistance," 55.

31. The examples are from James A. Weisheipl, "The Principle *Omne quod movetur ab alio movetur* in Medieval Physics," *Isis* 56 (1965): 28.

32. Lindberg, *Beginnings of Western Science*, 302.

33. Aristotle, *Physics*, bk. 7, trans. W. D. Ross (Oxford: Clarendon Press, 1936), 7.243^{a}37–243^{a}40.

34. Lindberg, *Beginnings of Western Science*, 302.

35. Grant, *Source Book*, 266.

36. Peter King, "Duns Scotus on the Reality of Self-Change," in *Self-Motion: From Aristotle to Newton*, ed. Mary Louise Gill and James G. Lennox (Princeton, NJ: Princeton University Press, 1994), 228.

37. Grant, *Source Book*, 266.

38. Aristotle, *Physics*, 8.10.266^{b}29–267^{a}15; quotation from Graham's commentary, 173.

39. Maier, "Causes, Forces, and Resistance," 51.

40. Rega Wood, "The Influence of Arabic Aristotelianism on Scholastic Natural Philosophy: Projectile Motion, the Place of the Universe, and Elemental Composition," in Pasnau and Van Dyke, *Cambridge History of Medieval Philosophy*, 1:249–54.

41. A discussion of the context of Francis de Marchia's discovery and an edition of the key passages of his *Sentences* appear in Chris Schabel, "Francis of Marchia's *Virtus derelicta* and the Context of Its Development," *Vivarium* 44 (2006): 41–80.

42. Fabio Zanin, "Francis of Marchia, *Virtus derelicta*, and Modifications of the Basic Principles of Aristotelian Physics," *Vivarium* 44 (2006): 85, 86, and see also 92.

43. Anneliese Maier, "The Significance of the Theory of Impetus for Scholastic Natural Philosophy," in Maier, *On the Threshold of the Exact Science*, 85.

44. Jean Buridan, *Physics, La physique de Bruges de Buridan et le traité du ciel d'Albert de Saxe: étude critique, textuelle et doctrinale*, ed. Benoît Patar, vol. 2 (Longueuil, Québec: Les Presses Philosophiques, 2001), *ultima lectura*, 8.12.35–43; *Questions on the Eight Books on the Heavens and the World of Aristotle*, trans. Clagett, 8.12.4, in Clagett, *Science of Mechanics*, 534–35. (Clagett translates from an early printed version of Buridan's *Physics*.) See also Ernest A. Moody, "Laws of Motion in Medieval Physics," in *Toward Modern Science: Studies in Ancient and Medieval Science*, vol. 1, ed. Robert Palter (New York: Noonday Press, 1961), 230–32; Moody, "Galileo and His Precursors," (1966), 397–98, and "Jean Buridan" (1969), especially 447, both in Moody, *Studies in Medieval Philosophy, Science, and Logic*; and Lang, *Aristotle's Physics*, 168–71.

45. Buridan, *Quaestiones Super Libris Quattuor de Caelo et Mundo*, 2.12; *Questions on the Four Books on the Heavens and the World of Aristotle*, trans. Clagett, in Clagett, *Science of Mechanics*, 560; Clagett observes of the passage that Buridan principally indicates the distance of the motion but also seems to suggest the time of the movement as causal factors in increasing impetus and therefore velocity in *Science of Mechanics*, 563. For Oresme's description of acceleration as a "degree of velocity" being "more intense or greater" depending on the time and distance traveled, see *Qualities and Motions*, 2.3.

46. Buridan, *Physics, ultima lectura*, 8.12.41–42; *Questions on the Eight Books*, trans. Clagett, in Clagett, *Science of Mechanics*, 535.

47. Bert Hansen, "An Overview," 119.

48. Clagett, *Science of Mechanics*, 339–40n12; *Nicole Oresme and the Medieval Geometry of Qualities and Motions*, 135–36.

49. Clagett, introduction, *Qualities and Motions*, 14, 73–125; Claire Richter Sherman, *Imaging Aristotle: Verbal and Visual Representation in Fourteenth-Century France* (Berkeley: University of California Press, 1995), 3–33; William J. Courtenay, "The Early Career of Nicole Oresme," *Isis* 91 (2000): 542–48.

50. *Nicole Oresme and the Marvels of Nature: A Study of His De Causis Mirabilium with Critical Edition, Translation, and Commentary*, trans. Bert Hansen (Toronto: Pontifical Institute of Mediaeval Studies, 1985), 298–301, 354–55; Clagett, *Science of Mechanics*, 552–53, 570; Clagett, introduction, *Qualities and Motions*, 6–10; Marshall Clagett, "Oresme, Nicole," in *Complete Dictionary of Scientific Biography*, vol. 10 (Detroit: Charles Scribner's Sons, 2008), 223–24, 226.

51. Wood, "Influence of Arabic Aristotelianism," 255.

52. Pierre Duhem is credited with discovering the "Parisian precursors to Galileo" in *Études sur Léonard de Vinci*, vol. 3 (Paris: A. Hermann, 1906), 582. See also Anneliese Maier, "Galileo and the Scholastic Theory of Impetus," in Maier, *On the Threshold of the Exact Science*, 103–13; Moody, "Galileo and His Precursors," 393–408; William A. Wallace, *Prelude to Galileo: Essays on Medieval and Sixteenth-Century Sources of Galileo's Thought* (Dordrecht: Reidel, 1981).

53. Moody, "Laws of Motion," 223.

54. *Velocitas instantanea* appears to be first used in William Heytesbury's *Regule solvendi sophismata*; see document 4.4 in Clagett, *Science of Mechanics*, 235–42.

55. J. D. North, "Natural Philosophy in Late Medieval Oxford," in Catto and Evans, *History of the University of Oxford*, 2:65–102.

56. Kaye, *History of Balance*, 401; Kaye discusses Bradwardine's ideas on 398–409.

57. Moody, "Laws of Motion," 223; Murdoch and Sylla, "The Science of Motion," 241.

58. *Thomas of Bradwardine: His Tractatus de proportionibus; Its Significance for the Development of Mathematical Physics*, trans. H. Lamar Crosby (Madison: University of Wisconsin Press, 1955), 118, 119. Clagett discusses the terminology in *Science of Mechanics*, 210.

59. *Tractatus de proportionibus*, chapter 3. The problem may be expressed as follows: if the proportion of force to resistance is the velocity, and the force equals 1 and the velocity equals 1, then the resistance must also equal 1. The force is therefore not greater than the resistance, yet the object has velocity; it moves. The problem with the formula is that if the force and the velocity are the same, then force will always equal the resistance. An additional problem is that even if the resistance is greater than the motive force, Aristotle's equation will still result in a movement. For example, if force equals 1 and resistance equals 3, then velocity equals 1/3; an object would take three times as long but would still travel a unit of distance. See also North, "Natural Philosophy in Late Medieval Oxford."

60. In addition to the Bradwardine cited above, see Anneliese Maier, "The Concept of the Function in Fourteenth-Century Physics," in Maier, *On the Threshold of the Exact Science*, 73; Murdoch and Sylla, "The Science of Motion," 224–27; Edith D. Sylla, "Medieval Dynamics," *Physics Today* 61 (2008): 53; and Nicole Oresme, *De proportionibus proportionum, and Ad pauca respicientes*, trans. Edward Grant (Madison: University of Wisconsin Press, 1966), 14–24. Bradwardine did not use exponential equations, but expressed in today's terms, if the rate is a number n, then $nV = (F/R)^n$. For example, if the rate of velocity is 3, then $3V = (F/R)^3$. In modern terms, velocity is the logarithm of F/R, or $V = \log(F/R)$. The logarithm of 1/1 is 0, which means an object with equivalent force and resistance does not move. Similarly, any object where the resistance exceeds the force means a negative logarithm, which for Bradwardine is impossible, and the object again does not move.

61. Edith D. Sylla, "The Oxford Calculators," in Kretzmann, Kenny, and Pinborg, *Cambridge History of Later Medieval Philosophy*, 540–42.

62. Clagett, *Science of Mechanics*, 255–329; James A. Weisheipl, "The Place of John Dumbleton in the Merton School," *Isis* 50 (1959): 452–53. See also Weisheipl, "Interpretation of Aristotle's *Physics* and the Science of Motion," 534–35.

63. Heytesbury, *Regule solvendi sophismata*, 241, lines 84–88, "William Heytesbury, Rules for Solving Sophisms," trans. Moody, in Clagett, *Science of Mechanics*, 236. On Heytesbury and another contemporary of Bradwardine's, Richard Kilvington, see Sylla, "Oxford Calculators."

64. Gilles Châtelet, *Figuring Space: Philosophy, Mathematics and Physics*, trans. Robert Shore and Muriel Zagha (Dordrecht, Netherlands: Kluwer Academic Publishers, 1999), 44.

65. Oresme, *Qualities and Motions*, 1.16.25.

6. Motion in Literature

1. Grant, *History of Natural Philosophy*, 200, 231–32. See also Gransden, "Realistic Observation in Twelfth-Century England," 29–51; King, "Mediaeval Thought-Experiments," 43–64.

2. *Oxford English Dictionary*, s.v. "errant."

3. Malpas, *Place and Experience*, 134–35.

4. *Mandeville's Travels*, 135–36.

5. *A Porteous of Noblenes and Ten Other Rare Tracts*, published by Walter Chepman and Andro Myllar in Edinburgh; Parkinson, introduction, *Complete Works*.

6. *Dictionary of the Scots Language* (Edinburgh: Scottish Language Dictionaries, 2004), s.v. "ȝed(e)," http://www.dsl.ac.uk/entry/dost/3ede; *Middle English Dictionary*, s.v. "yēde."

7. Jill Mann, "The Planetary Gods in Chaucer and Henryson," in *Chaucer Traditions: Studies in Honour of Derek Brewer*, ed. Ruth Morse and Barry Windeatt (Cambridge: Cambridge University Press, 1990), 96.

8. Henryson, *Orpheus and Eurydice*, in *Complete Works*, 154–58.

9. *Middle English Dictionary*, s.v. "wil" and "wilsome."

10. Gavin Douglas, *Virgil's Aeneid Translated into Scottish Verse*, ed. David F. C. Coldwell, 4 vols., Scottish Text Society (Edinburgh: William Blackwood and Sons, 1957–1964), 1.122–24.

11. Ian Johnson, "Hellish Complexity in Henryson's *Orpheus*," *Forum for Modern Language Studies* 38 (2002): 416.

12. "Wilsum" also at lines 245 and 290. See also the note to lines 180–85 of the *Fables*, page 166 in Henryson, *Complete Works*.

13. See John MacQueen's exploration of Platonic influences on the poem in *Complete and Full with Numbers: The Narrative Poetry of Robert Henryson* (New York: Rodopi, 2006), 251–72.

14. *Orpheus and Eurydice*, 194.

15. *Orpheus and Eurydice*, 223. Note, incidentally, that Henryson uses the compound *mappamound* not in the sense of a map, nor does he restrict the "mound" to the earth in this scene.

16. *Orpheus and Eurydice*, 245. In Jill Mann's reading, Henryson's *Testament of Cresseid* is similarly constrictive and misleading to the point of frustration in contrast to Chaucer's *Troilus and Criseyde*: "Chaucer's creations of a cosmic perspective all take the form of a movement upwards and outwards. The imagination soars to the outermost limits of the universe and thence turns to gaze back on the 'litel spot of erthe' far below," "aerial flights" that imitate those in Boccaccio, Boethius, Dante, and others. "In all these cases, the aerial flight represents a liberation from earthly concerns; the elevation to a higher plane of being or understanding." In Henryson, in contrast, the *"Testament* makes a striking contrast to this venerable tradition. For here the order in which the planets appear suggests a movement *downwards and inwards.* . . . This downward motion is emphasized by the fact that the planetary gods themselves descend from their spheres to appear in assembly before Cresseid. This simple inversion of the normal upwards and outwards movement creates a sinister effect of claustrophobia; the cosmos seems to be bearing down on Cresseid." Mann, "Planetary Gods," 96. My reading of the *House of Fame* differs from Mann's interpretation of Chaucer and aerial flights.

17. *Orpheus and Eurydice*, 249.

18. *Orpheus and Eurydice*, 248.

19. *Orpheus and Eurydice*, 275–309.

20. *Orpheus and Eurydice*, 429–30. Comparisons and contrasts among Henryson, Boethius, and Trivet may be found in Rita Copeland, *Rhetoric, Hermeneutics, and Translation in the Middle Ages: Academic Traditions and Vernacular Texts* (Cambridge: Cambridge University Press, 1995), 228–29; and in Johnson's critique of Copeland in "Hellish Complexity," 412–19.

21. *Orpheus and Eurydice*, 431–40.

22. *Orpheus and Eurydice*, 606.

23. *Orpheus and Eurydice*, 600–604.

24. See the survey of criticism on the moralization in Henryson, *Complete Works*, 215–16.

25. Malpas, *Place and Experience*, 35.

26. *The Early English Version of the Gesta Romanorum*, ed. Sidney J. H. Herrtage, EETS, e.s. 33 (1878; repr., London: N. Trübner, 1962), 247.

27. On the production and reception of Latin and English copies of *De Proprietatibus Rerum*, see the summary of research in Elizabeth Keen, *The Journey of a Book: Bartholomew the Englishman and the Properties of Things* (Canberra: Australian National University E Press, 2011), 4–5.

28. Bartholomaeus, *On the Properties of Things*, vols. 1 and 2, 121a/b.

29. *The Earliest English Translation of Vegetius' De Re Militari*, ed. Geoffrey Lester (Heidelberg: Winter, 1988), 65.

30. *Middle English Dictionary*, s.v. "as(s)aut."

31. James A. Weisheipl, "Ockham and Some Mertonians," *Mediaeval Studies* 30 (1968): 192.

32. Clagett, *Science of Mechanics*, 632–33 and, for more discussion of influences, see also 440–43 and 633–35. On the spread of Bradwardine within and beyond Merton College, see Weisheipl, "Place of John Dumbleton," 440–41; Courtenay, *Schools and Scholars in Fourteenth-Century England*, 240–49; and Joel Kaye, *Economy and Nature in the Fourteenth Century: Money, Market Exchange, and the Emergence of Scientific Thought*

(Cambridge: Cambridge University Press, 1998), 2–4. On connections among Oxford, Strode, Robert Holcot, and Chaucer, see the *Riverside Chaucer*, 939, 1058; Rodney Delasanta, "Chaucer and Strode," *Chaucer Review* 26 (1991): 205–18; Richard J. Utz, "Negotiating the Paradigm: Literary Nominalism and the Theory and Practice of Re-reading Late Medieval Texts," in *Literary Nominalism and the Theory of Rereading Late Medieval Texts: A New Research Paradigm*, ed. Richard J. Utz (New York: Edwin Mellen, 1995), 1–30; Richard J. Utz, "'As writ myn auctour called Lollius': Divine and Authorial Omnipotence in Chaucer's Troilus and Criseyde," in *Nominalism and Literary Discourse: New Perspectives*, ed. Christoph Bode, Hugo Keiper, and Richard J. Utz (Amsterdam: Rodopi, 1997), 128–29, 148–55; Kathryn L. Lynch, *Chaucer's Philosophical Visions* (Woodbridge, Suffolk: Boydell and Brewer, 2000), 18–25; Neil Cartlidge, "Ripples on the Water? The Acoustics of Geoffrey Chaucer's *House of Fame* and the Influence of Robert Holcot," *Studies in the Age of Chaucer* 39 (2017): 73.

33. For instance, the circulation of Bradwardine's *Tractatus de proportionibus* is discussed in Crosby's introduction to the edition (4), and the influence of Oresme's ideas is described in Clagett, introduction, *Qualities and Motions*, by Oresme, 73–111.

34. Nun's Priest's Tale, 7.3242; Knighton, *Chronicon Henrici Knighton, vel Cnitthon, monachi Leycestrensis*, ed. Joseph Rawson Lumby, Vol. 2, Rolls Series 92 (London: Eyre and Spottiswoode, 1896), 63; *Knighton's Chronicle 1337–1396*, trans. G. H. Martin (Oxford: Clarendon Press, 1996), 103.

35. Kaye, *Economy and Nature*, 2. Kaye discusses Bradwardine, Oresme, and the Merton School in chapter 8 of *History of Balance*.

36. Pierre Souffrin, "La quantification du mouvement chez les scolastiques: La vitesse instantanée chez Nicole Oresme," in *Autour de Nicole Oresme: Actes du Colloque Oresme*, ed. Jeannine Quillet (Paris: Vrin, 1990), 66, in Kaye, *Economy and Nature*, 205n19; Lynn Staley, *Languages of Power in the Age of Richard II* (University Park: Pennsylvania State University Press, 2005). See especially her discussion of Oresme and the court of Charles V, 87–93.

37. Sheila Delany, *Chaucer's House of Fame: The Poetics of Skeptical Fideism* (Chicago: University of Chicago Press, 1972), 69.

38. Delany, *Chaucer's House of Fame*, 76.

39. Delany, *Chaucer's House of Fame*, 83.

40. Cartlidge, "Ripples on the Water?," 93.

41. Lynch, *Chaucer's Philosophical Visions*, 64. Lynch cites John M. Fyler, *Chaucer and Ovid* (New Haven, CT: Yale University Press, 1979), 43.

42. Robertson, *Nature Speaks*, 9, 12, 27.

43. Robertson, *Nature Speaks*, 25, 28–29.

44. *House of Fame*, 1–110.

45. For a discussion of the difficulty of understanding whether December 10 has significance, see the notes to the *Riverside* edition, 979.

46. *House of Fame*, 114–18.

47. *The Complete Works of Geoffrey Chaucer*, ed. Walter W. Skeat, vol. 3 (Oxford: Clarendon Press, 1894), note 1.115 (p. 248); H. M. Smyser, "Chaucer's Two-Mile Pilgrimage," *Modern Language Notes* 56 (1941): 205–7; Tony Davenport, "Chaucer's *House of Fame*, 111–18: A Windsor Joke?," *Notes and Queries* 48 (2001): 222–24. Davenport locates the shrine elsewhere, at a hermitage west of Windsor Castle.

48. Even the *Book of the Duchess* does not say more about the dreamer's location. Also compare Chaucer's translation, *The Romaunt of the Rose*, an important influence on the *House of Fame*, which has no location until the dreamer is asleep, and there it is only that the dreamer leaves a town.

49. Malpas, *Place and Experience*, 8. Valerie Allen says, "Where one tarries is where one is, existentially speaking, and where one is is where one thinks." "Road," *postmedieval: A Journal of Medieval Cultural Studies* 4 (2013): 27; also in "Introduction: Roads and Writing," *Roadworks*, 9.

50. *House of Fame*, 121–31. On glass in Chaucer's poems, including the *House of Fame*, see David K. Coley, "'Withyn a temple ymad of glas': Glazing, Glossing, and Patronage in Chaucer's *House of Fame*," *Chaucer Review* 45 (2010): 59–84.

51. *House of Fame*, 140.

52. *Middle English Dictionary*, s.v. "romen."

53. *House of Fame*, 142. Jesse Simon notes several interesting uses of a word like *table* in Greek texts that show representations of the *oikoumene*, including a brass one in Herodotus's *Histories*, which was presented by Aristagoras of Miletus. "Chorography Reconsidered," 33.

54. *House of Fame*, 197, 237.

55. *House of Fame*, 474–75.

56. *House of Fame*, 647–50.

57. *House of Fame*, 889, 896–903.

58. *House of Fame*, 907.

59. *House of Fame*, 1641, 1867.

60. Ruth Evans, "Chaucer in Cyberspace," *Studies in the Age of Chaucer* 23 (2001): 44.

61. *House of Fame*, 151, 162, 174; 193, 209, 212, 219, 253.

62. *House of Fame*, 474–75.

63. Chaucer, *Romaunt of the Rose*, 509–30. *The Book of the Duchess*, 321–34, similarly finds the dreamer waking up with walls and windows painted with images of Troy and from the *Romance of the Rose*, and January enters a garden by means of a "wiket" in the Merchant's Tale, IV.2118.

64. *House of Fame*, 480–88.

65. Steven F. Kruger, "Imagination and the Complex Movement of Chaucer's *House of Fame*," *Chaucer Review* 28 (1993): 122.

66. Delany, among others, notes the unusual nature of the desert in *Chaucer's House of Fame*, 58.

67. *House of Fame*, 478–79.

68. *The Wars of Alexander: An Alliterative Romance, Translated Chiefly from the Historia Alexandri Magni de Preliis*, ed. Walter W. Skeat, EETS, e.s. 47 (London: N. Trübner, 1886), 5563, 5565; *The N-Town Plays*, ed. Douglas Sugano (Kalamazoo, MI: Medieval Institute, 2007), Play 22: Baptism, 123, 127.

69. *House of Fame*, 493.

70. *House of Fame*, 1116–23.

71. *House of Fame*, 714–20, 841–42.

72. Roads appear earlier in the *House of Fame*, as when after the vertiginous scene of looking down, the eagle bids Chaucer look up, and he views the Milky Way, which is likened to Watling Street (likely a borrowing from Bartholomeus Anglicus).

Furthermore, when Geffrey arrives at the House of Fame and the eagle sets him down, even though he is not sure whether he arrives there "in body or in gost," he nevertheless describes himself as being in a street and as near the place as someone might throw a spear. *House of Fame*, 939, 981, 1046–48.

73. *House of Fame*, 745. The physics of this resistance seems to lie behind the idea that the "way" to Fame's house is "so overt" (open or available) that all sound goes there. *House of Fame*, 718. Chaucer has added this description; it is not in Ovid.

74. Gabrovsky, *Chaucer the Alchemist*, 20; and see his chapter "Thought Experiments in Geffrey's Dream: The Poetics of *Motus Localis*, Measurement, and Relativity in the *House of Fame*," 27–64.

75. Valerie Allen, *On Farting: Language and Laughter in the Middle Ages* (New York: Palgrave Macmillan, 2007), 38–42.

76. *House of Fame*, 765–81.

77. *House of Fame*, 811–13.

78. Cartlidge, "Ripples on the Water?," 79–80.

79. *House of Fame*, 817–18. On the particular physics, see Maier, "The Significance of the Theory of Impetus," 79.

80. *House of Fame*, 847.

81. *House of Fame*, 854–57, 878.

82. *House of Fame*, 1034–42, 1071–83.

83. *House of Fame*, 1289–91, 1368–76, 1493–96; *Boece*, bk. 1, prosa 1, 12–17.

84. *House of Fame*, 1883–85.

85. *House of Fame*, 1886–95, 1915.

86. *House of Fame*, 1918.

87. John Kleiner, *Mismapping the Underworld: Daring and Error in Dante's Comedy* (Stanford, CA: Stanford University Press, 1994), 38–39, 47.

88. Kleiner, *Mismapping the Underworld*, 54.

89. Rebecca Davis, "Fugitive Poetics in Chaucer's *House of Fame*," *Studies in the Age of Chaucer* 37 (2015): 113.

90. *House of Fame*, 1927–30.

91. *House of Fame*, 1938, 1959, 1979.

92. *House of Fame*, 1996, 1998.

93. *House of Fame*, 2134–36.

94. *House of Fame*, 2138.

95. *House of Fame*, 2141.

96. Davis, "Fugitive Poetics," 102.

97. Davis, "Fugitive Poetics," 103, 126.

98. *House of Fame*, 2158.

7. Intense Proximate Affect

1. Herodotus, *The Histories*, trans. A. D. Godley, Loeb Classical Library (Cambridge, MA: Harvard University Press, 1920), 2.108–9. Godley identifies Sesostris as Rameses II (called Sesostris and Ozymandias by the Greeks, 2.102 n1), but the *Oxford Encyclopedia of Ancient Egypt* identifies him as Senwosret III. Robert Delia, "Senwosret III," in

The Oxford Encyclopedia of Ancient Egypt (Oxford University Press, 2001), http://www.oxfordreference.com.i.ezproxy.nypl.org/view/10.1093/acref/9780195102345.001.0001/acref-9780195102345-e-0653.

2. Aristophanes, *The Clouds*, trans. John Claughton and Judith Affleck (Cambridge: Cambridge University Press, 2012), 203–5. For discussions of Aristophanes and Herodotus on geometry, see A. F. Shore, "Egyptian Cartography," in Harley and Woodward, *History of Cartography*, 1:124–25, 1:138–39.

3. Hugh of St. Victor, *Didascalicon*, 2.9.

4. "A Treatise on the Mensuration of Heights and Distances," in Halliwell, *Rara Mathematica*, 56–57.

5. "The Babees Book," in *The Babees Book*, ed. Furnivall, 78–87, 132–34, 196; "How the Good Wiff Tauȝte Hir Douȝtir," in *The Babees Book*, ed. Furnivall, 35, 90.

6. Julian of Norwich, *The Writings of Julian of Norwich*, 281.

7. Jacques Rossiaud, "The City-Dweller and Life in Cities and Towns," in *Medieval Callings*, ed. Jacques Le Goff (Chicago: University of Chicago Press, 1995), 156. See also C. M. Woolgar, *The Senses in Late Medieval England* (New Haven, CT: Yale University Press, 2006), 35–36.

8. Reginald Pecock, *The Folewer to the Donet*, ed. Elsie Vaughan Hitchcock, EETS, o.s. 164 (Oxford: Oxford University Press, 1924), 195–96.

9. Bartholomaeus, *On the Properties of Things*, 465.

10. *Middle English Translation of Guy de Chauliac's Treatise on "Apostemes,"* 151.

11. Bartholomaeus, *On the Properties of Things*, 396, 487.

12. For example, Joseph Taylor, "Chaucer's Uncanny Regionalism: Rereading the North in The Reeve's Tale," *Journal of English and Germanic Philology* 109 (2010): 468–89; Elliot Kendall, "Family, *Familia*, and the Uncanny in *Sir Orfeo*," *Studies in the Age of Chaucer* 35 (2013): 289–327.

13. Keith Wrightson, "The 'Decline of Neighbourliness' Revisited," in *Local Identities in Late Medieval and Early Modern England*, ed. Norman L. Jones and Daniel Woolf (New York: Palgrave, 2007), 19–49; George Edmondson, *The Neighboring Text: Chaucer, Boccaccio, Henryson* (Notre Dame, IN: University of Notre Dame Press, 2011).

14. For example, Carolyn Dinshaw, "Chaucer's Queer Touches / A Queer Touches Chaucer," *Exemplaria* 7 (1995): 75–92; Glenn Burger, *Chaucer's Queer Nation* (Minneapolis: University of Minnesota Press, 2003), 142–49, 218n10; Lara Farina, "Wondrous Skins and Tactile Affection: The Blemmye's Touch," and Elizabeth Robertson, "*Noli me tangere*: The Enigma of Touch in Middle English Religious Literature and Art for and about Women," both in *Reading Skin in Medieval Literature and Culture*, ed. Katie L. Walter (New York: Palgrave Macmillan, 2013), 11–28, 29–55.

15. Clagget, *Science of Mechanics*, 206.

16. John Wyclif, *The English Works of Wyclif*, ed. F. D. Matthew, EETS, o.s. 74 (London: Kegan Paul, Trench, Trübner, 1880), 27.19.

17. Pecock, *Folewer to the Donet*, 39.

18. Aristotle, *Physics*, $8.5.256^{b}19$ and $8.5.258^{a}17$–21, and see the commentary on $8.4.10.267^{b}7$ for a discussion of problems with Aristotle's theory and its implications for the first unmoved mover in Aristotle.

19. Aquinas, *Commentaria in octo libros Physicorum*, http://www.corpusthomisticum.org/cpy011.html; Aquinas, *Commentary on Aristotle's Physics*, 7, lectio 3, chap. 2, par. 897, http://www.dhspriory.org/thomas/Physics1.htm.

20. Aristotle, *Physics*, 4.4.212^{a}5–6.

21. Edith D. Sylla, "Godfrey of Fontaines on Motion with Respect to Quantity of the Eucharist," in *Studi sul xiv secolo in memoria di Anneliese Maier*, ed. Maierù and Paravicini Bagliani (Rome: Storia e Letteratura, 1981), 124. Sylla cites Anneliese Maier, *Zwei Grundprobleme der scholastischen Naturphilosophie: Das Problem der intensiven Grösse: Die Impetustheorie*, 2nd ed. (Rome: Storia e Letteratura, 1951), 36–43.

22. On these topics, in addition to the Sylla cited previously, see also Sylla, "Medieval Quantifications of Qualities: The 'Merton School,'" *Archive for History of Exact Sciences* 8 (1971), 15n17; Marshall Clagett, "Richard Swineshead and Late Medieval Physics," *Osiris* 9 (1950): 132–39; Clagett, *Science of Mechanics*, 206; Stephen Dumont, "Godfrey of Fontaines and the Succession Theory of Forms at Paris in the Early Fourteenth Century," in *Philosophical Debates at Paris in the Early Fourteenth Century*, ed. Stephen F. Brown, Thomas Dewender, and Theo Kobusch (Leiden: Brill, 2009), 39–126; Gabrovsky, *Chaucer the Alchemist*, 34–35, 37.

23. *Qualities and Motions*, Proemium.1.0.4–5.

24. *Qualities and Motions*, 1.24.1–12.

25. *Qualities and Motions*, 1.1.1–8, 2.13.18–19; see also Clagett's discussion of these ideas in his introduction, 36–37, and notes, 468–70.

26. *Qualities and Motions*, 2.1.6, 2.1.24–31.

27. *Qualities and Motions*, 2.14.

28. Clagett, introduction, *Qualities and Motions*, 112. Clagett cites Duhem, who discusses the works of those philosophers in *Le système du monde*, 576–82.

29. Clagett, introduction, *Qualities and Motions*, 15.

30. *Qualities and Motions*, 1.25.3. Clagett translates the phrase as "natural bodies ... mutually compared," but the sense of physical proximity is clear in the remainder of the chapter.

31. *Qualities and Motions*, 1.25.9–14, 1.25.21.

32. Mary Beth Mader, "Whence Intensity? Deleuze and the Revival of a Concept," in *Gilles Deleuze and Metaphysics*, ed. Alain Beaulieu, Edward Kazarian, and Julia Sushytska (Lanham, MD: Lexington Books-Rowman and Littlefield, 2014), 225–48, with a discussion of Oresme on 236–37.

33. Brian Massumi, "Concrete Is as Concrete Doesn't," introduction to *Parables for the Virtual: Movement, Affect, Sensation* (Durham, NC: Duke University Press, 2002), 1.

34. Melissa Gregg and Gregory J. Seigworth, "An Inventory of Shimmers," in *The Affect Theory Reader* (Durham, NC: Duke University Press, 2009), 1.

35. Holly Crocker, "Medieval Affects Now," *Exemplaria* 29 (2017): 82–98.

36. *Qualities and Motions*, 1.23.3–10.

37. *Qualities and Motions*, 2.37.3–7.

38. *Qualities and Motions*, 1.31.2–21.

39. *Qualities and Motions*, 1.32.13–15, 1.32.19–22.

40. Clagett, introduction, *Qualities and Motions*, 5.

41. Oresme's arguments against magic are discussed, among other places, in Lynn Thorndike, *A History of Magic and Experimental Science*, Vol. 3–4 in Vol. 3 (New York: Columbia University Press, 1934), 398–517; G. W. Coopland, *Nicole Oresme and the Astrologers: A Study of His Livre de Divinacions* (Liverpool: Liverpool University Press, 1952).

42. *Qualities and Motions*, 2.38.33–37.

43. *Qualities and Motions*, 2.35.34–37.

44. *Qualities and Motions*, 1.27.

45. *Qualities and Motions*, 1.27.18–20.

46. *Qualities and Motions*, 1.29.3–28. Joan Cadden discusses Oresme but not his *Qualities and Motions* in *Nothing Natural Is Shameful: Sodomy and Science in Late Medieval Europe* (Philadelphia: University of Pennsylvania Press, 2013), 156–62.

47. *Qualities and Motions*, 1.27.27–35.

48. *Qualities and Motions*, 1.24.3.

49. *Qualities and Motions*, 1.26.20–21. On Oresme and species in his *Quotlibeta* or *Quodlibeta*, see Thorndike, *History of Magic and Experimental Science*, 446–48.

50. *Qualities and Motions*, 1.26.13–16.

51. *Qualities and Motions*, 1.27.13–14.

52. *Nicole Oresme and the Kinematics of Circular Motion: Tractatus de commensurabilitate vel incommensurabilitate motuum celi*, trans. Edward Grant (Madison: University of Wisconsin Press, 1971), 3.324–59.

53. *Qualities and Motions*, 1.30.3–17.

54. *Qualities and Motions*, 2.39.42–43.

55. *Qualities and Motions*, 2.40.13–14.

56. *Qualities and Motions*, 2.37.17–25.

57. *Qualities and Motions*, 2.38.1–2.

58. *Qualities and Motions*, 2.38.4–7.

59. *Qualities and Motions*, 2.38.24–29.

60. *Qualities and Motions*, 2.38.29–39.

61. *Qualities and Motions*, 2.38.62–65.

62. Clagett, introduction, *Qualities and Motions*, 6.

8. Proximal Literature

1. Jessica Brantley, "Venus and Christ in Chaucer's *Complaint of Mars*: The Fairfax 16 Frontispiece," *Studies in the Age of Chaucer* 30 (2008): 178.

2. Canon's Yeoman's Prologue, 8.585.

3. Foucault, "Des espaces autres," 752, 754; Foucault, "Of Other Spaces," 22, 23.

4. Massumi, *Parables for the Virtual*, 1.

5. Massumi, *Parables for the Virtual*, 7.

6. Holly Crocker outlines the critical history of medieval feelings in the "intellectual soul," and Stephanie Trigg defines *affect* as encompassing the "preconscious, non-discursive, [and] non-narrated." Crocker, "Medieval Affects Now," 83–84; Trigg, "Introduction: Emotional Histories—Beyond the Personalization of the Past and the Abstraction of Affect Theory," *Exemplaria* 26 (2014): 8.

7. *Sir Gawain the Green Knight*, ed. J. R. R. Tolkien and E. V. Gordon, 2nd ed., rev. Norman Davis (Oxford: Oxford University Press, 1925), 2312.

8. *Book of the Duchess*, 195.

9. *Troilus and Criseyde*, 1.276.

10. *Oxford English Dictionary*, s.v. "won, wone" (v.); *Middle English Dictionary*, s.v. "wonen."

11. *Troilus and Criseyde*, 1.282–87.

12. Sarah Stanbury, *The Visual Object of Desire in Late Medieval England* (Philadelphia: University of Pennsylvania Press, 2015), 109. Stanbury's observation on this point also applies to the earlier passage where Troilus was told that Criseyde's "goodly lokyng gladed al the prees. / Nas nevere yet seyn thyng to ben preysed derre [more highly]," 1.173–74.

13. *Troilus and Criseyde*, 1.273.

14. *Troilus and Criseyde*, 1.290–98.

15. *Qualities and Motions*, 2.38.40–43.

16. *Qualities and Motions*, 2.38.4–7.

17. *Troilus and Criseyde*, 1.304–7.

18. *Testament of Cresseid*, *The Complete Works*, 496–97, http://d.lib.rochester.edu/teams/text/parkinson-henryson-complete-works-testament-of-cresseid.

19. *Testament of Cresseid*, 499, 503.

20. *Testament of Cresseid*, 505–511.

21. A summary of this criticism may be found in the introductory notes to the *Legend* in the *Riverside Chaucer*, 1059–60.

22. *Legend of Good Women*, 16–21. Line numbers refer to the F version unless otherwise noted.

23. *Legend of Good Women*, 40–49, 197–211.

24. See Laura Howes, "Chaucer's Forests, Parks, and Groves," *Chaucer Review* 49 (2014): 125–33.

25. *Legend of Good Women*, 239–40.

26. *Legend of Good Women*, F.305, and by "degre" in the G version, G.231.

27. *Legend of Good Women*, 315–18.

28. "The Babees Book"; "How the Good Wiff Tau3te Hir Dou3tir"; Geoffrey de la Tour Landry, *The Book of the Knight of La Tour-Landry, Compiled for the Instruction of His Daughters*, ed. Thomas Wright, EETS, o.s. 33. 1868 (Oxford: Oxford University Press, 1973).

29. *Legend of Good Women*, 319–27.

30. *Legend of Good Women*, 688–97.

31. *Legend of Good Women*, 163, 342.

32. Betsy McCormick, Leah Schwebel, and Lynn Shutters, "Introduction: Looking Forward, Looking Back on the *Legend of Good Women*," *Chaucer Review* 52 (2017): 5.

33. Sheila Delany, *The Naked Text: Chaucer's Legend of Good Women* (Berkeley: University of California Press, 1994), 191–93; Kara Doyle, "Thisbe out of Context: Chaucer's Female Readers and the Findern Manuscript," *Chaucer Review* 40 (2006): 254.

34. Steven Kruger, "Passion and Order in Chaucer's *Legend of Good Women*," *Chaucer Review* 23 (1989): 229–30.

35. Glenn Burger, "'Pite renneth soone in gentil herte: Ugly Feelings and Gendered Conduct in Chaucer's *Legend of Good Women*," *Chaucer Review* 52 (2017): 67.

36. Lynn Shutters, "The Thought and Feel of Virtuous Wifehood: Recovering Emotion in the *Legend of Good Women*," *Chaucer Review* 52 (2017): 95, 104.

37. Gower, *Confessio amantis*, 3.1492–93.

38. R. Allen Shoaf, "'A Pregnant Argument': Dante's Comedy, Chaucer's *Troilus*, Henryson's *Testament*," in *Fleshly Things and Spiritual Matters: Studies on the Medieval Body in Honour of Margaret Bridges*, ed. Nicole Nyffenegger and Katrin Rupp (Cambridge: Cambridge Scholars, 2011), 196.

39. Gower, *Confessio amantis*, marginal gloss, page 262.

40. *Legend of Good Women*, 708–9, 713, 719–20.

41. *Legend of Good Women*, 736, 738–39.

42. *Legend of Good Women*, 740. Compare Christine in the *Epistle of Othea to Hector; or, The Boke of Knyghthode*, trans. Stephen Scrope, ed. George F. Warner (London: J. B. Nichols, 1904), 52, and Metham in *Amoryus and Cleopes*, 1161–62.

43. *Legend of Good Women*, 745–46.

44. *Legend of Good Women*, 758, 763–65.

45. *Legend of Good Women*, 782, 784.

46. *Legend of Good Women*, 894–95.

47. *Legend of Good Women*, 894–99.

48. Gower, *Confessio amantis*, 3.1496–501, 1613–17; Metham's hermit revives Piramus and Thisbe in the "last book" of *Amoryus and Cleopes*, beginning line 1807.

49. *Legend of Good Women*, 900–904.

50. *Middle English Dictionary*, s.v. "in-fere," "in-feere," "yfere," "yfeere."

51. *Legend of Good Women*, 921.

52. Pardoner's Tale, 6.966.

53. *Legend of Good Women*, 1229–30.

54. *Legend of Good Women*, 1643.

55. *Legend of Good Women*, 1908–9.

56. *Legend of Good Women*, 1970–72.

57. *Legend of Good Women*, 2162.

58. *Legend of Good Women*, 2211–14.

59. *Legend of Good Women*, 2690.

60. *Legend of Good Women*, 2711–16. The phrase about motion, "a ful good pas," occurs only one other time in Chaucer's work, when Thisbe hurries to the tree to meet Pyramus at 802.

61. *Legend of Good Women*, 2723.

62. Critiques and analyses of the "unfinished" and "bored" arguments include Robert Worth Frank, *Chaucer and the Legend of Good Women* (Cambridge, MA: Harvard University Press, 1972), 189–210; Elaine Tuttle Hansen, "Irony and the Antifeminist Narrator in Chaucer's *Legend of Good Women*," *Journal of English and Germanic Philology* 82 (1983): 29–31; William Quinn, "The *Legend of Good Women*: Performance, Performativity, and Presentation," in *The Legend of Good Women: Context and Reception*, ed. Carolyn Collette (Cambridge: D. S. Brewer, 2006), 27–31.

Afterword

1. A further underlying guide has been to assemble the evidence and to take a modest "creative" risk in the spirit of Alfred North Whitehead, Bruno Latour, and others. See Bruno Latour, "Why Has Critique Run Out of Steam? From Matters of Fact to

Matters of Concern," *Critical Inquiry* 30 (2004): 245–46. Latour draws inspiration from Alfred North Whitehead's *The Concept of Nature* (Cambridge: Cambridge University Press, 1920). Latour also discusses creativity in "An Attempt at a "Compositionist Manifesto," *New Literary History* 41 (2010): 471–90. There he addresses Whitehead's *Process and Reality: An Essay in Cosmology* (1929; New York: Free Press, 1978).

2. Aristotle, *Physics*, 4.1.208^{a}27–31; Aquinas, *Commentaria in octo libros Physicorum*, http://www.corpusthomisticum.org/cpy011.html; Aquinas, *Commentary on Aristotle's Physics*, 4.1.407, http://www.dhspriory.org/thomas/Physics1.htm.

3. Martin Heidegger, "Building Dwelling Thinking," in *Basic Writings: Second Edition, Revised and Expanded* (New York: Harper Collins, 1993), 359.

4. Merleau-Ponty, *Phenomenology of Perception*, 253.

5. Malpas, *Place and Experience*, 15. Significant ideas about place and being may also be found in Heidegger, *Ontology: Hermeneutics of Facticity*, 5–16; and Zumthor, *La mésure du monde*, 51–68. Malpas provides analysis of Heidegger and Merleau-Ponty in *Place and Experience*, 15. See also Holgar Zaborowski, "Heidegger's Hermeneutics: Towards a New Practice of Understanding," in *Interpreting Heidegger: Critical Essays*, ed. Daniel O. Dahlstrom (Cambridge: Cambridge University Press, 2011), 15–41; and Linn Miller and Jeff Malpas, "On the Beach: Between the Cosmopolitan and the Parochial," in *Ocean to Outback: Cosmopolitanism in Contemporary Australia*, ed. Keith Jacobs and Jeff Malpas (Crawley, Western Australia: University of Western Australia Press, 2011), 123–49.

6. The *Oxford English Dictionary* registers the appellations "ubiquiter," "ubiquitist," and "ubiquitarian," the latter as a "person, esp. a Lutheran, who believes that Christ in his human nature is present everywhere." *Oxford English Dictionary*, s.v. "ubiquiter," "ubiquitist," "ubiquitarian."

7. Aristotle, *Physics*, on time: 4.10.218^{b}13, 4.11.220^{b}5, and elsewhere; on the six "dimensions," 4.1.208^{b}8. This, incidentally, seems to me the notion of direction that Merleau-Ponty takes up and considers in *Phenomenology of Perception*, 101–2.

8. Jammer, *Concepts of Space*, 30.

9. Aquinas, *Summa theologiae, Opera Omnia*, pt. 1, question 8, art. 1, http://www.corpusthomisticum.org/sth1003.html.

10. Man of Law's Tale, 2.534; Clerk's Tale, 4.167; Second Nun's Tale, 8.209; *Troilus and Criseyde*, 4.61–62.

11. *Middle English Dictionary*, s.v. "thik(ke)" (adv. 3c.).

12. Hugh of St. Victor, *Didascalicon*, 3.19.

13. Erich Auerbach, "Philology and *Weltliteratur*," trans. Maire Said and Edward Said, *Centennial Review* 13 (1969): 1–17. Auerbach quotes Hugh on 17.

14. Edward Said quotes Auerbach on Hugh in "Reflections on Exile," in *Reflections on Exile and Other Essays* (Cambridge, MA: Harvard University Press, 2000), 185.

15. On distinguishing globalism from cosmopolitanism, see, for example, Gerard Delanty, "Introduction: The Emerging Field of Cosmopolitanism Studies," in *The Routledge Handbook of Cosmopolitanism Studies*, ed. Gerard Delanty (New York: Routledge, 2012), 2.

16. Jacques Derrida, *On Cosmopolitanism and Forgiveness*, trans. Mark Dooley and Michael Hughes (New York: Routledge, 2001).

17. Man of Law's Tale, 2.492–93.

18. Patricia Clare Ingham, "Contrapuntal Histories," in *Postcolonial Moves: Medieval through Modern*, ed. Patricia Clare Ingham and Michelle R. Warren (New York: Palgrave Macmillan, 2003), 60. Discussions of the Man of Law's Tale in terms of its structure include Paul M. Clogan, "The Narrative Style of the Man of Law's Tale," *Medievalia et Humanistica*, n.s. 8 (1977): 217–33; Ann W. Astell, "Apostrophe, Prayer and the Structure of Satire in the Man of Law's Tale," *Studies in the Age of Chaucer* 13 (1991): 81–97; and Helen Cooper, *Oxford Guides to Chaucer: The Canterbury Tales* (Oxford: Oxford University Press, 1996), 129–30.

19. Pardoner's Tale, 6.672, 6.676, 6.675–79. It is, to a degree but a degree only, an understandable mistake: *contree* could mean a space of almost any size, for example, indicating Europe, a nation-like area, and a whole region all the way down to a smaller region; Chaucer refers to the "contree" of Holderness and of Saluzzo, for example (Summoner's Tale, 3.1710, Clerk's Tale, 4.44).

20. Pardoner's Prologue, 6.345, 6.394, 6.443, 6.453.

21. Pardoner's Prologue, 6.518–20.

22. Massimo Cacciari explores the lexico-philosophical roots of the role of boundaries in making place: "The *limen* is the threshold, which the god Limentinus guards, the *passage* through which one accesses a domain or through which one exits from it. Through the threshold we are received, or otherwise *e-liminated*. It can direct us to the 'center' or open onto the *un-limited*, to that which does not have form or measure, 'where' we fatally disappear." He continues, indicating the significance of "the line (*lyra*) that encloses," one which "must represent a *finis* strong enough to condemn the one who comes to be *e-liminated* into the *de-lirium*. . . . Delirium comes to the one who does not acknowledge the limit or who cannot be accepted by it." "Place and Limit," 13.

23. Or he *has* walked to India: the phrasing is slightly ambiguous in that it hovers between the subjunctive and the indicative: "I ne kan fynde / A man, though that I walked into Ynde, / Neither in citee ne in no village." 6.721–23.

24. Wife of Bath's Prologue, 3.824; *Boece*, 3.m5.4–7.

25. Foucault, "Des espaces autres," 753; Foucault, "Of Other Spaces," 22–23.

26. Pardoner's Tale, 6.731, 6.749.

27. Pardoner's Tale, 6.966.

28. For example, Dinshaw, "Chaucer's Queer Touches"; Burger, *Chaucer's Queer Nation*; Walter, *Reading Skin in Medieval Literature and Culture*.

29. Carolyn Dinshaw, *Getting Medieval: Sexualities and Communities, Pre- and Postmodern* (Durham, NC: Duke University Press, 1999); Carolyn Dinshaw, *How Soon Is Now? Medieval Texts, Amateur Readers, and the Queerness of Time* (Durham, NC: Duke University Press, 2012). See also D. Vance Smith, who discusses Dinshaw and affect theory: "The Application of Thought to Medieval Studies: The Twenty-First Century," *Exemplaria* 22 (2010): 85–94.

30. Malpas, "Thinking Topographically."

31. Pardoner's Tale, 6.732.

Bibliography

Primary Sources

Anglo-Norman Dictionary. The Anglo-Norman On-Line Hub, Universities of Aberystwyth and Swansea. http://www.anglo-norman.net/.

The Anonimalle Chronicle, 1333 to 1381. Edited by V. H. Galbraith. Manchester: Manchester University Press, 1927.

Aquinas, Thomas. *Commentaria in octo libros Physicorum.* Turin: Textum Leoninum, 1954. http://www.corpusthomisticum.org/.

———. *Commentary on Aristotle's Physics.* Books 1–2. Translated by Richard J. Blackwell, Richard J. Spath, and W. Edmund Thirlkel. New Haven, CT: Yale University Press, 1963. Html edition by Joseph Kenny, http://www.dhspriory.org/thomas/Physics.htm.

———. *Commentary on Aristotle's Physics.* Books 3–8. Translated by Pierre H. Conway. Columbus, OH: College of St. Mary of the Springs, 1958–1962. Html edition by Joseph Kenny, http://www.dhspriory.org/thomas/Physics.htm.

———. *In libros Aristotelis De caelo et mundo expositio: The Heavens.* Translated by Fabian R. Larcher and Pierre H. Conway. 2 vols. Columbus, OH: College of St. Mary of the Springs, 1964. http://dhspriory.org/thomas/DeCoelo.htm.

———. *Summa theologiae, Opera Omnia.* 1888. Pamplona: University of Navarra, 2000. http://www.corpusthomisticum.org/.

Aristophanes. *The Clouds.* Translated by John Claughton and Judith Affleck. Cambridge: Cambridge University Press, 2012.

Aristotle. *De caelo.* Translated by J. L. Stocks. Oxford: Clarendon Press, 1922. http://classics.mit.edu/Aristotle/heavens.html.

———. *Physics.* Books 1–2. Translated by W. Charlton. Clarendon Aristotle Series. Oxford: Oxford University Press, 1970.

———. *Physics.* Books 1–4. Translated by Philip H. Wicksteed and Francis M. Cornford. Vol. 1. Cambridge, MA: Harvard University Press, 1963.

———. *Physics.* Books 3–4. Translated by Edward Hussey. Clarendon Aristotle Series. Oxford: Oxford University Press, 1983.

———. *Physics.* Book 7. Translated by W. D. Ross. Oxford: Clarendon Press, 1936.

———. *Physics.* Book 8. Translated by Daniel W. Graham. Clarendon Aristotle Series. Oxford: Oxford University Press, 1999.

"The Babees Book." In *The Babees Book*, ed. Furnivall, 1–9.

The Babees Book. Edited by Frederick J. Furnivall. Early English Text Society, o.s. 91. London: N. Trübner, 1868. Corpus of Middle English Prose and Verse. Ann

Arbor: University of Michigan Library, 2006. http://name.umdl.umich.edu/AHA6127.0001.001.

Bartholomaeus Anglicus. *On the Properties of Things: John Trevisa's Translation of De Proprietatibus Rerum, a Critical Text*. Edited by M. C. Seymour. 3 vols. Oxford: Clarendon Press, 1975–1988.

Boccaccio, Giovanni. *Decameron*. Edited by Vittore Branca. Torino: Einaudi, 1980.

Bradwardine, Thomas. *Thomas of Bradwardine: His Tractatus de proportionibus; Its Significance for the Development of Mathematical Physics*. Translated by H. Lamar Crosby. Madison: University of Wisconsin Press, 1955.

Buridan, Jean. *Physics. La physique de Bruges de Buridan et le traité du ciel d'Albert de Saxe: étude critique, textuelle et doctrinale*. Edited by Benoît Patar. 2 vols. Longueuil, Québec: Les Presses Philosophiques, 2001.

———. *Quaestiones Super Libris Quattuor de Caelo et Mundo*. Edited by Ernest A. Moody. Cambridge, MA: Medieval Academy of America, 1942.

———. *Quaestiones super octo libros Physicorum Aristotelis (secundum ultimam lecturam)*. Books 1–2. Edited by Michiel Streijger and Paul J. J. M. Bakker. History of Science and Medicine Library 50. Leiden: Brill, 2015.

———. *Questions on the Eight Books on the Heavens and the World of Aristotle*. Translated by Clagett. In Clagett, *Science of Mechanics*, 532–40.

———. *Questions on the Four Books on the Heavens and the World of Aristotle*. Translated by Clagett. In Clagett, *Science of Mechanics*, 557–64.

Chaucer, Geoffrey. *The Complete Works of Geoffrey Chaucer*. Edited by Walter W. Skeat. Vol. 3. Oxford: Clarendon Press, 1894.

———. *The Riverside Chaucer*. Edited by Larry D. Benson. 3rd ed. Boston: Houghton Mifflin, 1987.

———. *A Treatise on the Astrolabe*. Edited by Sigmund Eisner. Variorum ed. Vol. 6. Pt. 1. Norman: University of Oklahoma Press, 2002.

Christine de Pizan. *Epistle of Othea to Hector; or, The Boke of Knyghthode*. Translated by Stephen Scrope. Edited by George F. Warner. London: J. B. Nichols, 1904.

Clagett, Marshall, ed. *The Science of Mechanics in the Middle Ages*. Madison: University of Wisconsin Press, 1959.

Dante Alighieri. *The Divine Comedy*. Edited and translated by Charles S. Singleton. 3 vols. Princeton, NJ: Princeton University Press, 1975.

Deguileville, Guillaume de. *Le pelerinage de vie humaine*. Edited by J. J. Stürzinger. Roxburghe Club. London: J. B. Nichols, 1893.

———. *The Pilgrimage of the Lyfe of the Manhode: From the French*. Edited by William Addis Wright. Roxburghe Club. London: J. B. Nichols, 1869.

A Dictionary of English Etymology. Edited by Hensleigh Wedgwood. Vol. 2. London: Trübner, 1862.

Dictionary of the Scots Language. Edinburgh: Scottish Language Dictionaries, 2004. http://www.dsl.ac.uk/.

Dobson, R. B., ed. *The Peasants' Revolt of 1381*. London: Macmillan, 1983.

Dominicus de Clavasio. *The Practica Geometriae of Dominicus de Clavasio*. Edited by H. L. L. Busard. *Archive for History of Exact Sciences* 2 (1965): 520–75.

Douglas, Gavin. *Virgil's Aeneid Translated into Scottish Verse*. Edited by David F. C. Coldwell. 4 vols. Scottish Text Society. Edinburgh: William Blackwood and Sons, 1957–1964.
The Earliest English Translation of Vegetius' De Re Militari. Edited by Geoffrey Lester. Heidelberg: Winter, 1988.
The Early English Version of the Gesta Romanorum. Edited by Sidney J. H. Herrtage. EETS, e.s. 33. London: N. Trübner,1878. Reprint 1962.
Geoffrey de la Tour Landry. *The Book of the Knight of La Tour-Landry, Compiled for the Instruction of His Daughters*. Edited by Thomas Wright. EETS, o.s. 33. 1868. Oxford: Oxford University Press, 1973.
Gerald of Wales. *The Historical Works of Giraldus Cambrensis Containing the Topography of Ireland, and the History of the Conquest of Ireland*. Translated by Thomas Forester. Edited by Thomas Wright. London: H. G. Bohn, 1863.
———. *De rebus a se Gestus. Giraldi Cambrensis opera*. Edited by J. S. Brewer. Vol. 1. Rerum Britannicarum medii aevi scriptores. London: Longman, Green, Longman, and Roberts, 1861.
———. *Topographia Hibernia. Giraldi Cambrensis Opera*. Edited by James F. Dimock. Vol. 5. Rerum Britannicarum medii aevi scriptores. London: Longman, Green, and Roberts, 1861.
Gower, John. *Confessio amantis. The Complete Works*. Edited by G. C. Macauley. Vols. 2 and 3. Oxford: Clarendon Press, 1901.
Grant, Edward, ed. *A Source Book in Medieval Science*. Cambridge, MA: Harvard University Press, 1974.
A Greek-English Lexicon. Edited by Henry George Liddell and Robert Scott. Oxford: Clarendon Press, 1940. Perseus Digital Library, http://www.perseus.tufts.edu/hopper/text?doc=Perseus%3atext%3a1999.04.0057.
Guy de Chauliac. *The Middle English Translation of Guy de Chauliac's Treatise on "Apostemes": Book II of the Great Surgery*. Edited by Björn Wallner. Publications of the New Society of Letters at Lund 80. Lund: Lund University Press, 1988.
Halliwell, James Orchard, ed. *Rara Mathematica; or, A Collection of Tretises on the Mathematics and Subjects Connected with Them*. London: J. W. Parker, 1839.
Henryson, Robert. *The Complete Works*. Edited by David J. Parkinson, TEAMS. Kalamazoo, MI; Medieval Institute Publications, 2010. http://d.lib.rochester.edu/teams/publication/parkinson-henryson-the-complete-works.
Ad C. Herennium. Translated by Harry Caplan. Loeb Classical Library. Cambridge, MA: Harvard University Press, 1954.
Herodotus. *The Histories*. Translated by A. D. Godley. Loeb Classical Library. Cambridge, MA: Harvard University Press, 1920.
Heytesbury, William. *Regule solvendi sophismata Guillelmi Heytesberi*. Translated by Ernest Moody. "William Heytesbury, *Rules for Solving Sophisms*." In Clagett, *Science of Mechanics*, 235–42.
Hilton, R. H., and T. H. Aston, eds. *The English Rising of 1381*. Cambridge: Cambridge University Press, 1984.
Hoccleve, Thomas. *The Regiment of Princes*. Edited by Charles R. Blyth, TEAMS. Kalamazoo, MI: Medieval Institute Publications, 1999.
"How the Good Wiff Tauȝte Hir Douȝtir." In *The Babees Book*, ed. Furnivall, 36–47.

Hugh of St. Victor. *De arca Noe morali*. In *Practical Geometry*, trans. Homann, 85–86.
———. *The Didascalicon of Hugh of St. Victor: A Medieval Guide to the Arts*. Translated by Jerome Taylor. New York: Columbia University Press, 1991.
———. *Hugonis de Sancto Victore Didascalicon de Studio Legendi: A Critical Text*. Edited by Charles Henry Buttimer. Studies in Medieval and Renaissance Latin 10. Washington, DC: Catholic University Press, 1939.
———. *Practica geometriae. Hugonis de Sancto Victore Opera Propaedeutica*. Edited by Roger Baron. Notre Dame, IN: University of Notre Dame Press, 1966.
———. *Practical Geometry (Practica Geometriae). Attributed to Hugh of St. Victor*. Translated by Frederick A. Homann. Milwaukee, WI: Marquette University Press, 1991.
Johannes de Muris. *De arte mensurandi*. Translated and introduction by H. L. L. Busard. Stuttgart: Franz Steiner Verlag, 1998.
Julian of Norwich. *The Writings of Julian of Norwich: A Vision Showed to a Devout Woman and A Revelation of Love*. Edited by Nicholas Watson and Jacqueline Jenkins. University Park: Pennsylvania State University Press, 2006.
Kempe, Margery. *The Book of Margery Kempe*. Edited by Barry Windeatt. Cambridge: D. S. Brewer, 2000.
Kilwardby, Robert. *De ortu scientiarum*. Edited by Albert G. Judy. Oxford: Clarendon Press, 1976.
Knighton, Henry. *Chronicon Henrici Knighton, vel Cnitthon, monachi Leycestrensis*. Edited by Joseph Rawson Lumby. Vol. 2. Rolls Series 92. London: Eyre and Spottiswoode, 1896.
———. *Knighton's Chronicle 1337–1396*. Translated by G. H. Martin. Oxford: Clarendon Press, 1996.
Langtoft, Peter. *Peter Langtoft's Chronicle (as Illustrated and Improv'd by Robert of Brunne)*. Part 2. Edited by Thomas Hearne. Oxford: Printed at the Theater, 1725. Corpus of Middle English Prose and Verse. Ann Arbor: University of Michigan Library, 2006. http://name.umdl.umich.edu/ABA2096.0001.001.
Le Livre Griseldis. "The Clerk's Tale." Edited by Thomas J. Farrell and Amy W. Goodwin. In *Sources and Analogues of the Canterbury Tales*, vol. 1, 141–67.
Lydgate, John. *Fall of Princes*. Edited by Henry Bergen. 4 vols. Washington, DC: Carnegie Institute of Washington, 1923–1927.
———. *The Siege of Thebes*. Edited by Robert R. Edwards, TEAMS. Kalamazoo, MI: Medieval Institute Publications, 2001.
Lyf of the Noble and Crysten Prynce, Charles the Grete. Translated by William Caxton. Edited by Sidney J. H. Herrtage. EETS, e.s., 37. London: N. Trübner, 1881.
Mandeville, Sir John. *Mandeville's Travels*. Edited by M. C. Seymour. Oxford: Clarendon Press, 1967.
Manning, Robert, of Brunne. *The Story of England*. Part 1. Edited by Frederick J. Furnivall. Rolls Series 87. London: Longman, 1887. Corpus of Middle English Prose and Verse. Ann Arbor: University of Michigan Library, 2006. http://name.umdl.umich.edu/AHB1379.0001.001.
Matthew of Vendôme. *Ars Versificatoria. Mathei Vindocinensis: Opera*. Edited by Franco Munari. Vol. 3. Rome: Edizioni di Storia e Letteratura, 1988.
———. *Ars Versificatoria: The Art of the Versemaker*. Translated by Roger Parr. Milwaukee, WI: Marquette University Press, 1981.

Metham, John. *Amoryus and Cleopes*. Edited by Stephen F. Page, TEAMS. Kalamazoo, MI: Medieval Institute Publications, 1999.

"A Method Used in England in the Fifteenth Century for Taking the Altitude of a Steeple or Inaccessible Object." In Halliwell, *Rara Mathematica*, 27–28.

Middle English Dictionary. Edited by Hans Kurath, Sherman M. Kuhn, and Robert E. Lewis. Ann Arbor: University of Michigan Press, 1952–2001. http://quod.lib.umich.edu/m/med/.

The N-Town Plays. Edited by Douglas Sugano. Kalamazoo, MI: Medieval Institute, 2007.

Newton, Sir Isaac. *Philosophiae Naturalis Principia Mathematica*. 1686. Project Gutenberg, 2009. http://www.gutenberg.org/files/28233/28233-h/28233-h.htm.

———. *The Principia: Mathematical Principles of Natural Philosophy*. Translated by I. Bernard Cohen and Ann Whitman. Berkeley: University of California Press, 2014.

Old French Online. Linguistics Research Center, University of Texas at Austin. https://lrc.la.utexas.edu/eieol_base_form_dictionary/ofrol/15.

Oresme, Nicole. *Nicole Oresme and the Kinematics of Circular Motion: Tractatus de commensurabilitate vel incommensurabilitate motuum celi*. Translated by Edward Grant. Madison: University of Wisconsin Press, 1971.

———. *Nicole Oresme and the Marvels of Nature: A Study of His De Causis Mirabilium with Critical Edition, Translation, and Commentary*. Translated by Bert Hansen. Toronto: Pontifical Institute of Mediaeval Studies, 1985.

———. *Nicole Oresme and the Medieval Geometry of Qualities and Motions: A Treatise on the Uniformity and Difformity of Intensities Known as Tractatus de configurationibus qualitatum et motuum*. Translated by Marshall Clagett. Madison: University of Wisconsin Press, 1968.

———. *De proportionibus proportionum, and Ad pauca respicientes*. Translated by Edward Grant. Madison: University of Wisconsin Press, 1966.

The Oxford English Dictionary. OED Online. Oxford: Oxford University Press, 2018. http://www.oed.com.

Pecock, Reginald. *The Folewer to the Donet*. Edited by Elsie Vaughan Hitchcock. EETS, o.s. 164. Oxford: Oxford University Press, 1924.

Petrarca, Francesco (Petrarch). *Historia Griseldis*. "The Clerk's Tale." Edited by Thomas J. Farrell and Amy W. Goodwin. In *Sources and Analogues of the Canterbury Tales*, vol. 1, 108–29.

Plato. *Plato's Cosmology: The Timaeus of Plato*. Translated and commentary by Francis M. Cornford. New York: Routledge, 1935.

Pliny the Elder. *Natural History*. Translated by H. Rackham. Cambridge, MA: Harvard University Press, 1938.

Practical Geometry in the High Middle Ages: Artis Cuiuslibet Consummatio and the Pratike De Geometrie. Translated by Stephen K. Victor. Philadelphia: American Philosophical Society, 1979.

Ptolemy, Claudius. *Ptolemy's Almagest*. Translated by G. J. Toomer. London: Gerald Duckworth, 1984.

———. *Ptolemy's Geography: An Annotated Translation of the Theoretical Chapters.* Translated by J. Lennart Berggren and Alexander Jones. Princeton, NJ: Princeton University Press, 2000.
Puttenham, George. *Arte of English Poesie.* Edited by Gladys Dodge Willcock and Alice Walker. Cambridge: Cambridge University Press, 1936.
Quintilian. *Insitutio oratoria.* Translated by H. E. Butler. Loeb Classical Library. Cambridge, MA: Harvard University Press, 1920–1922.
"The Richard II Quadrant." 1860,0519.1. www.britishmuseum.org/collection. British Museum. Accessed June 23, 2018.
Sir Gawain the Green Knight. Edited by J. R. R. Tolkien and E. V. Gordon. 2nd ed. rev. Norman Davis. Oxford: Oxford University Press, 1925.
Sources and Analogues of the Canterbury Tales. Edited by Robert M. Correale and Mary Hamel. 2 vols. Cambridge: D. S. Brewer, 2002, 2005.
Thomas of Elmham. *Historia Monasterii S. Augustini Cantuariensis.* Edited by Charles Hardwick. Rolls Series 8. London: Longman, 1858.
"A Treatise on the Mensuration of Heights and Distances." In Halliwell, *Rara Mathematica,* 56–71.
The Wars of Alexander: An Alliterative Romance, Translated Chiefly from the Historia Alexandri Magni de Preliis. Edited by Walter W. Skeat. EETS, e.s. 47. London: N. Trübner, 1886.
Wyclif, John. *The English Works of Wyclif.* Edited by F. D. Matthew. EETS, o.s. 74. London: Kegan Paul, Trench, Trübner, 1880.

Secondary Sources

Ackermann, Silke, and John Cherry. "Richard II, John Holland and Three Medieval Quadrants." *Annals of Science* 56 (1999): 3–23.
Agnew, John A. *Place and Politics: The Geographical Mediation of State and Society.* Boston: Allen and Unwin, 1987.
Allen, Valerie. *On Farting: Language and Laughter in the Middle Ages.* New York: Palgrave Macmillan, 2007.
———. "Road." *postmedieval: A Journal of Medieval Cultural Studies* 4 (2013): 18–29.
———. "When Things Break: Mending Roads, Being Social." In Allen and Evans, *Roadworks,* 74–96.
Allen, Valerie, and Ruth Evans, eds. *Roadworks: Medieval Britain, Medieval Roads.* Manchester: Manchester University Press, 2016.
Armstrong, Dorsey, and Kenneth Hodges. *Mapping Malory: Regional Identities and National Geographies in Le Morte Darthur.* New York: Palgrave Macmillan, 2014.
Astell, Ann W. "Apostrophe, Prayer and the Structure of Satire in the Man of Law's Tale." *Studies in the Age of Chaucer* 13 (1991): 81–97.
Auerbach, Erich. "Philology and *Weltliteratur.*" Translated by Maire Said and Edward Said. *Centennial Review* 13 (1969): 1–17.
Bale, Anthony. *Feeling Persecuted: Christians, Jews and Images of Violence in the Middle Ages.* London: Reaktion, 2010.

———. "'ut legi': Sir John Mandeville's Audience and Three Late Medieval English Travelers to Italy and Jerusalem." *Studies in the Age of Chaucer* 38 (2016): 201–37.
Barley, M. W. "Sherwood Forest, Nottinghamshire." In Skelton and Harvey, *Local Maps and Plans*, 131–39.
Barthes, Roland. *Empire of Signs*. Translated by Richard Howard. New York: Hill and Wang, 1982.
Bartlett, Robert. *Gerald of Wales, 1146–1223*. Oxford: Clarendon Press, 1982.
———. *The Natural and the Supernatural in the Middle Ages*. Cambridge: Cambridge University Press, 2008.
Beer, Gillian. *Darwin's Plots: Evolutionary Narrative in Darwin, George Eliot and Nineteenth-Century Fiction*. 1983. Rev. ed. Cambridge: Cambridge University Press, 2000.
Bennett, J. A. W. *Chaucer at Oxford and at Cambridge*. Oxford: Oxford University Press, 1974.
Bennett, Michael J. "Mandeville's Travels and the Anglo-French Moment." *Medium Aevum* 75 (2006): 273–92.
Beresford, M. W. "Inclesmoor, West Riding of Yorkshire." In Skelton and Harvey, *Local Maps and Plans*, 147–61.
Bettridge, William Ellen, and Francis Lee Utley. "New Light on the Origin of the Griselda Story." *Texas Studies in Literature and Language* 13 (1971): 153–208.
Birkholz, Daniel. "Hereford Maps, Hereford Lives: Biography and Cartography in an English Cathedral City." In Lilley, *Mapping Medieval Geographies*, 225–49.
———. *The King's Two Maps: Cartography and Culture in Thirteenth-Century England*. New York: Routledge, 2004.
Blurton, Heather, and Hannah Johnson. *The Critics and the Prioress: Antisemitism, Criticism, and Chaucer's Prioress's Tale*. Ann Arbor: University of Michigan Press, 2017.
Bostock, David. *Space, Time, Matter, and Form*. Oxford: Oxford University Press, 2006.
Bourdieu, Pierre. *The Field of Cultural Production: Essays on Art and Literature*. Edited by Randal Johnson. New York: Columbia University Press, 1993.
Brantley, Jessica. "Venus and Christ in Chaucer's *Complaint of Mars*: The Fairfax 16 Frontispiece." *Studies in the Age of Chaucer* 30 (2008): 171–204.
Breen, Katharine. *Imagining an English Reading Public, 1150–1400*. Cambridge: Cambridge University Press, 2010.
Brosseau, Marc. "Geography's Literature." *Progress in Human Geography* 18 (1994): 333–53.
Brown, Peter. *Chaucer and the Making of Optical Space*. Bern: Peter Lang, 2007.
———. "The Containment of Symkyn: The Function of Space in the 'Reeve's Tale,'" *Chaucer Review* 14 (1980): 225–36.
Burger, Glenn. *Chaucer's Queer Nation*. Minneapolis: University of Minnesota Press, 2003.
———. "'Pite renneth soone in gentil herte: Ugly Feelings and Gendered Conduct in Chaucer's *Legend of Good Women*." *Chaucer Review* 52 (2017): 66–84.

Burnett, Charles. "The Introduction of Aristotle's Natural Philosophy into Great Britain: A Preliminary Survey of the Manuscript Evidence." In Marenbon, *Aristotle in Britain during the Middle Ages*, 21–50.

Butterfield, Ardis. "Nationhood." In *Chaucer: An Oxford Guide*, 50–65. Oxford: Oxford University Press, 2005.

Cacciari, Massimo. "Place and Limit." In Malpas, *Intelligence of Place*, 13–22.

Cadden, Joan. *Nothing Natural Is Shameful: Sodomy and Science in Late Medieval Europe*. Philadelphia: University of Pennsylvania Press, 2013.

———. "The Organization of Knowledge." In Lindberg and Shank, *The Cambridge History of Science*, 2:240–67.

Callus, D. A. "The Introduction of Aristotelian Learning in Oxford." *Proceedings of the British Academy* 29 (1943): 229–81.

Campbell, Mary B. *The Witness and the Other World: Exotic European Writing, 400–1600*. Ithaca, NY: Cornell University Press, 1988.

Campbell, Tony. "Portolan Charts from the Late Thirteenth Century to 1500." In Harley and Woodward, *History of Cartography*, 1:371–463.

Carruthers, Mary J. *The Book of Memory: A Study of Memory in Medieval Culture*. Cambridge: Cambridge University Press, 1990.

Cartlidge, Neil. "Ripples on the Water? The Acoustics of Geoffrey Chaucer's *House of Fame* and the Influence of Robert Holcot." *Studies in the Age of Chaucer* 39 (2017): 57–98.

Casey, Edward S. "Do Places Have Edges? A Geo-Philosophical Inquiry." In *Envisioning Landscapes, Making Worlds: Geography and the Humanities*, edited by Stephen Daniels, Dydia DeLyser, J. Nicholas Entrikin, and Doug Richardson, 65–73. Abingdon, Oxon: Routledge, 2011.

———. *The Fate of Place: A Philosophical History*. Berkeley: University of California Press, 2013.

———. *Getting Back Into Place: Toward a Renewed Understanding of the Place-World*. Bloomington: Indiana University Press, 1993.

———. "How to Get from Space to Place in a Fairly Short Stretch of Time: Phenomenological Prolegomena." In *Senses of Place*, edited by Steven Feld and Keith H. Basso, 13–52. Santa Fe, NM: School of American Research Press, 1996.

———. "Place and Edge." In Malpas, *Intelligence of Place*, 23–38.

———. *Representing Place: Landscape Painting and Maps*. Minneapolis: University of Minnesota Press, 2002.

Catto, J. I., and T. A. R. Evans, eds. *The History of the University of Oxford*. Vol. 2, *Late Medieval Oxford*. Oxford: Oxford University Press, 1992.

Certeau, Michel de. *The Practice of Everyday Life*. Translated by Steven F. Rendall. Berkeley: University of California Press, 1984.

Châtelet, Gilles. *Figuring Space: Philosophy, Mathematics and Physics*. Translated by Robert Shore and Muriel Zagha. Dordrecht, Netherlands: Kluwer Academic Publishers, 1999.

Clagett, Marshall. "Oresme, Nicole." In *Complete Dictionary of Scientific Biography*, 223–30. Vol. 10. Detroit: Charles Scribner's Sons, 2008.

———. "Richard Swineshead and Late Medieval Physics." *Osiris* 9 (1950): 131–61.

Classen, Albrecht. *East Meets West in the Middle Ages and Early Modern Times: Transcultural Experiences in the Premodern World*. Berlin: Walter de Gruyter, 2013.

Clogan, Paul M. "The Narrative Style of the Man of Law's Tale." *Medievalia et Humanistica*, n.s. 8 (1977): 217–33.

Cohen, I. Bernard. *The Birth of a New Physics*. Rev. ed. New York: W. W. Norton, 1985.

Cohen, Jeffrey J. *Hybridity, Identity, and Monstrosity in Medieval Britain: On Difficult Middles*. New York: Palgrave Macmillan, 2006.

———. *Medieval Identity Machines*. Minneapolis: University of Minnesota Press, 2003.

Coleman, Joyce. "Illuminations in Gower's Manuscripts." In *The Routledge Research Companion to John Gower*, edited by Ana Saez-Hidalgo, Brian Gastle, and R. F. Yeager, 117–31. New York: Routledge, 2017.

Coley, David K. "'Withyn a temple ymad of glas': Glazing, Glossing, and Patronage in Chaucer's *House of Fame*." *Chaucer Review* 45 (2010): 59–84.

Connolly, Daniel K. *The Maps of Matthew Paris: Medieval Journeys through Space, Time and Liturgy*. Woodbridge, Suffolk: Boydell and Brewer, 2009.

Cooper, Helen. *The English Romance in Time: Transforming Motifs from Geoffrey of Monmouth to the Death of Shakespeare*. Oxford: Oxford University Press, 2004.

———. *Oxford Guides to Chaucer: The Canterbury Tales*. Oxford: Oxford University Press, 1996.

Cooper, Lisa H. *Artisans and Narrative Craft in Late Medieval England*. Cambridge: Cambridge University Press, 2011.

Coopland, G. W. *Nicole Oresme and the Astrologers: A Study of His Livre De Divinacions*. Liverpool: Liverpool University Press, 1952.

Copeland, Rita. *Rhetoric, Hermeneutics, and Translation in the Middle Ages: Academic Traditions and Vernacular Texts*. Cambridge: Cambridge University Press, 1995.

Copeland, Rita, and Ineke Sluiter, eds. *Medieval Grammar and Rhetoric: Language Arts and Literary Theory, AD 300–1475*. Oxford: Oxford University Press, 2009.

Cosgrove, Denis. *Geography and Vision: Seeing, Imagining and Representing the World*. London: Tauris, 2008.

———. "Introduction: Mapping Meaning." In *Mappings*, edited by Denis Cosgrove, 1–23. London: Reaktion, 1999.

Courtenay, William J. "The Early Career of Nicole Oresme." *Isis* 91 (2000): 542–48.

———. *Schools and Scholars in Fourteenth-Century England*. Princeton, NJ: Princeton University Press, 1987.

Cresswell, Tim. *Place: An Introduction*. 2nd ed. Malden, MA: Wiley-Blackwell, 2014.

Cripps, Judith A. "Barholm, Greatford, and Stowe, Lincolnshire." In Skelton and Harvey, *Local Maps and Plans*, 263–88.

Crocker, Holly. "Medieval Affects Now." *Exemplaria* 29 (2017): 82–98.

Curry, Michael. "Toward a Geography of a World without Maps: Lessons from Ptolemy and Postal Codes." *Annals of the Association of American Geographers* 95 (2005): 680–91.

Curtius, Ernst Robert. *European Literature and the Latin Middle Ages*. Princeton, NJ: Princeton University Press, 1952.

Darby, H. C. "The Agrarian Contribution to Surveying in England." *Geographical Journal* 82 (1933): 529–35.
Davenport, Tony. "Chaucer's *House of Fame*, 111–18: A Windsor Joke?" *Notes and Queries* 48 (2001): 222–24.
Davis, Rebecca. "Fugitive Poetics in Chaucer's *House of Fame*." *Studies in the Age of Chaucer* 37 (2015): 101–32.
Dekker, Elly. "'With his sharp lok perseth the sonne': A New Quadrant from Canterbury." *Annals of Science* 65 (2008): 201–20.
Delano-Smith, Catherine, and Roger J. P. Kain. *English Maps: A History*. Toronto: University of Toronto Press, 1999.
Delanty, Gerard. "Introduction: The Emerging Field of Cosmopolitanism Studies." In *The Routledge Handbook of Cosmopolitanism Studies*, edited by Gerard Delanty, 1–8. New York: Routledge, 2012.
Delany, Sheila. *Chaucer's House of Fame: The Poetics of Skeptical Fideism*. Chicago: University of Chicago Press, 1972.
———. *The Naked Text: Chaucer's Legend of Good Women*. Berkeley: University of California Press, 1994.
Delasanta, Rodney. "Chaucer and Strode." *Chaucer Review* 26 (1991): 205–18.
Delia, Robert D. "Senwosret III." In *The Oxford Encyclopedia of Ancient Egypt*. Oxford University Press, 2001. http://www.oxfordreference.com.i.ezproxy.nypl.org/view/10.1093/acref/9780195102345.001.0001/acref-9780195102345-e-0653.
Derrida, Jacques. *On Cosmopolitanism and Forgiveness*. Translated by Mark Dooley and Michael Hughes. New York: Routledge, 2001.
Dilke, O. A. W. "The Culmination of Greek Cartography in Ptolemy." In Harley and Woodward, *History of Cartography*, 1:177–200.
Dinshaw, Carolyn. "Chaucer's Queer Touches / A Queer Touches Chaucer." *Exemplaria* 7 (1995): 75–92.
———. *Getting Medieval: Sexualities and Communities, Pre- and Postmodern*. Durham, NC: Duke University Press, 1999.
———. *How Soon Is Now? Medieval Texts, Amateur Readers, and the Queerness of Time*. Durham, NC: Duke University Press, 2012.
Doyle, Kara A. "Thisbe out of Context: Chaucer's Female Readers and the Findern Manuscript." *Chaucer Review* 40 (2006): 231–61.
Duhem, Pierre. *Études sur Léonard de Vinci*. Vol. 3. Paris: A. Hermann, 1906.
———. *Medieval Cosmology: Theories of Infinity, Place, Time, Void, and the Plurality of Worlds*. Edited and translated by Roger Ariew. Chicago: University of Chicago Press, 1985.
———. *Le système du monde: Histoire des doctrines cosmologiques de Platon à Copernic*. Vol. 7. Paris: Hermann, 1956.
Dumont, Stephen D. "Godfrey of Fontaines and the Succession Theory of Forms at Paris in the Early Fourteenth Century." In *Philosophical Debates at Paris in the Early Fourteenth Century*, edited by Stephen F. Brown, Thomas Dewender, and Theo Kobusch, 39–126. Leiden: Brill, 2009.
Dyas, Dee. *Pilgrimage in Medieval English Literature, 700–1500*. Cambridge: D. S. Brewer, 2001.

Edgerton, Samuel Y. *The Renaissance Rediscovery of Linear Perspective*. New York: Basic Books, 1975.
Edmondson, George. *The Neighboring Text: Chaucer, Boccaccio, Henryson*. Notre Dame, IN: University of Notre Dame Press, 2011.
Elden, Stuart. *The Birth of Territory*. Chicago: University of Chicago Press, 2013.
Escalona, Julio. "The Early Middle Ages: A Scale-Based Approach." In *Scale and Scale Change in the Early Middle Ages: Exploring Landscape, Local Society, and the World Beyond*, edited by Julio Escalona and Andrew Reynolds, 1–22. Turnhout: Brepols, 2011.
Evans, Ruth. "Chaucer in Cyberspace." *Studies in the Age of Chaucer* 23 (2001): 43–69.
———. "Getting There: Wayfinding in the Middle Ages." In Allen and Evans, *Roadworks*, 127–56.
Fanous, Samuel. "Measuring the Pilgrim's Progress: Internal Emphases in *The Book of Margery Kempe*." In *Writing Religious Women: Female Spiritual and Religious Practices in Late Medieval England*, edited by Denis Renevey and Christiania Whitehead, 157–76. Toronto: University of Toronto Press, 2000.
Farina, Lara. "Wondrous Skins and Tactile Affection: The Blemmye's Touch." In Walter, *Reading Skin in Medieval Literature and Culture*, 11–28.
Farrell, Thomas J., and Amy W. Goodwin. "The Clerk's Tale." In *Sources and Analogues of the Canterbury Tales*, 101–67.
Febvre, Lucien, and Henri-Jean Martin. *The Coming of the Book: The Impact of Printing 1450–1800*. Translated by David Gerard. London: Verso, 1976.
Flannery, Mary C., and Carrie Griffin, eds. *Spaces for Reading in Later Medieval England*. New York: Palgrave Macmillan, 2016.
Fletcher, J. M. "Developments in the Faculty of Arts, 1370–1520." In Catto and Evans, *The History of the University of Oxford*, 2:315–45.
Flint, Valerie I. J. "The Hereford Map: Its Author(s), Two Scenes and a Border." *Transactions of the Royal Historical Society*, 6th ser., 8 (1998): 19–44.
Foucault, Michel. "Des espaces autres." In *Dits et écrits, 1954–1988*, tome 4, 1980–1988, 752–62. Paris: Gallimard, 1994. 752–62.
———. "Of Other Spaces." Translated by Jay Miskowiec. *Diacritics* 16 (1986): 22–27.
Fox, H. S. A. "Exeter, Devonshire." In Skelton and Harvey, *Local Maps and Plans*, 163–69.
Fradenburg, Louise O. "Criticism, Anti-Semitism, and the Prioress's Tale." *Exemplaria* 1 (1989): 69–115.
———. "Sacrificial Desire in Chaucer's *Knight's Tale*." *Journal of Medieval and Early Modern Studies* 27 (1997): 47–75.
Frank, Robert Worth. *Chaucer and the Legend of Good Women*. Cambridge, MA: Harvard University Press, 1972.
French, Roger. *Canonical Medicine: Gentile Da Foligno and Scholasticism*. Leiden: Brill, 2001.
Fyler, John M. *Chaucer and Ovid*. New Haven, CT: Yale University Press, 1979.
Gabrovsky, Alexander N. *Chaucer the Alchemist: Physics, Mutability, and the Medieval Imagination*. New York: Palgrave Macmillan, 2015.
Gautier Dalché, Patrick. *La Géographie de Ptolémée en Occident (IVe–XVIe siècle)*. Turnhout: Brepols, 2009.

———. "The Reception of Ptolemy's *Geography* (End of the Fourteenth to Beginning of the Sixteenth Century)." In *The History of Cartography*. Vol. 3, *Cartography in the European Renaissance*, edited by David Woodward, 285–364. Chicago: University of Chicago Press, 2007.

Germann, Nadja. "Natural Philosophy in Earlier Latin Thought." In Pasnau and van Dyke, *Cambridge History of Medieval Philosophy*, 1:219–31.

Gibson, James J. *The Ecological Approach to Visual Perception*. New York: Houghton Mifflin, 1979.

———. *The Perception of the Visual World*. Cambridge, MA: Riverside, 1950.

———. *The Senses Considered as Perceptual Systems*. Boston: Houghton Mifflin, 1966.

Gibson, Margaret, T. A. Heslop, and Richard W. Pfaff, eds. *The Eadwine Psalter: Text, Image, and Monastic Culture in Twelfth-Century Canterbury*. University Park: Pennsylvania State University Press, 1992.

Givens, Jean A. *Observation and Image-Making in Gothic Art*. Cambridge: Cambridge University Press, 2005.

Goehring, Margaret. *Space, Place, and Ornament: The Function of Landscape in Medieval Manuscript Illumination*. Turnhout: Brepols, 2014.

Goldie, Matthew Boyd. "An Early English Rutter: The Sea and Spatial Hermeneutics in the Fourteenth and Fifteenth Centuries." *Speculum: A Journal of Medieval Studies* 90 (2015): 701–27.

———. *The Idea of the Antipodes: Place, People, and Voices*. New York: Routledge, 2010.

Goodman, Anthony. *Margery Kempe and Her World*. Harlow, Essex: Pearson, 2002.

Gransden, Antonia. *Historical Writing in England*. Vol. 1, *c. 500 to c. 1307*. 1974; New York: Routledge, 1996.

———. *Historical Writing in England*. Vol. 2, *1307 to the Early Sixteenth Century*. 1974; New York: Routledge, 1996.

———. "Realistic Observation in Twelfth-Century England." *Speculum* 47 (1972): 29–51.

Grant, Edward. "The Concept of *Ubi* in Medieval and Renaissance Discussions of Place." *Manuscripta: A Journal for Manuscript Research* 20 (1976): 71–80.

———. *A History of Natural Philosophy: From the Ancient World to the Nineteenth Century*. New York: Cambridge University Press, 2007.

———. "The Medieval Doctrine of Place: Some Fundamental Problems and Solutions." In *Studi sul xiv secolo in memoria di Anneliese Maier*, edited by Maierù and Paravicini Bagliani, 57–79. Rome: Storia e Letteratura, 1981.

———. *Much Ado about Nothing: Theories of Space and Vacuum from the Middle Ages to the Scientific Revolution*. Cambridge: Cambridge University Press, 2008.

———. "Place and Space in Medieval Physical Thought." In *Motion and Time, Space and Matter: Interrelations in the History of Science and Philosophy*, edited by Peter K. Machamer and Robert G. Turnbull, 137–67. Columbus: Ohio State University Press, 1976.

Green, Richard Firth. "Griselda in Siena." *Studies in the Age of Chaucer* 33 (2011): 3–38.

Gregg, Melissa, and Gregory J. Seigworth. "An Inventory of Shimmers." In *The Affect Theory Reader*, 1–25. Durham, NC: Duke University Press, 2009.

Günther, Robert T. *The Astrolabes of the World*. 2 vols. Oxford: Oxford University Press, 1932.
Gust, Geoffrey W. *Constructing Chaucer: Author and Autofiction in the Critical Tradition*. New York: Palgrave MacMillan, 2009.
Hagen, Margareth, Randi Koppen, and Margery Vibe Skagen, eds. *The Art of Discovery: Encounters in Literature and Science*. Aarhus, Denmark: Aarhus University Press, 2010.
Hagen, Margareth, and Margery Vibe Skagen, eds. *Literature and Chemistry: Elective Affinities*. Aarhus, Denmark: Aarhus University Press, 2014.
Hahn, Nan L. "Medieval Mensuration: '*Quadrans Vetus*' and '*Geometrie Due Sunt Partes Principales*.'" *Transactions of the American Philosophical Society* 72 (1982): i–204.
Hanawalt, Barbara A., and Michal Kobialka, eds. *Medieval Practices of Space*. Minneapolis: University of Minnesota Press, 2000.
Hansen, Bert. "An Overview." In Oresme, *Nicole Oresme and the Marvels of Nature*, 3–16.
Hansen, Elaine Tuttle. "Irony and the Antifeminist Narrator in Chaucer's *Legend of Good Women*." *Journal of English and Germanic Philology* 82 (1983): 11–31.
Hardman, Phillipa. "Lydgate's Uneasy Syntax." In Scanlon and Simpson, *John Lydgate: Poetry, Culture, and Lancastrian England*, 12–35.
Harley, J. B., and David Woodward. "Concluding Remarks." In Harley and Woodward, *The History of Cartography*, 1:502–9.
———. "Greek Cartography in the Early Roman World." In Harley and Woodward, *The History of Cartography*, 1:161–76.
———, eds. *The History of Cartography*. Vol. 1, *Cartography in Prehistoric, Ancient, and Medieval Europe and the Mediterranean*. Chicago: University of Chicago Press, 1987.
———, eds. *The History of Cartography*. Vol. 2, *Cartography in the Traditional Islamic and South Asian Societies*. Chicago: University of Chicago Press, 1992.
Harrison, Dick. *Medieval Space: The Extent of Microspatial Knowledge in Western Europe during the Middle Ages*. Lund: Lund University Press, 1996.
Harvey, P. D. A. "Boarstall, Buckinghamshire." In Skelton and Harvey, *Local Maps and Plans*, 211–19.
———. *The History of Topographical Maps: Symbols, Pictures and Surveys*. London: Thames and Hudson, 1980.
———. "Local and Regional Cartography in Medieval Europe." In Harley and Woodward, *History of Cartography*, 1:464–501.
Heidegger, Martin. *Basic Writings: Second Edition, Revised and Expanded*. New York: Harper Collins, 1993.
———. *Country Path Conversations*. Translated by Bret W. Davis. Bloomington: Indiana University Press, 2010.
———. *Ontology: The Hermeneutics of Facticity*. Translated by John van Buren. Bloomington: Indiana University Press, 1999.
Heng, Geraldine. *Empire of Magic: Medieval Romance and the Politics of Cultural Fantasy*. New York: Columbia University Press, 2003.
Heslop, T. A. "Eadwine and His Portrait." In Gibson, Heslop, and Pfaff, *The Eadwine Psalter: Text, Image, and Monastic Culture in Twelfth-Century Canterbury*, 178–85.

Hiatt, Alfred. "Beowulf Off the Map." *Anglo-Saxon England* 38 (2009): 11–40.
Higgins, Iain Macleod. "Introduction." *The Book of John Mandeville: With Related Texts*. Edited and translated by Higgins (Indianapolis, In: Hackett, 2011).
———. *Writing East: The "Travels" of Sir John Mandeville*. Philadelphia: University of Pennsylvania Press, 1997.
Hilton, Rodney. *Bond Men Made Free: Medieval Peasant Movements and the English Rising of 1381*. New York: Viking, 1973.
Holley, Linda Tarte. *Chaucer's Measuring Eye*. Houston, TX: Rice University Press, 1990.
Hope, William St John. *The History of the London Charterhouse from Its Foundation until the Suppression of the Monastery*. London: Society for Promoting Christian Knowledge, 1925.
Howarth, William L. "Imagined Territory: The Writing of Wetlands." *New Literary History* 30 (1999): 509–39.
Howe, Nicholas. *Writing the Map of Anglo-Saxon England: Essays in Cultural Geography*. New Haven, CT: Yale University Press, 2008.
Howes, Laura L. "Chaucer's Forests, Parks, and Groves." *Chaucer Review* 49 (2014): 125–33.
Hull, F. "Cliffe, Kent." In Skelton and Harvey, *Local Maps and Plans*, 99–105.
———. "Isle of Thanet, Kent." In Skelton and Harvey, *Local Maps and Plans*, 119–26.
Ingham, Patricia Clare. "Contrapuntal Histories." In *Postcolonial Moves: Medieval through Modern*, edited by Patricia Clare Ingham and Michelle R. Warren, 47–70. New York: Palgrave Macmillan, 2003.
———. *The Medieval New: Ambivalence in an Age of Innovation*. Philadelphia: University of Pennsylvania Press, 2015.
———. *Sovereign Fantasies: Arthurian Romance and the Making of Britain*. Philadelphia: University of Pennsylvania Press, 2001.
Jammer, Max. *Concepts of Space: The History of Theories of Space in Physics*. Cambridge, MA: Harvard University Press, 1954.
Janni, Pietro. *La mappa e il periplo: Cartografia antica e spazio odologico*. Rome: Giorgio Bretschneider, 1984.
Jenkins, Alice. *Space and the 'March of Mind': Literature and the Physical Sciences in Britain 1815–1850*. Oxford: Oxford University Press, 2007.
Johnson, Ian. "Hellish Complexity in Henryson's *Orpheus*." *Forum for Modern Language Studies* 38 (2002): 412–19.
Justice, Steven. *Writing and Rebellion: England in 1381*. Berkeley: University of California Press, 1994.
Kay, Sarah. *The Place of Thought: The Complexity of One in Late Medieval French Didactic Poetry*. Philadelphia: University of Pennsylvania Press, 2007.
Kaye, Joel. *Economy and Nature in the Fourteenth Century: Money, Market Exchange, and the Emergence of Scientific Thought*. Cambridge: Cambridge University Press, 1998.
———. *A History of Balance, 1250–1375: The Emergence of a New Model of Equilibrium and Its Impact on Thought*. Cambridge: Cambridge University Press, 2014.
Keen, Elizabeth. *The Journey of a Book: Bartholomew the Englishman and the Properties of Things*. Canberra: Australian National University E Press, 2011.

Kendall, Elliot. "Family, *Familia*, and the Uncanny in *Sir Orfeo*." *Studies in the Age of Chaucer* 35 (2013): 289–327.
Kendrick, T. D. *British Antiquity*. London: Methuen, 1950.
Khanmohamadi, Shirin A. *In Light of Another's Word: European Ethnography in the Middle Ages*. Philadelphia: University of Pennsylvania Press, 2013.
Kimble, George H. T. *Geography in the Middle Ages*. London: Methuen, 1938.
King, Peter. "Duns Scotus on the Reality of Self-Change." In *Self-Motion: From Aristotle to Newton*, edited by Mary Louise Gill and James G. Lennox, 227–90. Princeton, NJ: Princeton University Press, 1994.
———. "Mediaeval Thought-Experiments: The Metamethodology of Mediæval Science." In *Thought Experiments in Science and Philosophy*, edited by Tamara Horowitz and Gerald J. Massey, 43–64. New York: Rowman and Littlefield, 1991.
Kleiner, John. *Mismapping the Underworld: Daring and Error in Dante's Comedy*. Stanford, CA: Stanford University Press, 1994.
Knorr, Wilbur R. "The Latin Sources of the *Quadrans vetus*, and What They Imply for Its Authorship and Date." In *Texts and Contexts in Ancient and Medieval Science: Studies on the Occasion of John E. Murdoch's Seventieth Birthday*, edited by Edith D. Sylla and Michael McVaugh, 23–67. Leiden: Brill, 1997.
Knowles, David, and W. F. Grimes. *Charterhouse, the Medieval Foundation in the Light of Recent Discoveries*. London: Longmans, Green, 1954.
Kretzmann, Norman, Anthony Kenny, and Jan Pinborg, eds. *The Cambridge History of Later Medieval Philosophy: From the Rediscovery of Aristotle to the Disintegration of Scholasticism, 1100–1600*. Cambridge: Cambridge University Press, 1982.
Kruger, Steven F. "Imagination and the Complex Movement of Chaucer's *House of Fame*." *Chaucer Review* 28 (1993): 117–34.
———. "Passion and Order in Chaucer's *Legend of Good Women*." *Chaucer Review* 23 (1989): 219–35.
Kymäläinen, Päivi, and Ari A. Lehtinen. "Chora in Current Geographical Thought: Places of Co-Design and Re-Membering." *Geografiska Annaler, Series B, Human Geography* 92 (2010): 251–61.
Lando, Fabio. "Fact and Fiction: Geography and Literature." *GeoJournal* 38 (1996): 3–18.
Lang, Helen S. *Aristotle's Physics and Its Medieval Varieties*. Albany: State University of New York Press, 1992.
Latour, Bruno. "An Attempt at a 'Compositionist Manifesto.'" *New Literary History* 41 (2010): 471–90.
———. "Why Has Critique Run Out of Steam? From Matters of Fact to Matters of Concern." *Critical Inquiry* 30 (2004): 225–48.
Lavezzo, Kathy. *The Accommodated Jew: English Antisemitism from Bede to Milton*. Ithaca, NY: Cornell University Press, 2016.
———. *Angels on the Edge of the World: Geography, Literature, and English Community, 1000–1534*. Ithaca, NY: Cornell University Press, 2006.
———, ed. *Imagining a Medieval English Nation*. Minneapolis: University of Minnesota Press, 2004.
Lee, A. D. *Information and Frontiers: Roman Foreign Relations in Late Antiquity*. Cambridge: Cambridge University Press, 1993.

Lees, Clare A., and Gillian R. Overing, eds. *A Place to Believe In: Locating Medieval Landscapes*. University Park: Pennsylvania State University Press, 2006.

Lefebvre, Henri. *The Production of Space*. Oxford: Blackwell, 1991.

Legassie, Shayne Aaron. *The Medieval Invention of Travel*. Chicago: University of Chicago Press, 2017.

Lilley, Keith D. "Geography's Medieval History: A Neglected Enterprise?" *Dialogues in Human Geography* 1 (2011): 147–62.

———, ed. *Mapping Medieval Geographies*. Cambridge: Cambridge University Press, 2013.

Lindberg, David C. *The Beginnings of Western Science: The European Scientific Tradition in Philosophical, Religious, and Institutional Context, 600 B.C. to A.D. 1450*. Chicago: University of Chicago Press, 1992.

Lindberg, David C., and Michael H. Shank, eds. *The Cambridge History of Science*. Vol. 2. Cambridge: Cambridge University Press, 2013.

Lukerman, F. "The Concept of Location in Classical Geography." *Annals of the Association of American Geographers* 51 (1961): 194–210.

Lynch, Kathryn L., ed. *Chaucer's Cultural Geography*. New York: Routledge, 2002.

———. *Chaucer's Philosophical Visions*. Woodbridge, Suffolk: Boydell and Brewer, 2000.

Machamer, Peter K. "Aristotle on Natural Place and Natural Motion." *Isis* 69 (1978): 377–87.

MacQueen, John. *Complete and Full with Numbers: The Narrative Poetry of Robert Henryson*. New York: Rodopi, 2006.

Macrae, Enya. "Geoffrey of Aspall's Commentaries on Aristotle." *Mediaeval and Renaissance Studies*, Warburg Institute 6 (1968): 94–134.

Mader, Mary Beth. "Whence Intensity? Deleuze and the Revival of a Concept." In *Gilles Deleuze and Metaphysics*, edited by Alain Beaulieu, Edward Kazarian, and Julia Sushytska, 225–48. Lanham, MD: Lexington Books-Rowman and Littlefield, 2014.

Magnusson, Roberta J. *Water Technology in the Middle Ages: Cities, Monasteries, and Waterworks after the Roman Empire*. Baltimore: Johns Hopkins University Press, 2003.

Maier, Anneliese. "The Achievements of Late Scholastic Natural Philosophy." 1964. In Maier, *On the Threshold of Exact Science*, 143–70.

———. "Causes, Forces, and Resistance." 1949. In Maier, *On the Threshold of the Exact Science*, 40–60.

———. "The Concept of the Function in Fourteenth-Century Physics." 1949. In Maier, *On the Threshold of the Exact Science*, 61–75.

———. "Galileo and the Scholastic Theory of Impetus." 1967. In Maier, *On the Threshold of the Exact Science*, 103–13.

———. "The Nature of Motion." 1944. In Maier, *On the Threshold of Exact Science*, 21–39.

———. *On the Threshold of Exact Science: Selected Writings of Anneliese Maier on Late Medieval Natural Philosophy*. Translated by Steven D. Sargent. Philadelphia: University of Pennsylvania Press, 1982.

———. "The Significance of the Theory of Impetus for Scholastic Natural Philosophy." 1955. In Maier, *On the Threshold of the Exact Science*, 76–102.

———. *Zwei Grundprobleme der scholastischen Naturphilosophie: Das Problem der intensiven Grösse: Die Impetustheorie*. 2nd ed. Rome: Storia e Letteratura, 1951.
Malpas, Jeff. *Heidegger and the Thinking of Place: Explorations in the Topology of Being*. Cambridge, MA: MIT Press, 2012.
———, ed. *The Intelligence of Place: Topographies and Poetics*. London: Bloomsbury, 2015.
———. *Place and Experience: A Philosophical Topography*. Cambridge: Cambridge University Press, 1999.
———. "Place and Singularity." In Malpas, *Intelligence of Place*, 65–92.
———. "Thinking Topographically: Place, Space, and Geography." http://jeffmalpas.com/downloadable-essays/. Accessed June 10, 2018.
Manly, John Matthews. "Chaucer and the Rhetoricians." 1926. In *Chaucer Criticism*, edited by Richard J. Schoeck and Jerome Taylor, 268–90. Vol. 1. Notre Dame, IN: University of Notre Dame Press, 1960.
Mann, Jill. "The Planetary Gods in Chaucer and Henryson." In *Chaucer Traditions: Studies in Honour of Derek Brewer*, edited by Ruth Morse and Barry Windeatt, 91–106. Cambridge: Cambridge University Press, 1990.
Manzotti, Riccardo. *The Spread Mind: Why Consciousness and the World Are One*. New York: OR Books, 2018.
Marenbon, John, ed. *Aristotle in Britain during the Middle Ages: Proceedings of the International Conference at Cambridge, 8–11 April 1994*. Société Internationale pour l'Etude de la Philosophie Médiévale 5. Turnhout: Brepols, 1996.
Massey, Doreen. *For Space*. London: Sage, 2005.
———. "Introduction: Geography Matters." In *Geography Matters! A Reader*, edited by Doreen Massey and John Allen, 1–11. Cambridge: Cambridge University Press, 1984.
Massumi, Brian. *Parables for the Virtual: Movement, Affect, Sensation*. Durham, NC: Duke University Press, 2002.
Matthews, David. "Laurence Minot, Edward III, and Nationalism." *Viator* 38 (2007): 269–88.
McCormick, Betsy, Leah Schwebel, and Lynn Shutters. "Introduction: Looking Forward, Looking Back on the *Legend of Good Women*." *Chaucer Review* 52 (2017): 3–11.
McIntyre, Ruth Summar. "Margery's 'Mixed Life': Place Pilgrimage and the Problem of Genre in *The Book of Margery Kempe*." *English Studies* 89 (2008): 643–61.
Merleau-Ponty, Maurice. *Phenomenology of Perception*. Translated by Colin Smith. London: Routledge and Kegan Paul, 1962.
Merrills, Andy. "Geography and Memory in Isidore's *Etymologies*." In Lilley, *Mapping Medieval Geographies*, 45–64.
Mileson, S. A. *Parks in Medieval England*. Oxford: Oxford University Press, 2009.
Miller, Linn, and Jeff Malpas. "On the Beach: Between the Cosmopolitan and the Parochial." In *Ocean to Outback: Cosmopolitanism in Contemporary Australia*, edited by Keith Jacobs and Jeff Malpas, 123–49. Crawley, Western Australia: University of Western Australia Press, 2011.
Moffitt, John F. "Medieval *Mappaemundi* and Ptolemy's *Chorographia*." *Gesta* 32 (1993): 59–68.

Moody, Ernest A. "Buridan, Jean." In *Complete Dictionary of Scientific Biography*, 603–8. Vol. 2. Detroit: Charles Scribner's Sons, 2008.
———. "Empiricism and Metaphysics in Medieval Philosophy." *Philosophical Review* 67 (1958): 145–63.
———. "Galileo and His Precursors." 1966. In Moody, *Studies in Medieval Philosophy, Science, and Logic*, 393–408.
———. "Jean Buridan." 1969. In Moody, *Studies in Medieval Philosophy, Science, and Logic*, 441–53.
———. "Laws of Motion in Medieval Physics." In *Toward Modern Science: Studies in Ancient and Medieval Science*, edited by Robert Palter, 220–34. Vol. 1. New York: Noonday Press, 1961.
———. *Studies in Medieval Philosophy, Science, and Logic: Collected Papers, 1933–1969*. Los Angeles: University of California Press, 1975.
Mooney, Linne R. "A Middle English Text on the Seven Liberal Arts." *Speculum* 68 (1993): 1027–52.
Morison, Benjamin. *On Location*. Oxford: Oxford University Press, 2002.
Murdoch, John E. "From Social into Intellectual Factors: An Aspect of the Unitary Character of Late Medieval Learning." In *The Cultural Context of Medieval Learning: Proceedings of the First International Colloquium on Philosophy, Science, and Theology in the Middle Ages, September 1973*, edited by John Emery Murdoch and Edith D. Sylla, 271–348. Dordrecht, Holland: Reidel, 1975.
Murdoch, John E., and Edith D. Sylla. "The Science of Motion." In *Science in the Middle Ages*, edited by David C. Lindberg, 206–64. Chicago: University of Chicago Press, 1978.
Myer-Lee, Robert J. "Lydgate's Laureate Prose." In Scanlon and Simpson, *John Lydgate: Poetry, Culture, and Lancastrian England*, 36–60.
Nakley, Susan. *Living in the Future: Sovereignty and Internationalism in the Canterbury Tales*. Ann Arbor: University of Michigan Press, 2017.
North, J. D. "Astronomy and Mathematics." In Catto and Evans, *History of the University of Oxford*, 2:103–74.
———. "Natural Philosophy in Late Medieval Oxford." In Catto and Evans, *History of the University of Oxford*, 2:65–102.
Olwig, Kenneth. "*Choros, Chora*, and the Question of Landscape." In *Envisioning Landscapes, Making Worlds: Geography and the Humanities*, edited by Stephen Daniels, Dydia DeLyser, J. Nicholas Entrikin, and Douglas Richardson, 44–54. Milton Park, Abingdon, Oxon: Routledge, 2011.
Oman, Charles. *The Great Revolt of 1381*. Oxford: Clarendon Press, 1906.
Ormrod, W. Mark. "John Mandeville, Edward III, and the King of Inde." *Chaucer Review* 46 (2012): 314–39.
Oswald, Al, John Goodall, Andrew Payne, and Tara-Jane Sutcliffe. *Thornton Abbey, North Lincolnshire: Historical, Archaeological and Architectural Investigations*. English Heritage Research Department Report 100–2010. Portsmouth: English Heritage, 2010.
Ovitt, George, Jr. "The Status of the Mechanical Arts in Medieval Classifications of Learning." *Viator* 14 (1983): 89–106.

Parker, Kate. "Lynn and the Making of a Mystic." In *A Companion to the Book of Margery Kempe*, edited by John H. Arnold and Katherine J. Lewis, 55–73. Cambridge: D. S. Brewer, 2004.

Pasnau, Robert, and Christina van Dyke, eds. *The Cambridge History of Medieval Philosophy*. 2 vols. Cambridge: Cambridge University Press, 2010.

Pearsall, Derek. "The Idea of Englishness in the Fifteenth Century." In *Nation, Court, and Culture: New Essays on Fifteenth-Century English Poetry*, edited by Helen Cooney, 15–27. Dublin: Four Courts, 2000.

———. "Rhetorical 'Descriptio' in 'Sir Gawain and the Green Knight.'" *Modern Language Review* 50 (1955): 129–34.

Pickwoad, Nicholas. "Codicology and Palaeography." In Gibson, Heslop, and Pfaff, *The Eadwine Psalter: Text, Image, and Monastic Culture in Twelfth-Century Canterbury*, 4–12.

Pocock, Douglas C. D. "Geography and Literature." *Progress in Human Geography* 12 (1988): 87–102.

Prestwich, Michael. "The Royal Itinerary and Roads in England under Edward I." In Allen and Evans, *Roadworks*, 177–97.

Price, Derek J. "Medieval Land Surveying and Topographical Maps." *Geographical Journal* 121 (1955): 1–7.

Punta, Francesco del, Silvia Donati, and Cecilia Trifogli. "Commentaries on Aristotle's *Physics* in Britain, ca. 1250–1270." In Marenbon, *Aristotle in Britain during the Middle Ages*, 265–83.

Quinn, William A. "The *Legend of Good Women*: Performance, Performativity, and Presentation." In *The Legend of Good Women: Context and Reception*, edited by Carolyn P. Collette, 1–32. Cambridge: D. S. Brewer, 2006.

Raguin, Virginia Chieffo. "Real and Imagined Bodies in Architectural Space: The Setting for Margery Kempe's *Book*." In Raguin and Stanbury, *Women's Space*, 105–40.

Raguin, Virginia Chieffo, and Sarah Stanbury, eds. *Women's Space: Patronage, Place, and Gender in the Medieval Church*. Albany: State University of New York Press, 2005.

Relph, Edward. "Disclosing the Ontological Depth of Place: *Heidegger's Topology* by Jeff Malpas." *Environmental and Architectural Phenomenology Newsletter* 19 (2008): 5–8.

Robertson, Elizabeth. "Modern Chaucer Criticism." In *Chaucer: An Oxford Guide*, edited by Steve Ellis, 355–68. Oxford: Oxford University Press, 2005.

———. "Noli me Tangere: The Enigma of Touch in Middle English Religious Literature and Art for and about Women." In Walter, *Reading Skin in Medieval Literature and Culture*, 29–55.

Robertson, Kellie. "Medieval Materialism: A Manifesto." *Exemplaria* 22 (2010): 99–118.

———. *Nature Speaks: Medieval Literature and Aristotelian Philosophy*. Philadelphia: University of Pennsylvania Press, 2017.

Rogers, Janine. *Unified Fields: Science and Literary Form*. Montreal, QC: McGill-Queen's University Press, 2015.

Roland, Meg. "'After poyetes and astronomyers': English Geographical Thought and Early English Print." In Lilley, *Mapping Medieval Geographies*, 127–51.
Rose, Paul Lawrence. "Humanist Culture and Renaissance Mathematics: The Italian Libraries of the Quattrocento." *Studies in the Renaissance* 20 (1973): 46–105.
Rossiaud, Jacques. "The City-Dweller and Life in Cities and Towns." In *Medieval Callings*, edited by Jacques Le Goff, 139–79. Chicago: University of Chicago Press, 1995.
Rothwell, W. "Anglo-French and English Society in Chaucer's 'The Reeve's Tale.'" *English Studies* 87 (2006): 511–38.
Rouse, Robert Allen. "What Lies Between? Thinking through Medieval Narrative Spatiality." In *Literary Cartographies: Spatiality, Representation, and Narrative*, edited by Robert T. Tally Jr., 13–29. New York: Palgrave Macmillan, 2014.
Rudd, Gillian. *Greenery: Ecocritical Readings of Late Medieval English Literature*. Manchester: Manchester University Press, 2010.
Ruddick, Andrea. *English Identity and Political Culture in the Fourteenth Century*. Cambridge: Cambridge University Press, 2013.
Rudy, Kathryn M. *Virtual Pilgrimages in the Convent: Imagining Jerusalem in the Late Middle Ages*. Turnhout: Brepols, 2011.
Said, Edward W. "Reflections on Exile." In *Reflections on Exile and Other Essays*, 173–86. Cambridge, MA: Harvard University Press, 2000.
Salih, Sarah. "Lydgate's Landscape History." In Weiss and Salih, *Locating the Middle Ages*, 83–92.
Scanlon, Larry, and James Simpson, eds. *John Lydgate: Poetry, Culture, and Lancastrian England*. Notre Dame, IN: University of Notre Dame Press, 2006.
Schabel, Chris. "Francis of Marchia's *Virtus derelicta* and the Context of Its Development." *Vivarium* 44 (2006): 41–80.
Severs, J. Burke. "The Clerk's Tale." In *Sources and Analogues of Chaucer's Canterbury Tales*, edited by W. F. Bryan and Germaine Dempster, 288–331. 1941. New York: Humanities Press, 1958.
——. *The Literary Relationships of Chaucer's Clerkes Tale*. New Haven, CT: Yale University Press, 1942.
Shank, Michael H. "Schools and Universities in Medieval Latin Science." In Lindberg and Shank, *Cambridge History of Science*, 2:207–39.
Sharp, Joanne P. "Towards a Critical Analysis of Fictive Geographies." *Area* 32 (2000): 327–34.
Shelby, Lon R. "The Geometrical Knowledge of Mediaeval Master Masons." *Speculum* 47 (1972): 395–421.
Sherman, Claire Richter. *Imaging Aristotle: Verbal and Visual Representation in Fourteenth-Century France*. Berkeley: University of California Press, 1995.
Shoaf, R. Allen. "'A Pregnant Argument': Dante's Comedy, Chaucer's *Troilus*, Henryson's *Testament*." In *Fleshly Things and Spiritual Matters: Studies on the Medieval Body in Honour of Margaret Bridges*, edited by Nicole Nyffenegger and Katrin Rupp, 103–208. Cambridge: Cambridge Scholars, 2011.
Shore, A. F. "Egyptian Cartography." In Harley and Woodward, *History of Cartography*, 1:117–29.

Shutters, Lynn. "The Thought and Feel of Virtuous Wifehood: Recovering Emotion in the *Legend of Good Women.*" *Chaucer Review* 52 (2017): 85–105.
Simon, Jesse. "Chorography Reconsidered: An Alternative Approach to the Ptolemaic Definition." In Lilley, *Mapping Medieval Geographies*, 23–44.
Skelton, R. A., and P. D. A. Harvey, eds. *Local Maps and Plans from Medieval England.* Oxford: Clarendon Press, 1986.
———. "Surveying in Medieval England." In Skelton and Harvey, *Local Maps and Plans*, 11–19.
Sleigh, Charlotte. *Literature and Science.* New York: Palgrave Macmillan, 2010.
Smith, D. Vance. "The Application of Thought to Medieval Studies: The Twenty-First Century." *Exemplaria* 22 (2010): 85–94.
Smith, Neal. "Contours of a Spatialized Politics: Homeless Vehicles and the Production of Geographical Scale." *Social Text* 33 (1992): 54–81.
Smyser, H. M. "Chaucer's Two-Mile Pilgrimage." *Modern Language Notes* 56 (1941): 205–7.
Snape, M. G. "Durham." In Skelton and Harvey, *Local Maps and Plans*, 189–94.
Sobecki, Sebastian. "'The writyng of this tretys': Margery Kempe's Son and the Authorship of Her Book." *Studies in the Age of Chaucer* 37 (2015): 257–83.
Souffrin, Pierre. "La quantification du mouvement chez les scolastiques: La vitesse instantanée chez Nicole Oresme." In *Autour de Nicole Oresme: Actes du Colloque Oresme*, edited by Jeannine Quillet, 63–83. Paris: Vrin, 1990.
Spearing, A. C. *The Medieval Poet as Voyeur: Looking and Listening in Medieval Love-Narratives.* Cambridge: Cambridge University Press, 1993.
Staley, Lynn. *Languages of Power in the Age of Richard II.* University Park: Pennsylvania State University Press, 2005.
———. *Margery Kempe's Dissenting Fictions.* University Park: Pennsylvania State University Press, 1994.
Stanbury, Sarah. "Margery Kempe and the Arts of Self-Patronage." In Raguin and Stanbury, *Women's Space*, 75–103.
———. *Seeing the Gawain-Poet: Description and the Act of Perception.* Philadelphia: University of Pennsylvania Press, 1991.
———. *The Visual Object of Desire in Late Medieval England.* Philadelphia: University of Pennsylvania Press, 2015.
Stewart, Susan. *On Longing: Narratives of the Miniature, the Gigantic, the Souvenir, the Collection.* Durham, NC: Duke University Press, 1993.
Sylla, Edith D. "Godfrey of Fontaines on Motion with Respect to Quantity of the Eucharist." In *Studi sul xiv secolo in memoria di Anneliese Maier*, edited by Maierù and Paravicini Bagliani, 105–41. Rome: Storia e Letteratura, 1981.
———. "A Guide to the Text." In Buridan, *Quaestiones super octo libros Physicorum Aristotelis*, xliii–clxxv.
———. "Medieval Dynamics." *Physics Today* 61 (2008): 51–56.
———. "Medieval Quantifications of Qualities: The 'Merton School.'" *Archive for History of Exact Sciences* 8 (1971): 9–39.
———. "The Oxford Calculators." In Kretzmann, Kenny, and Pinborg, *Cambridge History of Later Medieval Philosophy*, 540–63.

Taylor, A. E. *A Commentary on Plato's Timaeus*. Oxford: Clarendon Press, 1962.
Taylor, E. G. R. *The Haven-Finding Art: A History of Navigation from Odysseus to Captain Cook*. London: Hollis and Carter, 1956.
———. "The Surveyor." *Economic History Review* 17 (1947): 121–24.
Taylor, Joseph. "Chaucer's Uncanny Regionalism: Rereading the North in The Reeve's Tale." *Journal of English and Germanic Philology* 109 (2010): 468–89.
Thijssen, Johannes M. M. H. "The Nature of Change." In Pasnau and van Dyke, *Cambridge History of Medieval Philosophy*, 1:279–90.
Thorndike, Lynn. *A History of Magic and Experimental Science*. Vol. 3–4 in Vol. 3. *Fourteenth and Fifteenth Centuries*. New York: Columbia University Press, 1934.
Trifogli, Cecilia. "Change, Time, and Place." In Pasnau and van Dyke, *Cambridge History of Medieval Philosophy*, 1:267–78.
———. *Oxford Physics in the Thirteenth Century (ca. 1250–1270): Motion, Infinity, Place, and Time*. Leiden: Brill, 2000.
Trigg, Stephanie. *Congenial Souls: Reading Chaucer from Medieval to Postmodern*. Minneapolis: University of Minnesota Press, 2002.
———. "Introduction: Emotional Histories—Beyond the Personalization of the Past and the Abstraction of Affect Theory." *Exemplaria* 26 (2014): 3–15.
Tuan, Yi-Fu. *Space and Place: The Perspective of Experience*. Minneapolis: University of Minnesota Press, 1977.
Turville-Petre, Thorlac. *England the Nation: Language, Literature, and National Identity, 1290–1340*. Oxford: Clarendon Press, 1996.
Urry, William. "Canterbury, Kent, circa 1153 × 1161." In Skelton and Harvey, *Local Maps and Plans*, 43–58.
———. "Canterbury, Kent, Late 14th Century x 1414." In Skelton and Harvey, *Local Maps and Plans*, 107–17.
Utz, Richard J. "'As writ myn auctour called Lollius': Divine and Authorial Omnipotence in Chaucer's Troilus and Criseyde." In *Nominalism and Literary Discourse: New Perspectives*, edited by Christoph Bode, Hugo Keiper, and Richard J. Utz, 123–44. Amsterdam: Rodopi, 1997.
———. "Negotiating the Paradigm: Literary Nominalism and the Theory and Practice of Re-reading Late Medieval Texts." In *Literary Nominalism and the Theory of Rereading Late Medieval Texts: A New Research Paradigm*, edited by Richard J. Utz, 1–30. New York: Edwin Mellen, 1995.
Varnam, Laura. *The Church as Sacred Space in Middle English Literature and Culture*. Manchester: Manchester University Press, 2018.
Voaden, Rosalynn. "Travels with Margery: Pilgrimage in Context." In *Eastward Bound: Travel and Travellers, 1050–1550*, edited by Rosamund Allen, 177–95. Manchester University Press, 2004.
Voigts, Linda Ehrsam. "Scientific and Medical Books." In *Book Production and Publishing in Britain 1375–1475*, edited by Jeremy Griffiths and Derek Pearsall, 345–402. Cambridge: Cambridge University Press, 1989.
Wallace, William A. *Prelude to Galileo: Essays on Medieval and Sixteenth-Century Sources of Galileo's Thought*. Dordrecht, Holland: Reidel, 1981.
Walter, Eugene Victor. *Placeways: A Theory of the Human Environment*. Chapel Hill: University of North Carolina Press, 1988.

Walter, Katie L., ed. *Reading Skin in Medieval Literature and Culture*. New York: Palgrave Macmillan, 2013.

Warren, Michelle R. *History on the Edge: Excalibur and the Borders of Britain, 1100–1300*. Minneapolis: University of Minnesota Press, 2000.

———. "Lydgate, Lovelich, and London Letters." In *Lydgate Matters: Poetry and Material Culture in the Fifteenth Century*, edited by Lisa H. Cooper and Andrea Denny-Brown, 113–38. New York: Palgrave Macmillan, 2007.

Watson, Nicholas. "The Making of *The Book of Margery Kempe*." In *Voices in Dialogue: Reading Women in the Middle Ages*, edited by Linda Olson and Kathryn Kerby-Fulton, 395–434. Notre Dame, IN: University of Notre Dame Press, 2005.

Watt, Diane. "Faith in the Landscape: Overseas Pilgrimages in *The Book of Margery Kempe*." In Lees and Overing, *A Place to Believe In*, 170–87.

Watt, Ian. *The Rise of the Novel: Studies in Defoe, Richardson and Fielding*. London: Chatto and Windus, 1957.

Weisheipl, James A. "The Interpretation of Aristotle's *Physics* and the Science of Motion." In Kretzmann, Kenny, and Pinborg, *Cambridge History of Later Medieval Philosophy*, 521–36.

———. "Ockham and Some Mertonians." *Mediaeval Studies* 30 (1968): 163–213.

———. "The Place of John Dumbleton in the Merton School." *Isis* 50 (1959): 439–54.

———. "The Principle *Omne quod movetur ab alio movetur* in Medieval Physics." *Isis* 56 (1965): 26–55.

———. "Science in the Thirteenth Century." In *The History of the University of Oxford*. Vol. 1, *The Early Oxford Schools*, 435–69.. Edited by J. I. Catto. Oxford: Clarendon Press, 1984.

Weiss, Julian, and Sarah Salih, eds. *Locating the Middle Ages: The Spaces and Places of Medieval Culture*. London: King's College, Centre for Late Antique and Medieval Studies, 2012.

Westphal, Bertrand. *The Plausible World: A Geocritical Approach to Space, Place, and Maps*. Translated by Amy D. Wells. New York: Palgrave Macmillan, 2013.

Whitehead, Alfred North. *The Concept of Nature*. Cambridge: Cambridge University Press, 1920.

———. *Process and Reality: An Essay in Cosmology*. 1929. New York: Free Press, 1978.

Whitney, Elspeth. "Paradise Restored: The Mechanical Arts from Antiquity through the Thirteenth Century." *Transactions of the American Philosophical Society* 80 (1990): 1–169.

Willis, Martin. *Literature and Science*. New York: Palgrave, 2014.

Willis, Robert. "The Architectural History of the Conventual Buildings of the Monastery of Christ Church in Canterbury." *Archaeologia Cantiana* 7 (1868): 1–206.

Withers, Charles W. J. "Place and the 'Spatial Turn' in Geography and in History." *Journal of the History of Ideas* 70 (2009): 637–58.

Wood, Rega. "The Influence of Arabic Aristotelianism on Scholastic Natural Philosophy: Projectile Motion, the Place of the Universe, and Elemental Composition." In Pasnau and van Dyke, *Cambridge History of Medieval Philosophy*, 1:247–66.

Woodman, Francis. "The Waterworks Drawings of the Eadwine Psalter." In Gibson, Heslop, and Pfaff, *The Eadwine Psalter: Text, Image, and Monastic Culture in Twelfth-Century Canterbury*, 168–77.
Woods, William. *Chaucerian Spaces: Spatial Poetics in Chaucer's Opening Tales*. Albany: State University of New York Press, 2008.
Woodward, David. "Maps and the Rationalization of Geographic Space." In *Circa 1492: Art in the Age of Exploration*, edited by Jay A. Levenson, 83–87. Washington, DC: National Gallery of Art, 1991.
——. "Medieval *Mappaemundi*." In Harley and Woodward, *History of Cartography*, 1:286–370.
Woolgar, C. M. *The Senses in Late Medieval England*. New Haven, CT: Yale University Press, 2006.
Wrightson, Keith. "The Decline of 'Neighbourliness' Revisited." In *Local Identities in Late Medieval and Early Modern England*, edited by Norman L. Jones and Daniel Woolf, 19–49. New York: Palgrave, 2007.
Young, Karl. "Chaucer's Aphorisms from Ptolemy." *Studies in Philology* 34 (1937): 1–7.
Zaborowski, Holgar. "Heidegger's Hermeneutics: Towards a New Practice of Understanding." In *Interpreting Heidegger: Critical Essays*, edited by Daniel O. Dahlstrom, 15–41. Cambridge: Cambridge University Press, 2011.
Zanin, Fabio. "Francis of Marchia, *Virtus derelicta*, and Modifications of the Basic Principles of Aristotelian Physics." *Vivarium* 44 (2006): 81–95.
Zumthor, Paul. *La mésure du monde: Représentation de l'espace au Moyen Âge*. Paris: Seuil, 1993.
Zupko, Jack. "John Buridan." *The Stanford Encyclopedia of Philosophy*. 2014. http://plato.stanford.edu/archives/spr2014/entries/buridan/.

Index

Page numbers followed by *f* refer to figures.

absolute space, versus relative space, 28
abstracted space, 7, 30, 79–86, 100, 101–2; *abstract*, 81; in *Book of Margery Kempe*, 102–3, 108, 110, 112, 116; in *Book of Sir John Mandeville*, 102–3, 118, 120–21; in Chaucer's *House of Fame*, 156–57; experiencing, 96, 225n31; in Henryson's *Orpheus and Eurydice*, 144
acceleration, 130
accidents of the soul, 179
aesthetics, 182
affect, 171–72, 177–87, 255n6; intensities and, 178, 191
affordance, in Gibson's spatial theory, 78
air, as medium of motion, 161
Albert of Saxony, 65
Albertus Magnus, 64, 126, 132
altimetry, 20, 89, 90, 93
amicitie et inimicitie naturalis, 181–82
Anglo-Saxon sense of space, 9
Anonimalle Chronicle, 98–100
Aquinas, Thomas: on being, 2–3, 57, 126–27, 130, 210, 211; *De caelo et mundo expositio*, 130; *Commentaria in octo libros Physicorum*, 2–3, 123, 127, 128–29, 173, 210, 220n7; on intellect, 220n7; on motion, 126–27, 128–29; on natural place, 124, 128–30, 220n7, 245n21; on place, 64; on proximity, 173; *Summa theologiae*, 211, 220n7
area: and geography, chorography, and topography, 23–24; use of term, 14
Aristophanes, *Clouds*, 168
Aristotle: on being, 2–3, 56–57, 130, 210, 211; *De caelo*, 130; on change and constancy in nature, 126; in Guillaume de Deguileville, 62–63; *Metaphysics*, 223–24n26; on motion and space, 11–12; on natural place, 124, 128–30, 131, 152; on place, 63–64, 79–80, 129; Paris condemnations of, and curricula, 223–24n26; *Physics*, 2, 11, 57, 63–64, 79, 123, 126, 128–29, 130, 131–32, 135–37, 150, 173, 201, 211, 223–24n26, 225n31, 253n18; theory of motion of, 135–36
art, and geography in relation to scientific discussions of spatial phenomena, 221n16
Artis cuiuslibet consummatio, 89–90, 92, 93, 95
astrolabes, 10, 79, 87–89, 92, 93, 94–96
astronomy, proximity in, 169–70
Auerbach, Erich, 212
Averroës (Ibn Rushd), 126
Avicenna (Ibn Sina), 132

"Babees Book," 169
Barthes, Roland, 39, 48
Bartholomaeus Anglicus, *De Proprietatibus Rerum*, 77, 100, 126, 150, 169–70, 235n3
being: Aquinas on, 2–3, 57, 126–27, 130, 210, 211; Aristotle on, 2–3, 56–57, 130, 210, 211; Chaucer's *Legend of Good Women* and, 216; Chaucer's Pardoner's Prologue and Tale and, 209–17; Malpas on, 217; Merleau-Ponty on, 29, 210; Oresme and, 137; place or space and, 2–3, 7, 9, 14, 56–57, 128, 149, 210–11, 217
Bennett, J. A. W., 94
bird's-eye perspective. *See* overhead view
Bishop's Lynn, 111–12
Bloch, Marc, 22
Boethius, *Consolation of Philosophy*, 4, 82, 127–28, 144
Bokenham, Osbern, 50–52
Book of Margery Kempe: abstracted space in, 102–3, 108, 110, 112, 116; Bishop's Lynn in, 111–12; *Book of Sir John*

285

Book of Margery Kempe (continued)
Mandeville and, 122; Chaucer's *House of Fame* and, 157, 159; chorography and, 116; distance in, 111–13, 116–17; Henryson's *Orpheus and Eurydice* and, 144–45; heterogeneous space in, 116–17; Holy Land in, 108–10, 112–13; homogeneous space in, 102–3, 117; horizonal space in, 11, 103, 107–17; Jerusalem in, 109–10, 112–13; lack of descriptive detail and attention to place in, 242n26; local space in, 111–12; London in, 115; measurement in, 117, 243n51; overview of space and objects in, 102–4, 107; pilgrimage in, 11, 108–9, 112–13, 122; re-meetings in, 113; reputation of Margery Kempe in, 107–8, 113–15; strength of networks in, 114–15; structure of, according to conventions of topography and chorography, 116; visions in, 109–11

Book of Sir John Mandeville: abstracted space in, 102–3, 118, 120–21; antipodes in, 121; *Book of Margery Kempe* and, 122; chorography and, 118–19; circumnavigation in, 120–21; critique of limited geographical understanding in, 121; "diverse" locales in, 118–19; Earthly Paradise in, 120, 122; Great Khan in, 120; Henryson's *Orpheus and Eurydice* and, 144–45; heterogeneous space in, 118; Holy Land in, 119; homogeneous space in, 102–3, 118, 120–21; horizonal space in, 11, 103, 117–22; overview of space and objects in, 102–4, 107; Prester John in, 120–21; realism in, 118; Sumatra ("Lamary") in, 120; topography in, 118

boundaries. *See* edges
Bourdieu, Pierre, 15
Bowet, Henry, Archbishop of York, 114
Bradwardine, Thomas, *Tractatus de proportionibus velocitatum in motibus*: influence of, 150–51, 152; on intensities, 136, 137; new theorem of velocity in, 12, 123, 125, 135, 136–37, 141, 247nn59–60
Brantley, Jessica, 188
Brosseau, Marc, 221n16
Burger, Glenn, 200
Buridan, Jean, 12, 123, 143, 152, 161, 162, 173; on impetus, 12, 124, 125, 131, 132, 133–34, 246n45; *Quaestiones Super Libris Quattuor de Caelo et Mundo*, 133–34; *Quaestiones super octo libros Physicorum Aristotelis*, 127, 133
Burley, Walter, 137, 174

Cambridge, Trinity College, manuscript R.17.1, 32f, 33f, 229–30n40
Cambridge, Trinity Hall, manuscript 1, 51f
Campbell, Mary, 119
Campbell, Tony, 83
Canterbury: in London, British Library, Royal 18 D 2, 70–73; maps of, 31–36, 32f, 33f, 49, 229–30n40
Canterbury, Records of the Dean and Chapter of Canterbury Cathedral, manuscript Charta: Antiqua C.295, 37f
Cartlidge, Neil, 152, 162
cartography. *See* maps and mapmaking
Casey, Edward, 15, 30, 38, 230n45
Caxton, William, 62
celestial motion, 126, 127
Certeau, Michel de, 15, 94, 238n51
Chapel of the Holy Sepulcher, 109–10
Chaucer, Geoffrey
—*Boece*, 4, 82, 127–28, 163, 215
—*Book of the Duchess*, 153, 251n48, 251n63; proximity in, 192
—*Canterbury Tales*
——Canon's Yeoman's Prologue and Tale, 213; pilgrimage in, 188; proximity in, 188–89
——Clerk's Prologue and Tale, 104–7, 211; chorography in, 104, 106; overhead view in, 104–5; topography in, 104–6
——Knight's Tale, 64–65
——and London, British Library, Royal 18 D 2, 56, 57, 67–74
——Man of Law's Tale, 211, 212–13; local space in, 212
——Pardoner's Prologue and Tale, 14, 114; being and, 209–17; horizonal space in, 215; local space in, 209, 213, 216–17; proximity in, 205, 215–16; ubiquity in, 209–17
——Prioress's Tale: edges in, 61; local space in, 58–62; realism in, 61
——Reeve's Tale, 65
——Wife of Bath's Prologue, 214–15
—"Complaint of Mars," 4; proximity in, 188
—*House of Fame*: abstracted space in, 156–57; and Chaucer's *Book of the*

Duchess, 157, 159; desert in, 158–60; history in, 155–58, 166; horizontal space in, 143, 157; impetus in, 143, 149–50, 152, 156, 162, 167; local space in, 166; motion in, 142–44, 151–67; natural place in, 143, 152–53, 160, 162, 165, 167; overhead view in, 155–57, 159–60, 166; pilgrimage in, 154; rumor in, 164–66; sleep in, 153–55; sound in, 160–63, 166; structure of, 13; unmotivated motion in, 151–66
—*Legend of Good Women*, 14, 189, 199–200
——being and, 216
——Legend of Ariadne, 205–7; wall(s) in, 205–6
——Legend of Cleopatra, 196–99
——Legend of Dido, 205
——Legend of Hypermnestra, 205, 207
——Legend of Thisbe, 199–204; wall(s) in, 199, 201–3, 204, 205, 206
——motion in, 198–99
——Prologue, 196–98
——proximity in, 189, 196–208
——structure of, 205–8
—Manly on development of, 53
—*Romaunt of the Rose*, 66, 97, 251n48
—science in, 4, 7
—scientific influence on, 151
—*Treatise on the Astrolabe*, 4, 87, 88, 90, 94, 95
—*Troilus and Criseyde*, 82, 154, 211–12, 256n12; overhead view in, 157; proximity in, 191, 192–94, 196
—use of *space* and *place*, 64–65
chora, 23, 227–28n13
chorography / *chōrographia*: and *Book of Margery Kempe*, 116; and *Book of Sir John Mandeville*, 118–19; in Chaucer's Clerk's Tale, 104, 106; defined, 23–27, 227n12; and local space, 54, 62–63; and Ptolemy's *Geography*, 8, 19, 23–28, 63, 96, 227n12
Cicero, 96, 97
circumnavigation, in *Book of Sir John Mandeville*, 120–21
Clagett, Marshall, 126, 130, 172, 177, 187, 246n45
Cliffe, Kent, map of, 36–38, 37f, 49, 52
Clouds (Aristophanes), 168
Cohen, Jeffrey, 97
communitas, 57, 68–70, 74–75, 213, 216
Connolly, Daniel, 26
constant scale, 24, 84, 85–86, 101, 103

Cosgrove, Denis, 20
cosmimetry, 89
cosmological space, 223n23, 249n16
Crocker, Holly, 178, 255n6
Curry, Michael, 27, 49

Dante, 12, 58, 81–82, 126, 128, 152, 163, 164, 166
Davis, Rebecca, 165, 166–67
Deguileville, Guillaume de, 62–63
Delany, Sheila, 152, 158, 166, 199
Deleuze, Gilles, and Félix Guattari, 177–78
Derrida, Jacques, 212
descriptio, 96–97
deserts, 146, 158–60
Dinshaw, Carolyn, 216
dispositio, 241n11
Dominicus de Clavasio, 92
Domne Eafe, 53, 75
Douglas, Gavin, 146
Doyle, Kara, 199–200
Dryden, John, 231n69
Duhem, Pierre, 16–17, 172, 176, 247n52
Duns Scotus, John, 125, 132, 174, 187
Durham, map of, 39–40
Dyas, Dee, 112

Eadwine Psalter, 31–34
Edgerton, Samuel, 16, 225n37
edges, 8, 76, 80, 81, 199, 216–17, 259n22; on local maps, 21, 30–38, 40, 41, 48, 52, 54, 55–56, 61; of local space, 30–38, 55; in Chaucer's Prioress's Tale, 61; in London, British Library, manuscript Royal 18 D 2, 66, 71–74, 75
elevation, spatial, 98–100, 108–9
Emilia-Romagna, 104–5
estral space. *See* local space
estre, 8–9, 65–66. *See also* local space
Evans, Ruth, 157
Exeter, map of, 39
exile, 212
eyes, and intensities, 193–94

falling objects, 133–34
Fibonacci (Leonardo of Pisa), 95
forests, 22, 40, 57, 146, 156–57, 165
"forms, intension and remission of," 171, 173–76
Foucault, Michel, 15, 189–90, 215
Fradenburg, Louise, 59, 65
Francis de Marchia, 125, 132, 133, 134, 152, 161, 162, 173

friendship, natural, 181–82
Fyler, John, 152–53

Gabrovsky, Alexander, 7, 161
Gascoigne, Richard, 42
Gautier Dalché, Patrick, 17
genealogies, 169
geography, 7, 10, 221n16; in Ptolemy, 17, 23–25, 96
geōmetria, 168
geometrical space, measurement of, 168
geometry: *geōmetria*, 168; and horizonal perspective, 168; and Hugh of St. Victor's *Practica geometriae*, 89, 92–93, 95; maps in discipline of, 20; and Oresme's *Tractatus de commensurabilitate vel incommensurabilitate motuum celi*, 182; as parallel to Middle English literature, 5; treatises on, 78. *See also* practical geometries
Gerald of Wales, 97–98, 239nn66–67
Gibson, James J., 96; *Ecological Approach to Visual Perception*, 77–79, 109; *Perception of the Visual World*, 76, 100; *Senses Considered as Perceptual Systems*, 15–16
Givens, Jean, 73
Godfrey of Fontaines, 174
Gower, John: *Confessio amantis*, 23, 50, 66, 200–1, 204; *Vox clamantis*, 83
Gransden, Antonia, 54, 232n70
Grant, Edward, 4, 141, 172
graticules on maps, 10, 84–85, 101, 236n26
gravity, 133–34
Groningen, Bibliotheek der Rijksuniversiteit, manuscript 103, 139*f*
Grossetese, Robert, 64, 132
Guy de Chauliac, 62, 170

Harley, Brian, 2, 28–29, 80
Harrison, Dick, 22
Harvey, P. D. A., 19, 26, 94–95, 231n53, 238n54
Heidegger, Martin, 210, 216
Henryson, Robert
—*Orpheus and Eurydice*, 12–13, 50, 144–48, 149, 153, 154, 159, 248n15; overhead view in, 144, 146–47, 148
—*Testament of Cresseid*, 194–96
Hereford map, 83
Herodotus, 168
heterogeneous space, 42, 96, 101; in *Book of Margery Kempe*, 102, 116–17; in *Book of Sir John Mandeville*, 102, 118; and contents of local space, 38–39; versus homogeneous space, 28–30, 38–39, 48, 54, 56, 80–81, 85–86, 101, 104; on maps, 39, 46, 48, 52, 54, 56, 71, 85–86. *See also* horizonal space
Heytesbury, William, 137
history: changes in space throughout, 1–2; in Chaucer's *House of Fame*, 155–58, 166
Hoccleve, Thomas, 4
Holcot, Robert, 152
Holderness, 259n19
Holy Land: in *Book of Margery Kempe*, 108–10, 112–13; in *Book of Sir John Mandeville*, 119
homogeneous space, 39–42, 62, 70–71; and abstract space, 80–81; in *Book of Margery Kempe*, 102, 108, 117; in *Book of Sir John Mandeville*, 102–3, 118–21; in Chaucer's *House of Fame*, 156, 160; in Chaucer's Pardoner's Prologue and Tale, 215; in Henryson's *Orpheus and Eurydice*, 145; versus heterogeneous space, 28–30, 38–39, 48, 54, 56, 80–81, 85–86, 101, 104; horizonal space as, 101–2; on maps, 45–46, 82–86. *See also* overhead view
horizonal space, 9–11, 77–79, 101, 125, 168; astrolabes, quadrants, and practical geometries and, 86–94; in *Book of Margery Kempe*, 11, 107–13, 116–17; in *Book of Sir John Mandeville*, 11, 103, 117–22; in Chaucer's Clerk's Prologue and Tale, 104, 106; in Chaucer's *House of Fame*, 143, 157; in Chaucer's Pardoner's Prologue and Tale, 215; elevation and, 98–100; versus overhead view, 10–11, 27, 76–77, 80–83, 86, 98, 121–22, 168; *topographia* and, 96–98; and viewer relations, 92–96; Westphal on, 79–80. *See also* estral space; heterogeneous space
hostility, natural, 181–82
Howarth, William, 47
Howe, Nicholas, 9
"How the Good Wiff Tau3te Hir Dou3tir," 169
Hugh of St. Victor: *De arca Noe morali*, 89; *Didascalicon*, 5, 20, 89, 168, 212; *Practica geometriae*, 89, 92–93, 95
Hussey, Edward, 130

al-Idrīsī, Abu ʿabd-Allah Muhammad, 236n26
Ilium. *See* Thebes (Ilium)

INDEX 289

impetus, 161; Aristotle and fourteenth-century philosophers on, 173; in Bartholomaeus Anglicus's *De Proprietatibus Rerum*, 150; in Buridan, 12, 124, 125, 131, 132, 133–34, 246n45; in Chaucer's *House of Fame*, 143, 149–50, 152, 156, 162, 167; defined, 124; and motion, 149–50; in Vegetius, 150
Inclesmoor, West Riding, Yorkshire, maps of, 42–48, 43f, 44f, 49, 71, 230–31n52
Ingham, Patricia, 212–13
instantaneous velocity. *See* mean speed theorem
intellect: Aquinas on, 220n7; Oresme on, 179–80
"intension and remission of forms," 171, 173–76
intensities: and affect, 178, 191; Bradwardine on, 136, 137; configurations of, 171–77, 182–84; Duns Scotus on, 174; effects of, in an alien body, 185–86; and eyes, 193–94; Godfrey of Fontaines on, 174; and "intension and remission of forms," 171, 173–76; and natural amity and enmity, 181–82; Oresme on, 13–14, 137–38, 172, 174–87, 193–94; and pain and joy, 184–85; and pleasure and displeasure, 183–84; and proximity, 172–77, 186, 190–91; relations among, 180
intrinsic motion, 132
"inwith," 232n10
Ireland, 97–98

Jammer, Max, 211
Jean de Meun, *Roman de la rose*, 158
Jerusalem, in *Book of Margery Kempe*, 109–10, 112–13
Jones, Alexander, 227n12
Julian of Norwich, 81, 82, 110, 169

Kaye, Joel, 3, 17, 136, 151
Kempe, Margery. *See Book of Margery Kempe*
khôros, 23, 227–28n13
Kilwardby, Robert, 5
King's Lynn 111–12
Kleiner, John, 164
Knighton, Henry, 151
Kruger, Steven, 158, 200

Langland, William, *Piers Plowman*, 155
Lavezzo, Kathy, 98, 239n71

Lefebvre, Henri, 15, 224n28
Le Goff, Jacques, 22
Lilley, Keith, 22–23
limbus, 10, 87–90, 95
literature: Aquinas on, 3; and geography in relation to scientific discussions of spatial phenomena, 221n16; points of similarity and difference between science and, 3–6
Livre Griseldis, 105
local space, 7–9, 18–19; in *Book of Margery Kempe*, 111–12; in Chaucer's *House of Fame*, 166; in Chaucer's Man of Law's Tale, 212; in Chaucer's Pardoner's Prologue and Tale, 209, 213, 216–17; in Chaucer's Prioress's Tale, 58–62; chorography and, 22–28; contents of, 38–48, 56; edges of, 30–38, 55; *estre* and, 65–66; *local* and, 8–9, 62–63, 149; on London, British Library, manuscript Royal 18 D 2, 56, 57, 70–75; in Lydgate's *Siege of Thebes*, 9, 56, 57, 69–70; maps of, 8, 19–28, 55–56, 80; Ptolemy's *Geography* on, 7, 22–28; semantics of, 62–66; temporality of, 48–53. *See also* horizonal space
locus naturalis. See natural place
Lombardy, 104–5
London: map of charterhouse, 35; reputation of Margery Kempe in, 115
London, British Library, Archives of the Duke of Rutland, Belvoir Castle, map 125, 41f
London, British Library, manuscript Lansdowne 762, 93
London, British Library, manuscript Royal 18 D 2, 9, 56, 57, 66–67, 67f, 68, 70–71, 74–75
London, British Library, manuscript Royal 12 G II, 126, 244n8
Lydgate, John: *Fall of Princes*, 68; *Siege of Thebes*, 56–57, 66–75, 67f, 78, 205
Lynch, Kathryn, 152–53

Macrobian maps, 10, 25, 82–83, 85
Mader, Mary Beth, 177
magic, 180–81, 186
Maier, Anneliese, 124, 125, 126, 130, 132, 172, 223n26
Malpas, Jeff, 6, 7, 9, 15, 29, 78, 143, 149, 154, 210–11, 216–17, 235n6
Mandeville, Sir John. *See Book of Sir John Mandeville*

Manly, John, 53, 231n69
Mann, Jill, 145, 249n16
mappa, 50–52
maps and mapmaking: of Canterbury, 31–36, 49, 229–30n40; Certeau on medieval, 238n51; of Cliffe, Kent, 36–38, 49, 52; constant versus objective scale in, 84; of Durham, 39–40; early modern, 238n51; edges of local maps and, 21, 30–38, 40, 41, 48, 52, 54, 55–56, 61; of Exeter, 39; in geometry, 20; graticules on, 10, 84–85, 101, 236n26; Hereford, 83; heterogeneous space on, 39, 46, 48, 52, 54, 56, 71, 85–86; homogeneous space on, 45–46, 82–86; of Inclesmoor, West Riding, Yorkshire, 42–48, 49, 71, 230–31n52; and local space, 8, 19–28, 55–56, 80; of London charterhouse, 35; Macrobian, 10, 25, 82–83, 85; *mappa*, 50–52; *mappaemundi*, 10, 18, 24–25, 26, 40, 49–50, 55, 82–83, 84, 85–86; mechanical arts and, 20, 48; overhead view on, 27, 29, 34, 35, 83–84; of Palestine, 85; Ptolemy's *Geography* and local, 24–28; realism in local maps, 26–27, 31, 34, 35–36, 52, 54; of Sherwood Forest, 40–42, 49, 71; sources for, 55–56; and temporality of local space, 48–53; of Thanet, Kent, 50–52, 75; and water and waterworks, 32–36
Margery Kempe. See *Book of Margery Kempe*
Massey, Doreen, 15
Massumi, Brian, 177–78, 191
Matthew of Vendôme, *Ars Versificatoria*, 97
mean speed theorem, 135–37, 149, 247n54
measurement, 223n26
mechanical arts: and horizonal space, 76, 77, 79, 88–92; maps and, 20, 48; similarities between literature and, 5–6; status of, in late Middle Ages, 5, 78, 89
medicine, proximity in, 169–70
Merleau-Ponty, Maurice, 28, 29, 80, 210, 258n7
Merton School (Oxford Calculators), 12, 94, 123, 125, 134, 135, 136–37, 141, 151, 174, 249n32
Metham, John, 82
Moody, Ernest, 135
motion, 123–25; Aristotle on, 129–30, 135–36, 173; in Chaucer's *House of Fame*, 142–44, 151–67; in Chaucer's *Legend of Good Women*, 198–99; celestial, 126, 127; fourteenth-century culture of, 148–51; in Henryson's *Orpheus and Eurydice*, 144–45; intrinsic, 132; and local space, 22; measurement of, 135–40; in medieval discussions of space, 11–14; natural, 125, 127, 129–30, 131, 133, 245n21; new science of, 141–42; and Oresme's physics, 187; proximity and, 171; relationship between object and, 131–35, 141; traditional ideas of natural place and, 126–30
Mount Quarantania, 108–9, 119, 159
Mount Vesulus, 105–6
Murdoch, John, 126

nationalism, 7, 18, 210
natural motion, 125, 127, 129–30, 131, 133, 245n21
natural place: Aquinas on, 124, 128–30; Aristotle on, 124, 128–30, 131, 152; Buridan on, 133; in Chaucer's *House of Fame*, 143, 152–53, 160, 162, 165, 167; diminished significance of, 12, 124–25, 135, 171; origins of, 152; traditional ideas of motion and, 126–30, 166–67; violent motion and, 129–30
networks, Margery Kempe and strength of, 114–19
Newton, Isaac, 28
North, J. D., 136
N-Town Plays, 159

objective scale, 20, 84, 85, 86
objects: change in, 175–76; motion and place of, 129–30; movement of, through medium, 160–62; relationship between motion and, 131–35, 141; relationship between place and, 86–93, 101. *See also* proximity
Ockham, William, 64, 133, 223n23
optics, 17, 161, 193, 195, 221n15
Oresme, Nicole, 13–14; on eyes, 193–94; and graphs of motion, 125, 135, 137–39, 151; on impetus, 134; on intellect, 179–80; on intensities, 13–14, 137–38, 172, 174–87, 191; on motion, 135, 137–39, 187; on proximity, 172, 174–87; *Tractatus de commensurabilitate vel incommensurabilitate motuum celi*, 182; *Tractatus de configurationibus qualitatum*

et motuum, 13, 134, 137–39, 139*f*, 169, 172, 174–82, 180, 184–87, 188, 193–94
Ospringe, 70–71, 73, 74, 75
overhead view, 96, 100; in *Book of Margery Kempe*, 10–11, 102–3, 107, 108–10, 116, 122, 160; in *Book of Sir John Mandeville*, 103, 107, 118, 119–22; in Chaucer's Clerk's Prologue and Tale, 104–5; in Chaucer's *House of Fame*, 155–57, 159–60, 166; in Chaucer's *Troilus and Criseyde*, 157; in Henryson's *Orpheus and Euridice*, 144, 146–47, 148; versus horizontal space, 10–11, 27, 76–77, 80–83, 86, 98, 121–22, 168; in Julian of Norwich, 82; on maps, 27, 29, 34, 35, 83–84. *See also* abstracted space; homogeneous space
Oxford, Bodleian Library, manuscript Ashmole 1522, 90–92, 91*f*
Oxford, Bodleian Library, manuscript Bodley 619, 95
Oxford, Bodleian Library, manuscript Merton College 272, 126, 244n8
Oxford, Museum of the History of Science, Quadrans Vetus, Inv. 52020, 88*f*
Oxford Calculators. *See* Merton School (Oxford Calculators)

Palestine, graticular map of, 85
Paris, Bibliothèque Nationale de France, manuscript Latin 4939, 85
Paris, condemnations of Aristotle in, and curricula, 223–24n26
Paris, Matthew, 26
Parker, Kate, 113
passions, 178–79, 180, 184
Pecock, Reginald, 169, 172–73
Petrarca, Francesco (Petrarch), 104–105
Piedmont, 104–5
pilgrimage, 9, 115; in *Book of Margery Kempe*, 11, 108–9, 112–13, 122; in Chaucer's Canon's Yeoman's Prologue and Tale, 188; in Chaucer's *Canterbury Tales*, 216; in Chaucer's *House of Fame*, 154; in Lydgate's *Siege of Thebes*, 57, 68, 71, 74
place: Aristotle on, 2–3, 63–64, 79; being and, 2–3, 7, 9, 14, 56–57, 128, 149; enfolding of, 216–17; sense of, 228n29; versus space, 15, 225n31; ubiquity and, 210–17. *See also* natural place; space
planimetry, 20, 89, 93–94
Pliny the Elder, 239n71
portolan charts, 83, 84, 85–86

practical geometries, 88–92, 94, 95–96, 239n60
proximity, 13–14, 168–72; and affect theory, 177–87; Aquinas on, 173; in Chaucer's *Book of the Duchess*, 192; in Chaucer's Canon's Yeoman's Prologue and Tale, 188–89; in Chaucer's "Complaint of Mars," 188; in Chaucer's *Legend of Good Women*, 189, 196–208; in Chaucer's Pardoner's Prologue and Tale, 205, 215–16; in Chaucer's *Troilus and Criseyde*, 191, 192–94, 196; Foucault on, 190; in Henryson's *Testament of Cresseid*, 194–96; and intensities, 172–77, 186, 190–91; in medicine, 169–70; and motion, 171; in Oresme's *Tractatus de configurationibus qualitatum et motuum*, 172, 174–87, 193–94; in *Sir Gawain and the Green Knight*, 191
Ptolemy, Claudius: *Almagest*, 23; *Geography*, 8, 17, 21, 22–27, 96
Puttenham, George, *Arte of English Poesie*, 104

quadrants, 10, 79, 87–92, 88*f*, 92, 91*f*, 92, 93, 94–96, 237n29
quadrivium, 20
Quintilian, 96

Raguin, Virginia, 110, 242n26
raised viewpoints, 98–100
realism: in *Book of Sir John Mandeville*, 118; in Chaucer's Prioress's Tale, 61; in illuminations, 73; in local maps, 26–27, 31, 34, 35–36, 52, 54; in Lydgate's *Siege of Thebes*, 9, 57, 67
Reed, William, 94
relative space, versus absolute space, 28
Repingdon, Philip, 114
representation, Quintilian on, 96
Revolt of 1381, 98–100
rhumb lines, 84, 85–86, 236n23
Richard II, 98–100
Richard Rufus of Cornwall, 132–33, 161
Richmond, The National Archives (TNA): Public Record Office (PRO), Duchy of Lancaster, manuscript 42/12, 43*f*
Richmond, The National Archives (TNA): Public Record Office (PRO), Maps and Plans C 1/56, 44*f*
Robertson, Kellie, 17, 124, 153, 157, 166, 226n42

Robertus (Johannes) Anglicus, *Quadrans vetus*, 89–92, 93, 237n41
Rochester Castle, 73
romance, 191–95
Roman sense of space, 9
rumor, in Chaucer's *House of Fame*, 164–66

Said, Edward, 212
Said, Maire, 212
Salih, Sarah, 234n41
Saluzzo (Saluce, Saluces), 104–6, 259n19
scale, 20, 84, 85, 86, 119–20
scale jumping, 25
science(s): Aquinas on natural, 220n7; fourteenth-century innovations in, 149–51; points of similarity and difference between literature and, 3–6; use of term, 219n1
sea, 12, 24, 50, 83, 89, 127, 161, 162, 192, 212
Sesostris, 168, 252n1
shadow square. See *umbra recta*
Sherwood Forest, map of, 40–42, 41*f*, 49, 71
Shoaf, Allen, 200–201
Shutters, Lynn, 200
Simon, Jesse, 23, 24, 27, 251n53
Sir Gawain and the Green Knight, 146, 191
Skelton, R. A., 19, 26, 94–95, 238n54
Sobecki, Sebastian, 116
Souffrin, Pierre, 151
soul: accidents of, 179; effect of, on other beings, 185
sound, in *House of Fame*, 160–63, 166
space: absolute, versus relative, 28; Anglo-Saxon sense of, 9; being and, 2–3, 7, 9, 14, 56–57, 128, 149, 210–11; changes in, throughout history, 1–2; and comparison between science and literature, 3–6; cosmological, 223n23, 249n16; historical parameters and periodization of innovation about, 16–17; versus place, 15, 225n31; qualities and medieval apprehension of, 7–8; Roman sense of, 9; social constructivist arguments about, 15–16. *See also* abstracted space; *estre*; heterogeneous space; homogeneous space; horizonal space; local space; natural place
spatial turn, 15, 224n28
speed, measurement of, 135–40. *See also* velocity
Staley, Lynn, 151
Stanbury, Sarah, 193, 242n26, 256n12

Strode, Ralph, 151
Sumatra ("Lamary"), 120
Sylla, Edith, 126, 136–37
synecdoche, 118, 119, 129

Taylor, Jerome, 5
temporality: of local space, 48–53; in Lydgate's *Siege of Thebes*, 69–70, 74–75
Thanet, Kent, map of, 50–52, 51*f*, 75
Thebes (Ilium), 70, 74–75, 155. *See also* Lydgate, John: *Siege of Thebes*
Thomas of Chobham, 241n11
Thomas of Elmham, *Historia Monasterii S. Augustini Cantuariensis*, 50–53
Thrace, 66, 153
topography / *topographia*: and abstraction from earth, 79–80; in *Book of Sir John Mandeville*, 118; in Chaucer's Clerk's Prologue and Tale, 104–6; and elevation, 96–98; Gerald of Wales on, 97–98, 239nn66–67; and horizonal space, 10–11; and Ptolemy's *Geography*, 23–24, 96; structure of Margery Kempe according to conventions of, 116
topos, 15, 64, 225n31
Trevisa, John, *On the Properties of Things*, 77, 100, 126, 150, 169–70, 235n3
Trigg, Stephanie, 57, 68, 255n6
Trivet, Nicholas, 144, 146, 147
Tuan, Yi-Fu, 111

ubi, 64
ubiquity, 14, 209–17, 258n6
ubity, 211
umbra recta, 10, 87–88, 90
University of Paris, 223–24n26

Vegetius, 150
velocitas instantanea. *See* mean speed theorem
velocity, 12, 133, 135–38, 162, 246n45, 247nn59–60. *See also* speed, measurement of
visionary overview, 102–3, 108–9
visions, of Margery Kempe, 109–11
Voigts, Linda, 17

wall(s): in Chaucer's Legend of Ariadne, 205–6; in Chaucer's Legend of Thisbe, 199, 201–3, 204, 205, 206
Walter, E. V., 15, 225n31
Wars of Alexander, 159

water and waterworks: in Canterbury map, 229–30n40; in Inclesmoor maps, 45–48; in maps, 32–36; as medium of motion, 161–62. *See also* sea
Waterton, Robert, 42
Westphal, Bertrand, 79–80, 100, 103
wetlands, Howarth's writings on, 47
wilderness, 47, 145–48, 149, 159
William of Moerbeke, 223n26

Willis, Robert, 34
Woods, William, 65
Woodward, David, 2, 28–29, 79, 80, 100, 103
Wyclif, John: *De officio pastorali*, 172–73; scientific influence on, 151
Wycliffite Bible, 150

Zanin, Fabio, 133

CPSIA information can be obtained
at www.ICGtesting.com
Printed in the USA
BVHW031055010219
539200BV00004B/19/P